MATHEMATICS WITH APPLICATIONS

1 2 3 4 5 6 7 8 9 0 DODO 7 8 3 2 1 0 9

This book was set in Times Roman by York Graphic Services. The editors were Carol Napier and James W. Bradley; the designer was Hermann Strohbach; the production supervisor was Charles Hess. The cover illustrations and chapter numbers were done by Hermann Strohbach; all other drawings were done by J & R Services, Inc.
R. R. Donnelley & Sons Company was printer and binder.

Library of Congress Cataloging in Publication Data

Hoffmann, Laurence D date
Mathematics with applications.

Includes index.
1. Mathematics—1961- I. Orkin, Michael.
II. Title.
QA37.2.H65 510 78-13261
ISBN 0-07-029301-5

MATHEM
WITH APPLIC

Laurence D. Hof

CLAREMONT MEN'S COLLEGE

Michael Orki

CALIFORNIA STATE UNIVERSITY, HAYWAR

McGRAW-HILL BOOK COMPANY • New York • St. Louis • San Francisco • Auc
Düsseldorf • Johannesburg • London • Madrid • Mexico • Montreal • New Delhi
São Paulo • Singapore • Sydney • Tokyo • Toronto

CONTENTS

PREFACE

If you are a student in the social, management, or life sciences and if you have taken high school algebra, then this book was written for you. The purpose of the book is to teach you the basic mathematical techniques you will need in your undergraduate courses in economics, business, sociology, political science, psychology, and biology. The book will also prepare you for more specialized quantitative courses that may be required for your major such as econometrics and statistics.

The text is applications-oriented. Each new concept you learn is applied to a variety of practical situations. Special emphasis is placed on the techniques you will need to solve practical problems.

Theory for its own sake has been avoided. However, the main results are stated carefully and completely, and most of them are explained or justified. Whenever possible, explanations are informal and intuitive.

You learn mathematics by doing it. Each section in this text is accompanied by a set of problems. Many involve routine computation and are designed to help you master new techniques. Others ask you to apply the new techniques to practical situations. You can find the answers to odd-numbered problems at the back of the book. A study guide with supplementary problems is also available.

The text covers three basic areas of mathematics: linear analysis (Chapters 1, 2, and 3); probability and statistics (Chapters 4, 5, 6, 7, and 8); and calculus (Chapters 10, 11, 12, and 13). There is also a chapter on the mathematics of finance (Chapter 9). If your course is more than one semester long, you will be able to cover most of the material in the book. You should not expect to complete all 13 chapters during a one-semester course. In such a course, your instructor might decide to cover selected topics from each area or to concentrate on just two of the areas.

Many people helped us prepare this book. We are especially grateful to the students of Claremont Men's College who used a preliminary edition of the text and who offered many valuable suggestions. We are also indebted to Professors Howard Bell of Shippensburg State College, John Ewing of Indiana University, Russell Lee of Allen Hancock College, Kenneth Perrin of Pepperdine University, and George Springer of Indiana University, who read various versions of the manuscript and whose perceptive comments influenced us considerably. And we thank Ann Cambra who typed the entire manuscript at least twice.

Our editors at McGraw-Hill have been especially helpful, encouraging, and patient. We are grateful to Tony Arthur, Jim Bradley, and most of all, Carol Napier, who has been involved with this project since its inception almost three years ago.

Laurence D. Hoffmann
Michael Orkin

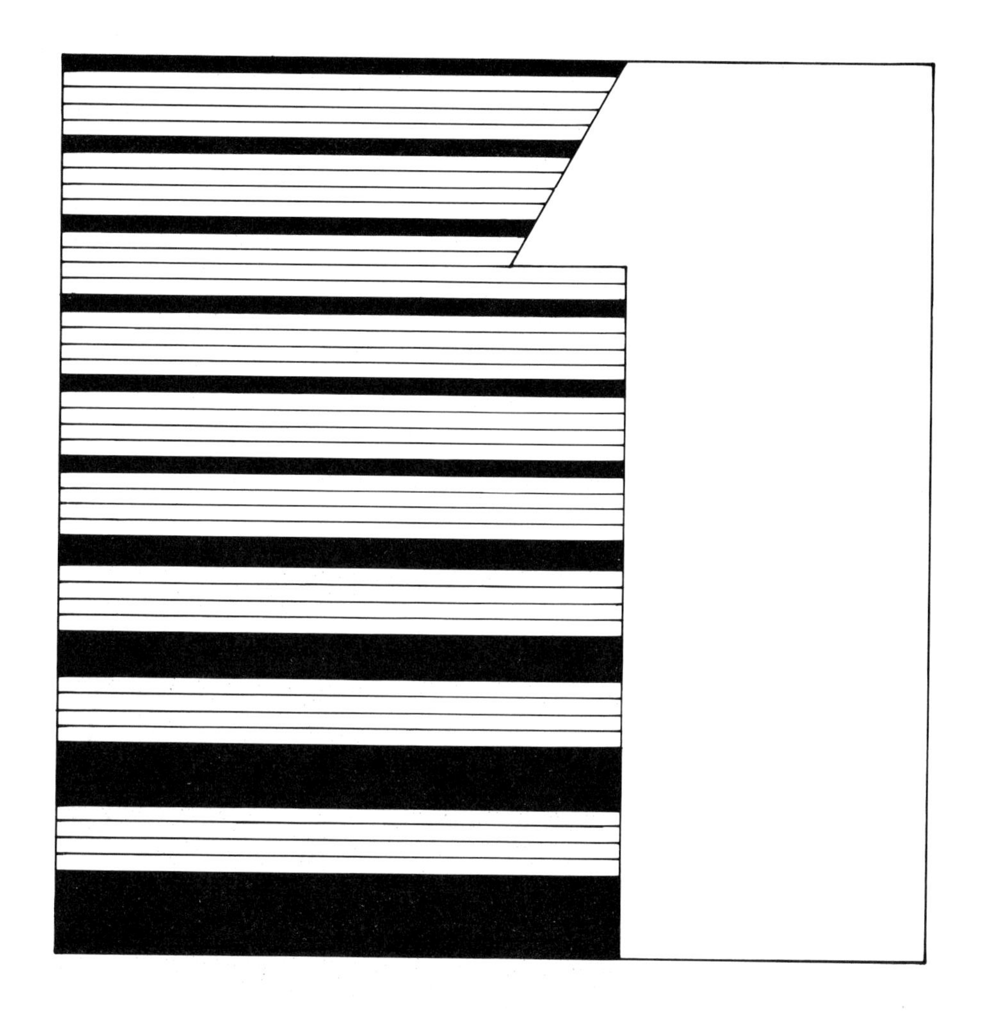

LINEAR FUNCTIONS AND STRAIGHT LINES

1 • FUNCTIONS AND LINES

In many practical situations, the value of one quantity may depend on the value of a second quantity. For example, the consumer demand for beef may depend on its current market price; the amount of air pollution in a metropolitan area may depend on the number of cars on the road; the value of a bottle of wine may depend on its age. Relationships of this sort can often be represented mathematically as **functions.**

Function
A *function* is a rule that associates with each object in a set A, one and only one object in a set B.

For the functions in this book, both A and B will be the set of all real numbers. For our purposes, then, it will be sufficient to think of a function as a rule according to which "new" numbers are associated with "old" numbers. Here is an example.

EXAMPLE 1.1 ● A certain function is defined by the rule that the new number is obtained by adding 1 to twice the old number. What number does this function associate with 3?

SOLUTION The number associated with 3 is $2(3) + 1$, or 7. ●

Variables

Often a function can be abbreviated conveniently by a mathematical formula. Traditionally, we let the letter x denote the old number and y the new number, and we write an equation relating x and y. For instance, the function in Example 1.1 can be expressed by the equation $y = 2x + 1$. The letters x and y that appear in such an equation are called **variables.** The numerical value of the variable y is determined by that of the variable x. For this reason, y is sometimes referred to as the **dependent variable** and x as the **independent variable.**

Functional notation

There is an alternative notation for functions that is widely used and somewhat more versatile. A letter such as f is chosen to stand for the function itself, and the value that the function associates with x is denoted by $f(x)$ instead of y. The symbol $f(x)$ is read "f of x." Using this **functional notation,** we can rewrite Example 1.1 as follows.

EXAMPLE 1.2 ● Find $f(3)$ if $f(x) = 2x + 1$.

SOLUTION

$$f(3) = 2(3) + 1 = 7$$ ●

Observe the convenience and simplicity of this notation. In Example 1.2, the compact formula $f(x) = 2x + 1$ completely defines the function, and the simple equation $f(3) = 7$ indicates that 7 is the number that the function associates with 3.

The next example illustrates how functional notation is used in a practical situation. Notice that in this example, as in many practical situations, letters other than f and x are used to represent the function and its independent variable.

EXAMPLE 1.3 ● Suppose the total cost in dollars of manufacturing q units of a certain commodity is given by the function $C(q) = 20q + 500$.

(a) Compute the cost of manufacturing 8 units of the commodity.
(b) Compute the cost of manufacturing the 8th unit of the commodity.

SOLUTION (a) The cost of manufacturing 8 units is the value of the total cost function when $q = 8$. That is,

$$\text{Cost of 8 units} = C(8) = 20(8) + 500 = \$660$$

(b) The cost of manufacturing the 8th unit is the difference between the cost of manufacturing 8 units and the cost of manufacturing 7 units. That is,

$$\text{Cost of 8th unit} = C(8) - C(7) = 660 - 640 = \$20$$ ●

The graph of a function Functions can be represented geometrically as graphs drawn on a rectangular coordinate system. Traditionally, the independent variable is represented on the horizontal axis and the dependent variable on the vertical axis.

The graph of a function
The *graph* of a function consists of all points (x, y) whose coordinates satisfy the equation that defines the function.

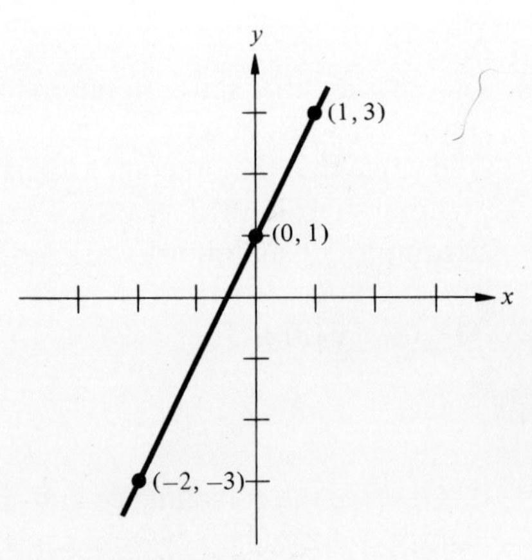

Figure 1.1 The graph of the function $y = 2x + 1$.

The graph of the function $y = 2x + 1$ is shown in Figure 1.1, and some of the points on the graph are labeled. Notice that the coordinates (x, y) of the labeled points satisfy the given equation $y = 2x + 1$.

Linear functions

The function $y = 2x + 1$ in Figure 1.1 is an example of a **linear function.** In general, any function of the form

$$y = mx + b$$

where m and b are constants is said to be linear. This is because the graph of such a function is a straight line. Linear functions arise frequently in the social sciences. You will see some typical applications in the next section. To work with linear functions, you will need to know the following facts about straight lines.

The slope of a line

An important geometric property of a line is its **slope.** This is the amount by which the y coordinate of a point on the line changes when the x coordinate is increased by 1. That is, the slope is the ratio of a change in y to the corresponding change in x.

The slope of a nonvertical line can be computed if any two points on the line are known. Suppose (x_1, y_1) and (x_2, y_2) lie on a line as indicated in Figure 1.2. Between these points, x changes by an amount $x_2 - x_1$ and y by an amount $y_2 - y_1$. The slope is the ratio

$$\text{Slope} = \frac{\text{change in } y}{\text{change in } x} = \frac{y_2 - y_1}{x_2 - x_1}$$

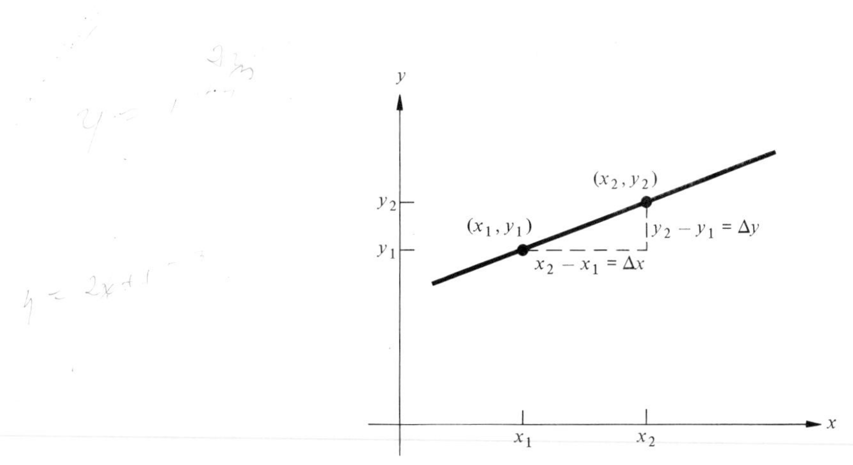

Figure 1.2 **Slope** $= \dfrac{y_2 - y_1}{x_2 - x_1} = \dfrac{\Delta y}{\Delta x}$.

It is sometimes convenient to use the symbol Δy instead of $y_2 - y_1$ to denote the change in y. The symbol Δy is read "delta y." Similarly, the symbol Δx is used to denote $x_2 - x_1$.

The slope of a line
The *slope* of the nonvertical line passing through the points (x_1, y_1) and (x_2, y_2) is given by the formula

$$\text{Slope} = \frac{\Delta y}{\Delta x} = \frac{y_2 - y_1}{x_2 - x_1}$$

The use of this formula is illustrated in the following example.

EXAMPLE 1.4 ● Find the slope of the line joining the points $(3, -1)$ and $(-2, 5)$.

SOLUTION

$$\text{Slope} = \frac{\Delta y}{\Delta x} = \frac{5 - (-1)}{-2 - 3} = -\frac{6}{5}$$ ●

The sign and magnitude of the slope of a line indicate the line's direction and steepness, respectively. The slope is positive if the height of the line increases as x increases and is negative if the height decreases as x increases. The absolute value of the slope is large if the slant of the line is severe and is small if the slant of the line is gradual. The situation is illustrated in Figure 1.3.

The slope-intercept form of the equation of a line

The constants m and b in the equation $y = mx + b$ have geometric interpretations. The constant m is the slope of the corresponding line. To see this, suppose that (x_1, y_1) and (x_2, y_2) are two points on the line $y = mx + b$. Then, $y_1 = mx_1 + b$ and $y_2 = mx_2 + b$ and so

$$\text{Slope} = \frac{y_2 - y_1}{x_2 - x_1} = \frac{(mx_2 + b) - (mx_1 + b)}{x_2 - x_1}$$

$$= \frac{mx_2 - mx_1}{x_2 - x_1} = \frac{m(x_2 - x_1)}{x_2 - x_1} = m$$

The constant b in the equation $y = mx + b$ is the value of y corresponding to $x = 0$. Hence, b is the height at which the line $y = mx + b$ crosses the y axis. The corresponding point $(0, b)$ is known as the **y intercept** of the line. The situation is illustrated in Figure 1.4.

Because the constants m and b in the equation $y = mx + b$ correspond to the slope and y intercept, respectively, this form of the equation of a line is known as the **slope-intercept form.**

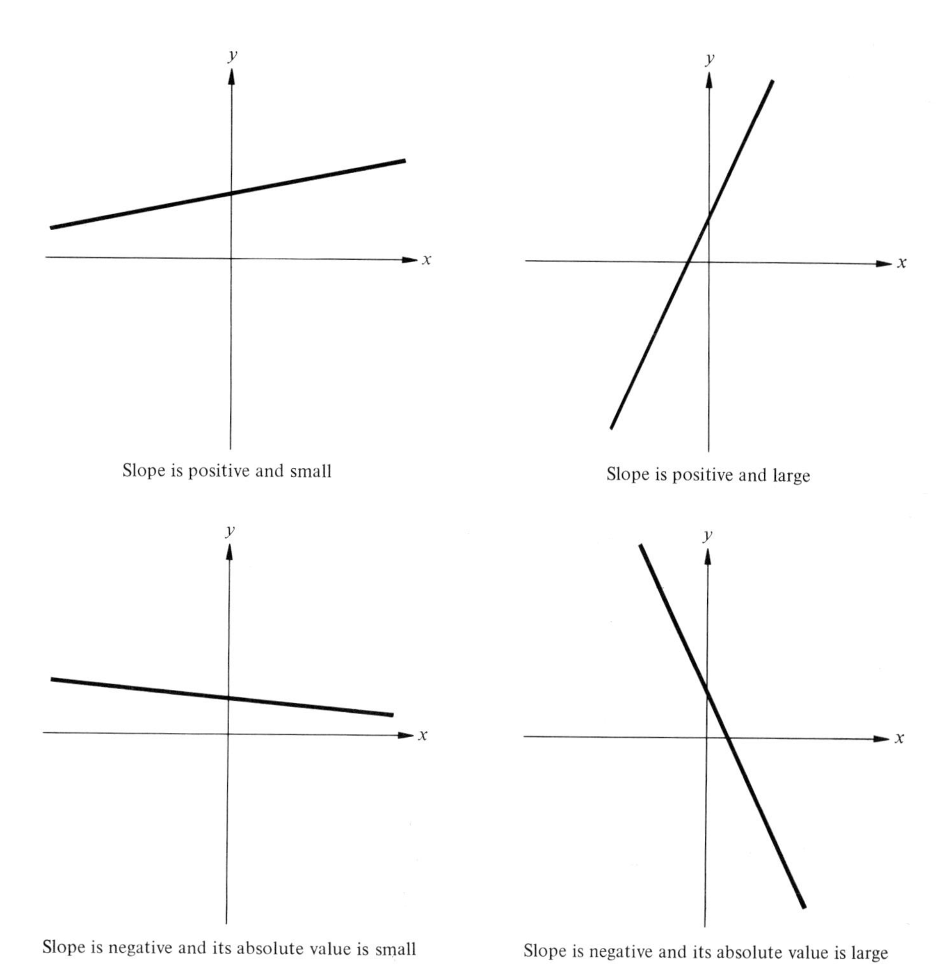

Figure 1.3 The direction and steepness of a line.

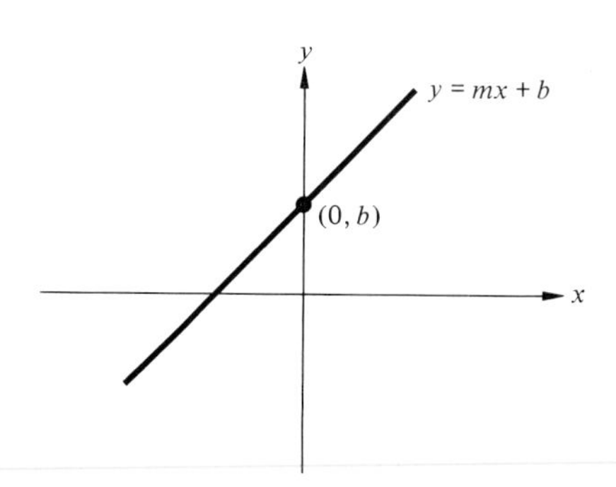

Figure 1.4 The y intercept of a line.

Slope-intercept form of the equation of a line
The equation

$$y = mx + b$$

is the equation of the line whose slope is m and whose y intercept is $(0, b)$.

The slope-intercept form of the equation of a line is particularly useful when geometric information about a line (such as its slope or y intercept) is to be determined from the line's algebraic representation. Here is a typical application.

EXAMPLE 1.5 ● Find the slope and y intercept of the line $3y + 2x = 6$ and draw the graph.

SOLUTION The first step is to put the equation $3y + 2x = 6$ in the slope-intercept form $y = mx + b$. To do this, we solve for y and get

$$y = -\tfrac{2}{3}x + 2$$

from which we conclude that the slope is $-\frac{2}{3}$ and the y intercept is $(0, 2)$.

To graph a linear function, we plot two of its points and draw a straight line through them. In this case, we already know one point, the y intercept $(0, 2)$. A convenient choice for the x coordinate of the second point is $x = 3$. The corresponding y coordinate is $y = -(\frac{2}{3})(3) + 2 = 0$. Drawing a line through the points $(0, 2)$ and $(3, 0)$ we obtain the graph shown in Figure 1.5. ●

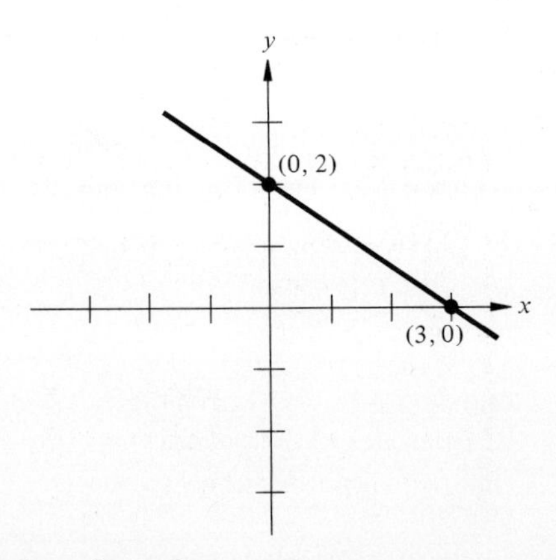

Figure 1.5 The line $3y + 2x = 6$.

Horizontal and vertical lines Horizontal and vertical lines (Figures 1.6*a* and 1.6*b*) have particularly simple equations. The y coordinates of all the points on a horizontal line are the same. Hence, a horizontal line is the graph of a linear function of the form $y = b$,

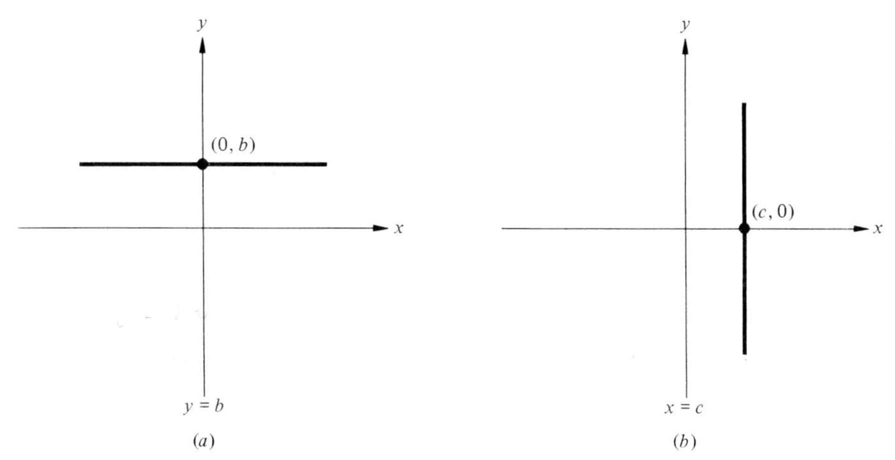

Figure 1.6 Horizontal and vertical lines.

where b is a constant. The slope of a horizontal line is zero, since changes in x produce no changes in y whatsoever.

The x coordinates of all the points on a vertical line are equal. Hence, vertical lines are characterized by equations of the form $x = c$, where c is a constant. The slope of a vertical line is undefined. This is because only the y coordinates of points on a vertical line can change, and so the denominator of the quotient (change in y)/(change in x) is zero.

The point-slope form of the equation of a line

Geometric information about a line can be obtained readily from the slope-intercept formula $y = mx + b$. There is another form of the equation of a line, however, that is usually more efficient for problems in which a line's geometric properties are known and the goal is to find the equation of the line.

Point-slope form of the equation of a line
The equation

$$y - y_0 = m(x - x_0)$$

is an equation of the line that passes through the point (x_0, y_0) and that has slope equal to m.

The point-slope form of the equation of a line is nothing more than the formula for slope in disguise. To see this, suppose that the point (x, y) lies on the line that passes through a given point (x_0, y_0) and that has slope m. Using the points (x, y) and (x_0, y_0) to compute the slope, we get

$$\frac{y - y_0}{x - x_0} = m$$

which we put in point-slope form

$$y - y_0 = m(x - x_0)$$

by multiplying both sides by $x - x_0$.

The use of the point-slope form of the equation of a line is illustrated in the next two examples.

EXAMPLE 1.6 ● Find an equation of the line that passes through the point (5, 1) and whose slope is equal to $\frac{1}{2}$.

SOLUTION Using the formula $y - y_0 = m(x - x_0)$ with $(x_0, y_0) = (5, 1)$ and $m = \frac{1}{2}$, we get

$$y - 1 = \tfrac{1}{2}(x - 5)$$

or, equivalently,

$$y = \tfrac{1}{2}x - \tfrac{3}{2}$$ ●

Instead of the point-slope form $y - y_0 = m(x - x_0)$, the slope-intercept form $y = mx + b$ could have been used to solve the problem in Example 1.6. For practice, solve the problem this way. Notice how much more efficient the solution based on the point-slope formula is.

The next example illustrates how the point-slope form of the equation of a line can be used to find an equation of a line that passes through two given points.

EXAMPLE 1.7 ● Find an equation of the line that passes through the points (3, −2) and (1, 6).

SOLUTION First we compute the slope:

$$m = \frac{6 - (-2)}{1 - 3} = \frac{8}{-2} = -4$$

Then, using (1, 6) as the given point (x_0, y_0) we get

$$y - 6 = -4(x - 1) \qquad \text{or} \qquad y = -4x + 10$$

Convince yourself that the resulting equation would have been the same if we had chosen (3, −2) as the given point (x_0, y_0). ●

Problems

1 The total cost in dollars of manufacturing q units of a certain commodity is $C(q) = 30q + 1{,}000$.
(a) Compute the cost of manufacturing 10 units of the commodity.
(b) Compute the cost of manufacturing the 10th unit.

2 It is estimated that t years from now, the average SAT score of incoming freshmen at a certain college will be $S(t) = -6t + 582$.
(a) What will the average SAT score be 5 years from now?
(b) What is the average SAT score of this year's freshman class?

3 It is estimated that x months from now, a certain reservoir will hold $V(x) = 200 - 4x$ million gallons of water.
(a) How much water will the reservoir hold 15 months from now?
(b) How much does the reservoir hold at the present time?
(c) How much water will the reservoir lose next month?
(d) How much water will the reservoir lose the month after next?

4 When a certain industrial machine is t years old, its trade-in value is $V(t) = 20{,}000 - 1{,}500t$ dollars.
(a) What is the value of the machine when it is 10 years old?
(b) What was the value of the machine when it was new?
(c) By how much does the value of the machine decrease each year? Is this annual depreciation constant or does it vary with time? Explain.

5 Determine which of the following functions are linear.
(a) $y = -\frac{1}{2}x + 3$ (b) $y = \frac{2}{x} + 3$
(c) $y = 2x^2 - 4$ (d) $y = -x$

In Problems 6 through 10, find the slope (if possible) of the line that passes through the given pair of points.

6 $(2, -3)$ and $(0, 4)$ **7** $(-1, 2)$ and $(2, 5)$
8 $(2, 0)$ and $(0, 2)$ **9** $(5, -1)$ and $(-2, -1)$
10 $(2, 6)$ and $(2, -4)$

In Problems 11 through 21, find the slope and y intercept (if possible) of the given line and draw a graph.

11 $y = 3x$ **12** $y = 5x - 6$
13 $y = x$ **14** $x + y = 2$
15 $3x + 2y = 6$ **16** $2x - 4y = 12$
17 $5y - 3x = 4$ **18** $4x = 2y + 6$
19 $(x/2) + (y/5) = 1$ **20** $y = 2$
21 $x = -3$

In Problems 22 through 32, write an equation for the line with the given properties.

22 Through $(2, 0)$ with slope 1
23 Through $(-1, 2)$ with slope $\frac{2}{3}$
24 Through $(5, -2)$ with slope $-\frac{1}{2}$
25 Through $(0, 0)$ with slope 5
26 Through $(2, 5)$ and parallel to the x axis
27 Through $(2, 5)$ and parallel to the y axis

28 Through $(1, 0)$ and $(0, 1)$
29 Through $(2, 5)$ and $(1, -2)$
30 Through $(-2, 3)$ and $(0, 5)$
31 Through $(1, 5)$ and $(3, 5)$
32 Through $(1, 5)$ and $(1, -4)$

2 • LINEAR MODELS

A mathematical representation of a real-life situation is called a **mathematical model.** In this section, you will see examples illustrating some of the techniques you can use to build models involving linear functions.

The rate at which the value of a linear function changes with respect to its independent variable is constant. This corresponds to the geometric fact that the height of a straight line changes at a constant rate, this rate being the slope of the line. Situations in which one quantity changes at a constant rate with respect to another can be represented mathematically by linear functions. Here are some examples.

EXAMPLE 2.1 ● Since the beginning of the year, the price of a loaf of whole-wheat bread at a local supermarket has been rising at a constant rate of 2 cents per month. By November first, the price had reached 64 cents per loaf.

(a) Express the price of the bread as a function of time.
(b) What was the price of the bread at the beginning of the year?

SOLUTION (a) Let x denote the number of months that have elapsed since the first of the year and y the price of a loaf of the bread. Since y changes at a constant rate with respect to x, the function relating y to x must be linear and its graph a straight line. The fact that the price y increases by 2 each time x increases by 1 implies that the slope of the line is 2. The fact that the price was 64 cents on November first (10 months after the first of the year) implies that the line passes through the point $(10, 64)$. To write an equation defining y as a function of x, we use the point-slope formula $y - y_0 = m(x - x_0)$ with $m = 2$ and $(x_0, y_0) = (10, 64)$ and get

$$y - 64 = 2(x - 10) \qquad \text{or} \qquad y = 2x + 44$$

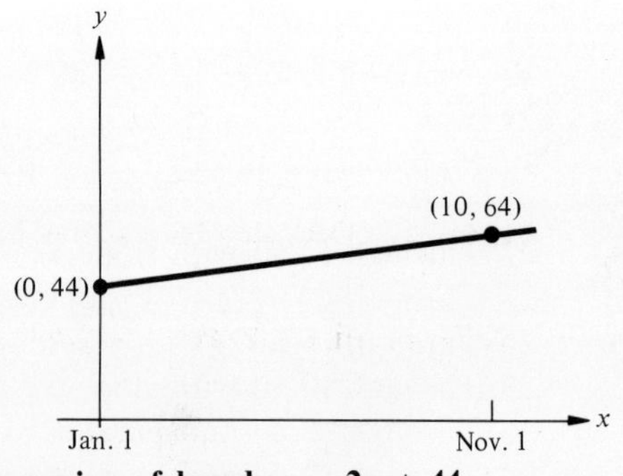

Figure 2.1 The rising price of bread: $y = 2x + 44$.

The corresponding line is shown in Figure 2.1.

(b) Since $y = 44$ when $x = 0$, we conclude that the price of the bread at the beginning of the year was 44 cents per loaf. ●

EXAMPLE 2.2 ● The average SAT scores of incoming freshmen at an eastern liberal arts college have been declining at a constant rate in recent years. In 1974, the average SAT score was 582 while in 1979 the average SAT score was only 552.

(a) Express the average SAT score as a function of time.
(b) If the trend continues, what will be the average SAT score of incoming freshmen in 1984?

SOLUTION (a) We let x denote the number of years since 1974 and y the average SAT score of incoming freshmen. Since y changes at a constant rate with respect to x, the function relating y to x must be linear. Since $y = 582$ when $x = 0$ and $y = 552$ when $x = 5$, the corresponding straight line must pass through the points (0, 582) and (5, 552). The slope of this line is

$$m = \frac{582 - 552}{0 - 5} = -6$$

Since one of the given points happens to be the y intercept (0, 582), we use the slope-intercept form and conclude immediately that

$$y = -6x + 582$$

The corresponding line is shown in Figure 2.2.

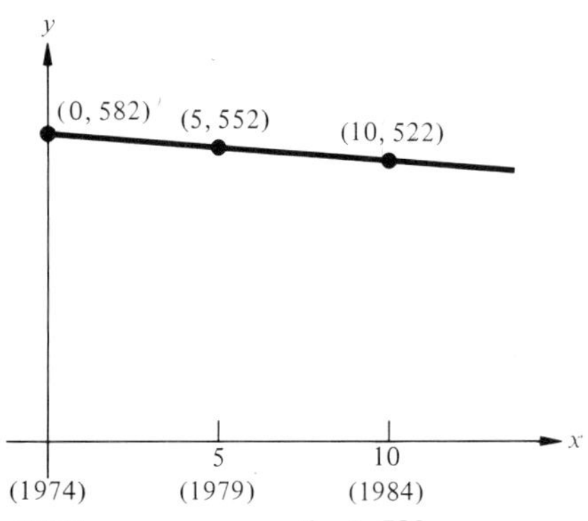

Figure 2.2 Declining SAT scores: $y = -6x + 582$.

(b) To predict the average SAT score in 1984, we compute y when $x = 10$ and get

$$y = -6(10) + 582 = 522$$ ●

In the next example, we will need three formulas to define the desired function.

EXAMPLE 2.3 ● During the 1977 drought, residents of Marin County, Calif., were faced with a severe water shortage. To discourage excessive use of water, the County Water District initiated drastic rate increases. The monthly rate for a family of four was \$1.22 per 100 cubic feet of water for the first 1,200 cubic feet, \$10 per 100 cubic feet for the next 1,200 cubic feet, and \$50 per 100 cubic feet thereafter. Express the monthly water bill for a family of four as a function of the amount of water used.

SOLUTION We let x denote the number of hundred-cubic-foot units of water used by the family during the month and $C(x)$ the corresponding cost in dollars. If $0 \leq x \leq 12$, the cost is simply the cost per unit times the number of units used:

$$C(x) = 1.22x$$

If $12 < x \leq 24$, each of the first 12 units costs \$1.22 and so the total cost of these 12 units is $1.22(12) = 14.64$ dollars. Each of the remaining $x - 12$ units costs \$10 and hence the total cost of these units is $10(x - 12)$ dollars. The cost of all x units is the sum

$$C(x) = 14.64 + 10(x - 12) = 10x - 105.36$$

If $x > 24$, the cost of the first 12 units is $1.22(12) = 14.64$ dollars, the cost of the next 12 units is $10(12) = 120$ dollars, and the cost of the remaining $x - 24$ units is $50(x - 24)$. The cost of all x units is the sum

$$C(x) = 14.64 + 120 + 50(x - 24) = 50x - 1{,}065.36$$

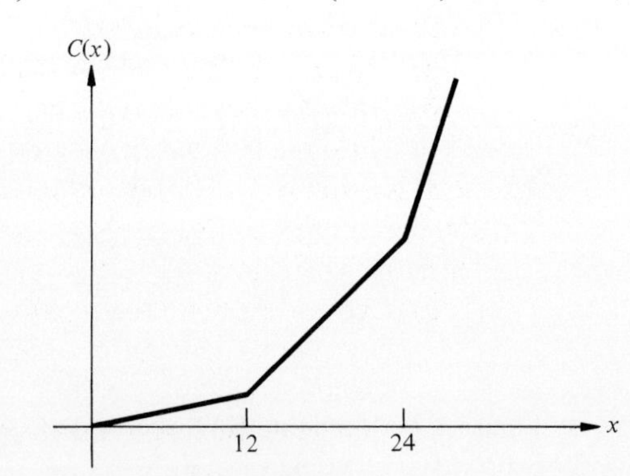

Figure 2.3 Cost of water in Marin County:

$$C(x) = \begin{cases} 1.22x & \text{if } 0 \leq x \leq 12 \\ 10x - 105.36 & \text{if } 12 < x \leq 24. \\ 50x - 1{,}065.36 & \text{if } x > 24 \end{cases}$$

Combining our three formulas, we conclude that

$$C(x) = \begin{cases} 1.22x & \text{if } 0 \leq x \leq 12 \\ 10x - 105.36 & \text{if } 12 < x \leq 24 \\ 50x - 1{,}065.36 & \text{if } x > 24 \end{cases}$$

The graph of this function is shown in Figure 2.3. Notice that the graph consists of three straight lines, each steeper than the preceding one. What aspect of the practical situation is reflected by the increasing steepness of the lines? ●

Problems

1 A certain oil well yields 400 barrels of crude oil a month that can be sold for \$15 per barrel.

(a) Derive a formula for the total revenue from the well over the next x months.

(b) How much revenue will the oil well generate over the next 6 months?

2 During the summer, a group of students builds kayaks in a converted warehouse. The rental cost of the warehouse is \$600 for the summer. The materials needed to build a kayak cost \$25. Express the group's total cost as a function of the number of kayaks built and graph this function.

3 A taxi fleet contains 30 cabs, each of which is driven approximately 200 miles per day and averages 15 miles per gallon. If the price of gasoline is 70 cents per gallon, derive a formula for the amount of money the taxi company can expect to spend on gasoline over the next x days.

4 Students at a state college may preregister for their fall classes by mail during the summer. Those who do not preregister must register in person in September. The registrar can process 35 students per hour during the September registration period. After 4 hours in September, a total of 360 students have been registered.

(a) Express the number of students registered as a function of time.

(b) How many students were registered after 3 hours?

(c) How many students preregistered during the summer?

5 Membership in a swimming club costs \$150 for the 12-week summer season. If a member joins after the start of the season, the fee is prorated; that is, the fee is reduced linearly.

(a) Express the membership fee as a function of the number of weeks that have elapsed by the time the membership is purchased and graph this function.

(b) Compute the cost of a membership that is purchased 5 weeks after the start of the season.

6 A manufacturer buys \$20,000 worth of machinery that depreciates linearly so that its trade-in value after 10 years will be \$1,000.

(a) Express the value of the machinery as a function of its age and graph this function.

(b) What is the value of the machinery after 4 years?

7 Since the beginning of the month, a local reservoir has been losing water at a constant rate. On the 12th of the month, the reservoir held 200 million gallons of water and on the 21st it held only 164 million gallons.

(a) Express the amount of water in the reservoir as a function of time.

(b) How much water was in the reservoir on the 8th of the month?

8 To encourage motorists to form car pools, the transit authority in a major metropolitan area has been offering a special reduced rate at toll bridges for vehicles containing four or more persons. When the program began 30 days ago, 157 vehicles qualified for the reduced rate during the morning rush hour. Since then, the number of vehicles qualifying for the reduced rate has been increasing at a constant rate, and today 247 vehicles qualified for the reduced rate during the morning rush hour.

(a) Express the number of vehicles qualifying each morning for the reduced rate as a function of time.

(b) If the trend continues, how many vehicles will qualify for the reduced rate during the morning rush hour 14 days from now?

9 (a) Temperature measured in degrees Fahrenheit is a linear function of temperature measured in degrees Celsius. Use the facts that 0°C is equal to 32°F and 100°C is equal to 212°F to write an equation for this linear function.

(b) Use the function you obtained in part (a) to convert 15°C to Fahrenheit.

(c) Convert 68°F to Celsius.

10 A local natural history museum charges admission to groups according to the following policy: Groups of fewer than 50 people are charged a rate of $1.50 per person, while groups of 50 people or more are charged a reduced rate of $1.00 per person.

(a) Express the amount a group will be charged for admission to the museum as a function of its size and graph this function.

(b) How much money will a group of 49 people save in admission costs if it can recruit one additional member?

11 In 1977, the rate for interstate telegrams was $4.75 for 15 words or less, plus 12 cents for each additional word.

(a) Express the cost of sending a telegram as a function of its length and graph this function.

(b) Compute the cost of sending a 60-word telegram.

12 The following table is taken from the 1972 federal income tax rate schedule for single taxpayers.

If taxable income is:		Income tax is:	
Over	But not over		Of the excess over
$8,000	$10,000	$1,590 + 25%	$8,000
$10,000	$12,000	$2,090 + 27%	$10,000
$12,000	$14,000	$2,630 + 29%	$12,000
$14,000	$16,000	$3,210 + 31%	$14,000

(a) Express an individual's income tax as a function of his taxable income x for $8{,}000 < x \leq 16{,}000$.
(b) Graph the function in part (a).
(c) Your graph in part (b) should consist of four line segments. Compute the slope of each segment. What happens to these slopes as the taxable income increases? Explain the behavior of the slopes in practical terms.

13 The value of a certain rare book doubles every 10 years. The book was originally worth \$3.00.
(a) How much is the book worth when it is 30 years old? When it is 40 years old?
(b) Is it possible to express the value of the book as a *linear* function of its age? Explain.

3 • INTERSECTIONS OF LINES

Sometimes it is necessary to determine when two functions are equal. This is the case, for example, when an economist computes the market price at which the demand for a certain commodity will be equal to its supply. It occurs when a manufacturer seeks to determine how many units have to be sold to make the total revenue equal to the total cost. And it occurs when a political analyst attempts to predict how long it will take for the popularity of a certain challenger to reach that of the incumbent.

In geometric terms, two functions $f(x)$ and $g(x)$ will be equal for those values of x for which their graphs intersect. To find these values of x algebraically, we set $f(x)$ equal to $g(x)$ and solve for x.

EXAMPLE 3.1 ● Where do the lines $y = 2x + 1$ and $y = -x + 4$ intersect?

SOLUTION We solve the equation

$$2x + 1 = -x + 4$$

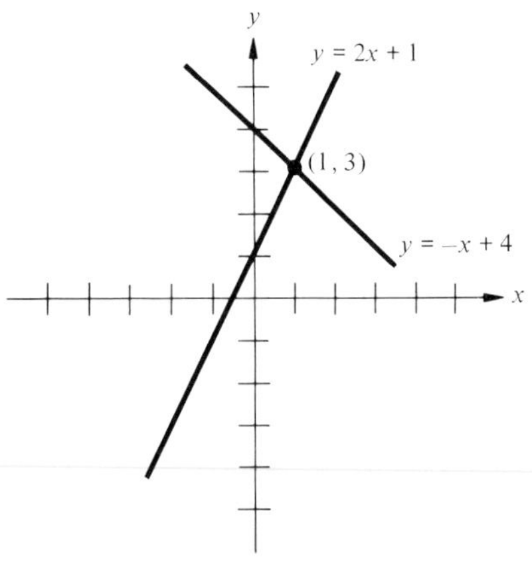

Figure 3.1 The lines $y = 2x + 1$ and $y = -x + 4$.

and find that $x = 1$. To get the corresponding value of y, we substitute $x = 1$ into either of the original equations $y = 2x + 1$ or $y = -x + 4$. We find that $y = 3$ and conclude that $(1, 3)$ is the point of intersection. The situation is illustrated in Figure 3.1. ●

EXAMPLE 3.2 ● Where do the lines $y = 3x - 5$ and $y = 3x + 2$ intersect?

SOLUTION If we try to solve the equation

$$3x - 5 = 3x + 2$$

we get $0 = 7$. This is clearly false and we conclude that the given lines do not intersect. This should come as no surprise since the lines have equal slopes and hence are parallel as shown in Figure 3.2. ●

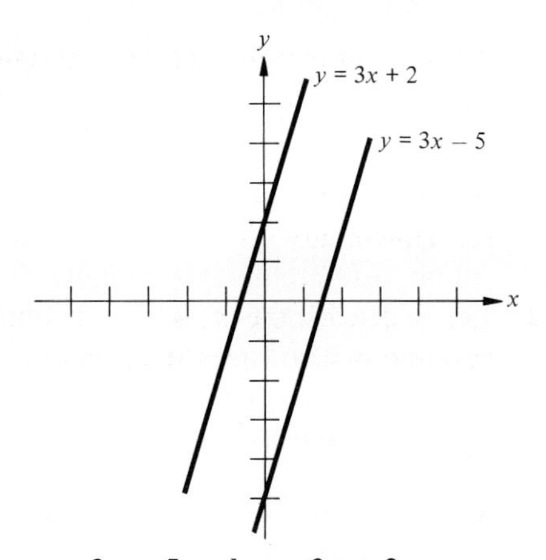

Figure 3.2 The lines $y = 3x - 5$ and $y = 3x + 2$.

Break-even analysis

Intersections of graphs arise in business in the context of **break-even analysis.** In a typical situation, a manufacturer wishes to determine how many units of a certain commodity have to be sold to make the total revenue equal to the total cost. Suppose that x denotes the number of units manufactured and sold, and let $C(x)$ and $R(x)$ be the corresponding total cost and total revenue, respectively. A pair of cost and revenue curves is sketched in Figure 3.3.

Because of fixed overhead costs, the total cost curve is initially higher than the total revenue curve. Hence, at low levels of production, the manufacturer suffers a loss. At higher levels of production, however, the total revenue curve is the higher one and the manufacturer realizes a profit. The point at which the two curves cross is called the **break-even point,** because when total revenue is equal to total cost, the manufacturer breaks even, experiencing neither a profit nor a loss.

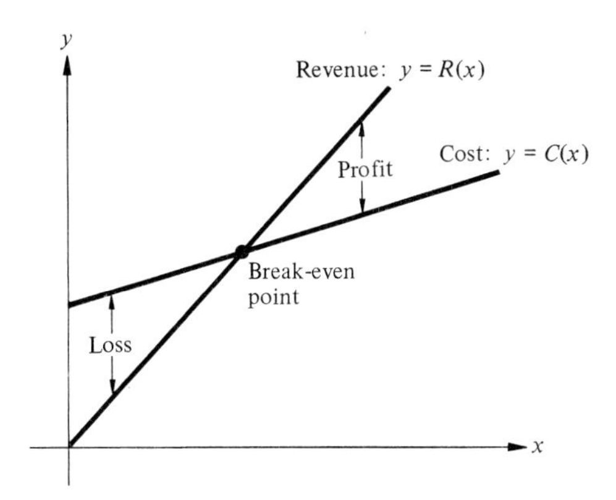

Figure 3.3 Linear cost and revenue functions.

EXAMPLE 3.3 ● A manufacturer can sell a certain product for $110 per unit. Costs consist of a fixed overhead of $7,500 and a production cost of $60 per unit.

(a) How many units must the manufacturer sell to break even?
(b) What is the manufacturer's profit or loss if 100 units are sold?
(c) How many units must the manufacturer sell to realize a profit of $1,250?

SOLUTION Let x denote the number of units manufactured and sold. Then the total revenue is given by the function

$$R(x) = 110x$$

and the total cost by the function

$$C(x) = 7{,}500 + 60x$$

(a) To find the break-even point, we set $R(x)$ equal to $C(x)$ and solve getting

$$\begin{aligned} 110x &= 7{,}500 + 60x \\ 50x &= 7{,}500 \\ x &= 150 \end{aligned}$$

We conclude that the manufacturer will have to sell 150 units to break even.

(b) The profit $P(x)$ is revenue minus cost. Hence,

$$P(x) = R(x) - C(x) = 110x - 7{,}500 - 60x = 50x - 7{,}500$$

Substituting $x = 100$ we get

$$P(100) = 5{,}000 - 7{,}500 = -2{,}500$$

The minus sign indicates a negative profit, or loss. We conclude that the manufacturer will lose $2,500 if only 100 units are sold.

(c) To determine the number of units that must be sold to produce a profit of $1,250, we set the formula for profit $P(x)$ equal to 1,250 and solve for x. We get

$$\begin{aligned} 50x - 7{,}500 &= 1{,}250 \\ 50x &= 8{,}750 \\ x &= 175 \end{aligned}$$

That is, 175 units must be sold to generate the desired profit. ●

The next example illustrates how break-even analysis can be used as a tool for decision making.

EXAMPLE 3.4 ● A leading car rental agency charges $14 plus 15 cents per kilometer. A second agency charges $20 plus 5 cents per kilometer. Which agency offers the better deal?

SOLUTION The answer depends on the number of kilometers the car is driven. For short trips, the first agency charges less than the second; but for long trips, the second charges less than the first. We use break-even analysis to determine the number of kilometers for which the two agencies charge the same amount.

Suppose a car is to be driven x kilometers. Then the first agency will charge

$$C_1(x) = 14 + 0.15x$$

dollars and the second will charge

$$C_2(x) = 20 + 0.05x$$

dollars. Setting these expressions equal to each other and solving we get

$$\begin{aligned} 14 + 0.15x &= 20 + 0.05x \\ 0.1x &= 6 \\ x &= 60 \end{aligned}$$

We conclude that for a 60-kilometer trip, the two agencies charge the same amount. For shorter trips, the first agency offers the better deal and for longer trips the second agency does. The situation is illustrated in Figure 3.4. ●

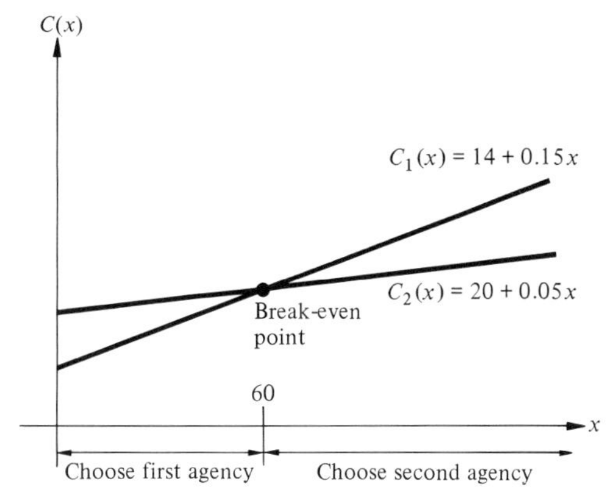

Figure 3.4 Car rental costs at competing agencies.

Market equilibrium

An important economic application involving intersections of graphs arises in connection with the **law of supply and demand.** In this context, we think of the market price p of a commodity as determining the number of units of the commodity that manufacturers are willing to supply as well as the number of units that consumers are willing to buy. In most cases, manufacturers' supply, $S(p)$, increases and consumers' demand, $D(p)$, decreases as the market price p increases. A pair of linear supply and demand curves is sketched in Figure 3.5. (The letter q used to label the vertical axis stands for "quantity.")

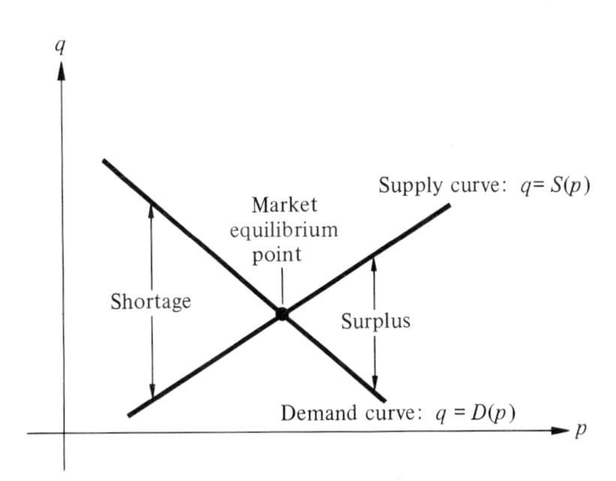

Figure 3.5 Market equilibrium: The intersection of supply and demand.

The point of intersection of the supply and demand curves is called the point of **market equilibrium.** The p coordinate of this point (the **equilibrium price**) is the market price at which supply equals demand; that is, the market price at which there will be neither a surplus nor a shortage of the commodity.

The law of supply and demand asserts that, in a situation of pure competition, a commodity will tend to be sold at its equilibrium price. If the commodity is sold for more than the equilibrium price, there will be an unsold surplus on the market, and retailers will tend to lower their prices. On the other hand, if the commodity is sold for less than the equilibrium price, the demand will exceed the supply, and retailers will be inclined to raise their prices.

Problems In Problems 1 through 5, find the point of intersection (if any) of the given pair of lines and draw the graph.

1 $y = 3x + 5$ and $y = -x + 3$

2 $y = x - 7$ and $y = 3 + x$

3 $y = -2x + 1$ and $y = 4x - 5$

4 $y = x + 5$ and $y = 9 - 3x$

5 $y = -5x + 5$ and $y = -5x - 2$

6 A furniture manufacturer can sell dining-room tables for \$70 apiece. The total cost consists of a fixed overhead of \$8,000 and a production cost of \$30 per table.

(a) Determine how many tables the manufacturer must sell to break even.

(b) Determine how many tables the manufacturer must sell to make a profit of \$6,000.

(c) Calculate the manufacturer's profit or loss if 150 tables are sold.

(d) On the same set of axes, graph the total-revenue and total-cost functions.

7 During the summer, a group of students builds kayaks in a converted warehouse. The rental for the warehouse is \$600 for the summer, and the materials needed to build a kayak cost \$25. The kayaks can be sold for \$175 apiece.

(a) How many kayaks must the students sell to break even?

(b) How many kayaks must the students sell to make a profit of \$450?

8 The charge for maintaining a checking account at a certain bank is \$2 per month plus 5 cents for each check that is written. A competing bank charges \$1 per month plus 9 cents per check. Find a criterion for deciding which bank offers the better deal.

9 Membership in a private tennis club costs \$500 per year and entitles the member to use the courts for a fee of \$1.00 per hour. At a competing club, membership costs \$440 per year and the charge for the use of the courts is \$1.75 per hour. If only financial considerations are to be taken into account, how should a tennis player choose which club to join?

10 When electric blenders are sold for p dollars apiece, manufacturers are willing to supply $3p - 20$ blenders to local retailers. On the other hand, the local consumers would buy a total of $60 - p$ blenders at this price. At what market price will the manufacturers' supply of electric blenders be equal to the consumers' demand for the blenders? How many blenders will be sold at this price?

11 The supply and demand functions for a certain commodity are given by $S(p) = p - 10$ and $D(p) = 80 - 2p$, respectively.

(a) Find the equilibrium price and the corresponding number of units supplied and demanded.

(b) Draw the supply and demand curves on the same set of axes.
(c) Where do the supply and demand curves cross the p axis? What is the economic significance of these points?

12 Suppose that the supply and demand functions for a certain commodity are $S(p) = ap + b$ and $D(p) = cp + d$, respectively.
(a) What can you say about the signs of the coefficients a, b, c, and d if the supply and demand curves are oriented as shown in the following diagram?

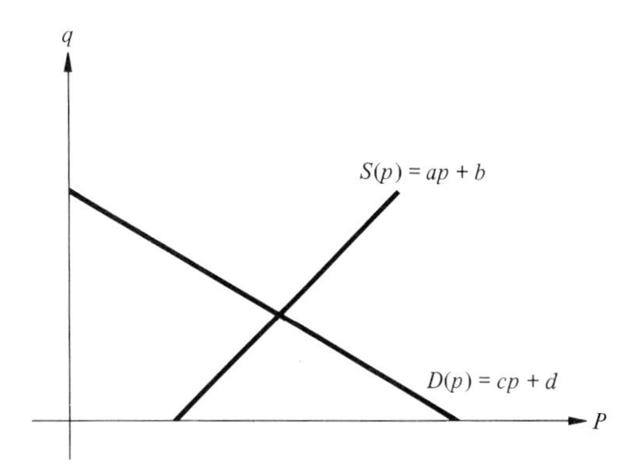

(b) Express the equilibrium price in terms of the coefficients a, b, c, and d.
(c) Use your answer in part (b) to determine what happens to the equilibrium price as a increases. Explain your answer in economic terms.
(d) Use your answer in part (b) to determine what happens to the equilibrium price as d increases. Explain your answer in economic terms.

13 The hero of a popular spy story has escaped from the headquarters of an international diamond-smuggling ring in the tiny Mediterranean country of Azusa. Our hero, driving a stolen milk truck at 72 kilometers per hour, has a 40-minute head start on his pursuers, who are chasing him in a Ferrari going 168 kilometers per hour. The distance from the smugglers' headquarters to the border, and freedom, is 83.8 kilometers. Will our hero make it?

14 Two jets bound for Los Angeles leave New York 30 minutes apart. The first travels 880 kilometers per hour, while the second goes 1,040 kilometers per hour. At what time will the second plane pass the first?

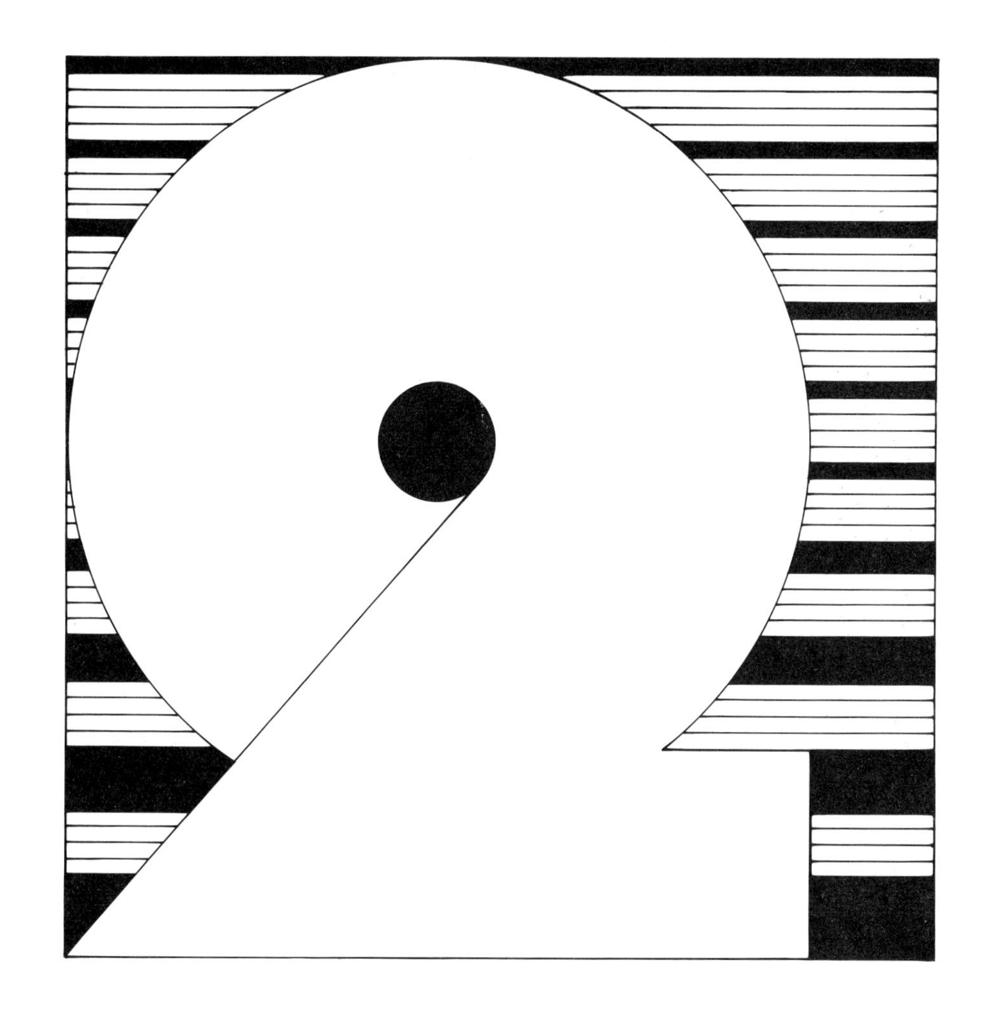

MATRIX ALGEBRA

1 • MATRICES

Rectangular arrays of numbers often simplify the organization and presentation of information. For example, the array

	Pizza	Beer	Salad
Pizza King	2.65	0.65	0.50
Campus Pizza	2.80	0.55	0.55

summarizes conveniently the prices of pizza, beer, and salad at two pizza parlors. Such an array is called a **matrix.** The numbers that form a matrix are called **entries,** and the position of each entry is specified by its **row** and **column.** For example, in the matrix summarizing the prices at the pizza parlors, the entry 2.80 is located in the second row and first column.

If a matrix has m rows and n columns, its **dimensions** are said to be $m \times n$ (read "m by n"). A matrix with an equal number of rows and columns is called a **square matrix.** Matrices that have only one row or one column are sometimes called **vectors.** A matrix in which every entry is zero is called a **zero matrix** and is denoted by the symbol 0.

It is customary to use capital letters to denote matrices and the corresponding lower case letters with appropriate subscripts to denote the entries. For example, if the letter A is used to denote a matrix, the symbol a_{ij} is used to denote the entry in row i and column j. Using this notation, a general 3×4 matrix is written as follows:

$$A = \begin{bmatrix} a_{11} & a_{12} & a_{13} & a_{14} \\ a_{21} & a_{22} & a_{23} & a_{24} \\ a_{31} & a_{32} & a_{33} & a_{34} \end{bmatrix}$$

EXAMPLE 1.1 ● Suppose

$$A = \begin{bmatrix} 7 & 0 & 1 & -5 \\ 6 & 1 & 2 & 10 \end{bmatrix}$$

(a) Specify the dimensions of A.
(b) Find a_{23}.
(c) Use subscript notation to denote the entry -5.

SOLUTION (a) The matrix has two rows and four columns. Hence, its dimensions are 2×4. It is customary to abbreviate this by writing $\dim(A) = 2 \times 4$.
(b) The symbol a_{23} stands for the entry in the second row and third column. Thus, $a_{23} = 2$.
(c) Since -5 is in the first row and fourth column, it follows that $-5 = a_{14}$. ●

Two matrices are said to be **equal** if they have the same dimensions and their corresponding entries are equal.

Two matrices that have the same dimensions can be added according to the following definition.

Matrix addition
If two matrices A and B have the same dimensions, their **sum** $A + B$ is the matrix whose entry in each position is the sum of the corresponding entries in A and B.

Matrix subtraction is defined analogously. Here is an example.

EXAMPLE 1.2 ● If

$$A = \begin{bmatrix} 1 & -2 \\ 0 & 2 \\ 5 & 3 \end{bmatrix} \quad B = \begin{bmatrix} -4 & 1 \\ 6 & 0 \\ -1 & 2 \end{bmatrix} \quad C = \begin{bmatrix} 2 & 0 & -1 \\ 3 & 8 & 5 \\ 6 & 0 & 2 \end{bmatrix}$$

compute (if possible)
(a) $A + B$ (b) $A + C$ (c) $B - A$

SOLUTION (a) Since the dimensions of both A and B are 3×2, addition is possible, and we add corresponding entries to get

$$A + B = \begin{bmatrix} -3 & -1 \\ 6 & 2 \\ 4 & 5 \end{bmatrix}$$

(b) The sum $A + C$ is not defined since A and C have different dimensions.
(c) Subtracting each entry in A from the corresponding entry in B we get

$$B - A = \begin{bmatrix} -5 & 3 \\ 6 & -2 \\ -6 & -1 \end{bmatrix}$$

●

In the context of matrix algebra, real numbers are usually referred to as **scalars.** A matrix can be multiplied by a scalar according to the following definition.

Multiplication of a matrix by a scalar
If A is a matrix and c is a scalar, the **product** cA is the matrix obtained by multiplying each entry of A by c.

EXAMPLE 1.3 ● Compute $-3A - 2B$ if

$$A = \begin{bmatrix} 0 & 3 & -2 & -1 \\ 2 & 0 & -3 & 5 \end{bmatrix} \quad \text{and} \quad B = \begin{bmatrix} 4 & -6 & 0 & 5 \\ 1 & 1 & -1 & 2 \end{bmatrix}$$

SOLUTION We first multiply the entries in A by -3 and the entries in B by 2 to get

$$-3A = \begin{bmatrix} 0 & -9 & 6 & 3 \\ -6 & 0 & 9 & -15 \end{bmatrix} \quad \text{and} \quad 2B = \begin{bmatrix} 8 & -12 & 0 & 10 \\ 2 & 2 & -2 & 4 \end{bmatrix}$$

and then subtract to get

$$-3A - 2B = \begin{bmatrix} -8 & 3 & 6 & -7 \\ -8 & -2 & 11 & -19 \end{bmatrix}$$ ●

Problems

1 Let

$$A = \begin{bmatrix} 7 & 0 & -1 & 1 \\ 2 & -2 & 9 & 6 \\ -3 & 8 & 5 & -4 \end{bmatrix}$$

(a) Specify the dimensions of A.
(b) Find a_{12}, a_{22}, a_{14}, and a_{34}.
(c) Use subscript notation to denote the entries 9, -3, 0, and 5.

2 Construct a 3×2 matrix A having the following entries: $a_{11} = 1$, $a_{21} = 2$, $a_{31} = 3$, $a_{12} = 4$, $a_{22} = 5$, and $a_{32} = 6$.

3 Construct the 3×4 matrix whose entries are defined by the rule $a_{ij} = i + j$.

4 Construct the 3×3 matrix whose entries are defined by the rule

$$a_{ij} = \begin{cases} 5 & \text{if } i = j \\ i - j & \text{if } i \neq j \end{cases}$$

5 The prices per unit for four items at three competing supermarkets are summarized in the following matrix:

	Market 1	Market 2	Market 3
Item A	1.52	1.48	1.43
Item B	0.28	0.31	0.32
Item C	2.58	2.60	2.65
Item D	0.52	0.50	0.47

(a) Which market has the lowest price on item C?

(b) What quantity is represented by the sum of all the entries in the second column?

(c) A shopper is planning to buy 2 units of item A, 9 units of item B, 1 unit of item C, and 12 units of item D. Calculate how much this would cost at each market.

6 The telephone company has received the following order for phones in two different styles and four different colors: 28 avocado desk phones, 41 beige desk phones, 34 black desk phones, 27 white desk phones, 10 beige wall phones, 15 black wall phones, and 23 white wall phones.

(a) Use a matrix to summarize this order.

(b) Use the matrix in part (a) to determine the total number of wall phones ordered.

(c) Use the matrix in part (a) to determine the total number of avocado phones ordered.

7 Three friends, Eric, Gabriel, and Kirk, write to one another from time to time. Last year, Eric received 1 letter from Gabriel and 5 from Kirk, Gabriel received 2 letters from Eric and 3 from Kirk, and Kirk received 1 letter from Eric and 6 from Gabriel.

(a) Summarize this information in a matrix.

(b) Use the matrix in part (a) to determine which of the friends wrote the most letters last year.

In Problems 8 through 13, perform (when possible) the indicated matrix operations if

$$A = \begin{bmatrix} 1 & -2 \\ 3 & 0 \\ -5 & -1 \end{bmatrix} \qquad B = \begin{bmatrix} 1 & -5 & 0 \\ 3 & 0 & 2 \end{bmatrix} \qquad C = \begin{bmatrix} -4 & 0 & 5 \\ 2 & -2 & -1 \end{bmatrix}$$

8 $B + C$ **9** $A + C$ **10** $B - C$

11 $C - B$ **12** $A + A$ **13** $3B - 2C$

In Problems 14 through 18, find the matrix C that makes the given equation true if

$$A = \begin{bmatrix} 1 & 0 & 3 \\ -2 & 1 & 5 \\ 0 & -4 & 2 \end{bmatrix} \quad \text{and} \quad B = \begin{bmatrix} -1 & 1 & 0 \\ 5 & -2 & 3 \\ 0 & 1 & 4 \end{bmatrix}$$

14 $A + C = B$ **15** $A - C = B$ **16** $A + 2C = A$

17 $C - 2B = A$ **18** $3C + A = 2B$

19 A music shop carries three brands of classical guitars. A brand X guitar lists for \$200 and its case for \$45. A brand Y guitar lists for \$260 and its case for \$40. And a brand Z guitar lists for \$320 and its case for \$50.

(a) Summarize this information in a 3×2 matrix.

(b) During a special clearance sale, the dealer offers a 15 percent discount on all merchandise. Modify the matrix in part (a) to obtain a matrix summarizing the sale prices. Express this matrix in terms of the matrix in part (a) using operations that were defined in this section.

20 An airline has two flights daily from Denver to Boston, flight #027 and flight #602. The number of passengers taking these flights on a particular day, and the corresponding fares, are summarized in the following two matrices.

$$A = \begin{array}{c} \\ \text{Flight \#027} \\ \text{Flight \#602} \end{array} \begin{array}{c} \begin{array}{ccc} \text{First class} & \text{Tourist class} & \text{Economy class} \end{array} \\ \begin{bmatrix} 22 & 60 & 25 \\ 15 & 41 & 20 \end{bmatrix} \end{array}$$

$$B = \begin{array}{l} \text{First class} \\ \text{Tourist class} \\ \text{Economy class} \end{array} \begin{array}{c} \text{Fare} \\ \begin{bmatrix} \$240 \\ \$200 \\ \$180 \end{bmatrix} \end{array}$$

(a) Calculate the airline's total receipts from ticket sales for flight #027.
(b) Calculate the airline's total receipts from ticket sales for flight #602.
(c) Summarize the information calculated in parts (a) and (b) in a 2×1 matrix.

21 A publisher with facilities in New York and San Francisco is preparing for the publication of several new texts and novels. The numbers of books of each type that will be produced in each facility are summarized in the matrix

$$\begin{array}{c} \\ \text{N.Y.} \\ \text{S.F.} \end{array} \begin{array}{c} \begin{array}{cc} \text{Text} & \text{Novel} \end{array} \\ \begin{bmatrix} 5 & 6 \\ 1 & 3 \end{bmatrix} \end{array}$$

The numbers of hours of work by various specialists that go into the production of each type of book are summarized in the matrix

$$\begin{array}{c} \\ \text{Text} \\ \text{Novel} \end{array} \begin{array}{c} \begin{array}{ccc} \text{Production editor} & \text{Designer} & \text{Copy editor} \end{array} \\ \begin{bmatrix} 50 & 8 & 30 \\ 30 & 2 & 20 \end{bmatrix} \end{array}$$

For each type of specialist, calculate the total number of hours that will be needed in each city to produce the books as planned, and summarize your results in a 2×3 matrix.

2 • MATRIX MULTIPLICATION

A useful operation in which the entries in a row of one matrix are multipled by the corresponding entries in a column of a second matrix and the resulting products then added is called **matrix multiplication** and is defined as follows.

> **Matrix multiplication**
> If A is an $m \times n$ matrix and B is an $n \times p$ matrix, the **product** AB is the $m \times p$ matrix whose entry in row i and column j is obtained by multiplying each entry in row i of A by the corresponding entry in column j of B and then adding the resulting products.

Notice that in matrix multiplication, each entry in the product matrix is obtained by multiplying the entries in a row from the first matrix by the entries in a column from the second matrix and adding. The row and column that are used to compute a particular entry in the product matrix are indicated in Figure 2.1. An example follows.

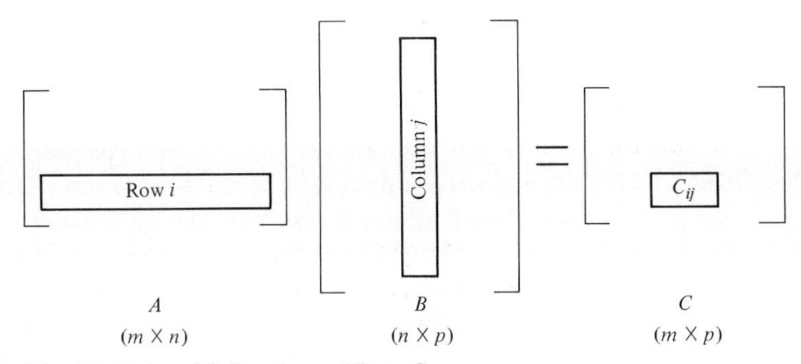

Figure 2.1 Matrix multiplication: $AB = C$.

EXAMPLE 2.1 ● Compute the product AB if

$$A = \begin{bmatrix} 3 & 1 \\ -1 & 2 \\ 1 & 5 \end{bmatrix} \quad \text{and} \quad B = \begin{bmatrix} -1 & -1 & 2 \\ 2 & -6 & 1 \end{bmatrix}$$

SOLUTION Since the dimensions of A are 3×2 and the dimensions of B are 2×3, the dimensions of the product AB will be 3×3.

To calculate the entry that goes in the first row and first column of the product matrix, we multiply each entry in the first row of A by the corresponding entry in the first column of B and add. We get $3(-1) + 1(2) = -1$ which we record in row 1, column 1 of the product matrix.

$$\begin{bmatrix} \boxed{3 \quad 1} \\ -1 \quad 2 \\ 1 \quad 5 \end{bmatrix} \begin{bmatrix} \boxed{\begin{matrix}-1 \\ 2\end{matrix}} & \begin{matrix}-1 \\ -6\end{matrix} & \begin{matrix}2 \\ 1\end{matrix} \end{bmatrix} = \begin{bmatrix} \boxed{-1} & & \\ & & \\ & & \end{bmatrix}$$

To calculate the entry that goes in the first row and second column of the product, we use the first row of A and the second column of B. We get $3(-1) + 1(-6) = -9$ which we record in row 1, column 2 of the product matrix.

$$\begin{bmatrix} \boxed{3 \quad 1} \\ -1 \quad 2 \\ 1 \quad 5 \end{bmatrix} \begin{bmatrix} -1 & \boxed{\begin{matrix} -1 \\ -6 \end{matrix}} & 2 \\ 2 & & 1 \end{bmatrix} = \begin{bmatrix} -1 & \boxed{-9} & \\ & & \\ & & \end{bmatrix}$$

Try to compute the remaining entries in the product matrix yourself. You should end up with the followng 3×3 matrix:

$$AB = \begin{bmatrix} -1 & -9 & 7 \\ 5 & -11 & 0 \\ 9 & -31 & 7 \end{bmatrix}$$

●

Algebraic properties of matrix multiplication

The order in which two matrices are multiplied can affect the answer. In Example 2.1, for instance, the product AB was a 3×3 matrix while the product BA of the same matrices in the opposite order is a 2×2 matrix. Even when both of the product matrices AB and BA have the same dimensions, they are rarely equal to each other.

One useful algebraic property of real numbers that is also valid for matrices is the **distributive property.**

The distributive property for matrices
If A and B are matrices,

$$A(B \pm C) = AB \pm AC \qquad \text{and} \qquad (A \pm B)C = AC \pm BC$$

whenever the indicated operations can be performed.

EXAMPLE 2.2 ● Compute $AC + BC$ if

$$A = \begin{bmatrix} 6 & -2 \\ 1 & 3 \\ 0 & 4 \end{bmatrix} \qquad B = \begin{bmatrix} 0 & 1 \\ 0 & 6 \\ 5 & -2 \end{bmatrix} \qquad C = \begin{bmatrix} 0 & 3 & 1 \\ -2 & 4 & -1 \end{bmatrix}$$

SOLUTION The most direct way to solve this problem is to compute the products AC and BC and then add. However, this involves the multiplication of two pairs of matrices and is slightly cumbersome.

A more efficient approach is to compute $(A + B)C$ which, according to the distributive property, is the same as $AC + BC$. (Notice that this approach involves the multiplication of only one pair of matrices.) Proceeding in this way we find that

$$A + B = \begin{bmatrix} 6 & -1 \\ 1 & 9 \\ 5 & 2 \end{bmatrix} \quad \text{and} \quad (A + B)C = \begin{bmatrix} 2 & 14 & 7 \\ -18 & 39 & -8 \\ -4 & 23 & 3 \end{bmatrix}$$

Check this answer by computing $AC + BC$ directly. ●

Identity matrix A square matrix of the form

$$I = \begin{bmatrix} 1 & 0 & \cdots & 0 \\ 0 & 1 & \cdots & 0 \\ \vdots & \vdots & & \vdots \\ 0 & 0 & \cdots & 1 \end{bmatrix}$$

whose entries are equal to one on the diagonal running from the upper left-hand corner to the lower right-hand corner and whose entries are zero off this diagonal is said to be an **identity matrix.** This is because multiplication of any matrix by the identity matrix with compatible dimensions leaves the matrix unchanged, just as multiplication of any real number by the multiplicative identity 1 leaves that number unchanged. (Do you see why?)

The identity property for matrix multiplication
For any matrix A,

$$AI = A \quad \text{and} \quad IA = A$$

where, in each case, I is the identity matrix of appropriate dimensions.

Practical application We conclude this section with a typical application of matrix multiplication.

EXAMPLE 2.3 ● A publisher with production facilities in New York and San Francisco is preparing for the publication of several new texts and novels. The numbers of books of each type that will be produced in each facility are summarized in matrix A.

$$A = \begin{matrix} \text{N.Y.} \\ \text{S.F.} \end{matrix} \overset{\begin{matrix} \text{Text} & \text{Novel} \end{matrix}}{\begin{bmatrix} 5 & 6 \\ 1 & 3 \end{bmatrix}}$$

Matrix B shows the numbers of hours of work by various specialists that go into the production of each type of book.

$$B = \begin{array}{l} \text{Text} \\ \text{Novel} \end{array} \overset{\begin{array}{ccc} \text{Production Editor} & \text{Designer} & \text{Copy Editor} \end{array}}{\begin{bmatrix} 50 & 8 & 30 \\ 30 & 2 & 20 \end{bmatrix}}$$

For each type of specialist, calculate the total number of hours that will be needed in each city to produce the books as planned and summarize your results in matrix form.

SOLUTION The information we are seeking can be summarized in a 2×3 matrix of the form

$$C = \begin{array}{l} \text{N.Y.} \\ \text{S.F.} \end{array} \overset{\begin{array}{ccc} \text{Production Editor} & \text{Designer} & \text{Copy Editor} \end{array}}{\begin{bmatrix} c_{11} & c_{12} & c_{13} \\ c_{21} & c_{22} & c_{23} \end{bmatrix}}$$

We compute c_{11} as follows.

$$\begin{aligned} c_{11} = \begin{array}{l} \text{production} \\ \text{editors' time} \\ \text{in New York} \end{array} &= \begin{pmatrix} \text{number of} \\ \text{texts in} \\ \text{New York} \end{pmatrix} \begin{pmatrix} \text{production} \\ \text{editors' time} \\ \text{per text} \end{pmatrix} \\ &\quad + \begin{pmatrix} \text{number of} \\ \text{novels} \\ \text{in New York} \end{pmatrix} \begin{pmatrix} \text{production} \\ \text{editors' time} \\ \text{per novel} \end{pmatrix} \\ &= 5(50) + 6(30) = 430 \text{ hours} \end{aligned}$$

Notice that the calculation of c_{11} is the same calculation we would perform to get the entry in row 1 and column 1 of the product matrix AB. Convince yourself that this correspondence is not accidental and that

$$C = AB = \begin{array}{l} \text{N.Y.} \\ \text{S.F.} \end{array} \overset{\begin{array}{ccc} \text{Production Editor} & \text{Designer} & \text{Copy Editor} \end{array}}{\begin{bmatrix} 430 & 52 & 270 \\ 140 & 14 & 90 \end{bmatrix}}$$

●

Look again at the matrices in Example 2.3. Do you see how the headings for the rows and columns of the product matrix AB are related to the headings for the original matrices A and B? Keep this relationship in mind. You will find it helpful whenever you have to interpret the information contained in a product matrix.

Problems

1 Suppose dim $(A) = 3 \times 2$, dim $(B) = 1 \times 4$, dim $(C) = 4 \times 3$, dim $(D) = 2 \times 1$, and dim $(E) = 2 \times 4$. In each of the following cases, determine whether the indicated matrix can be formed and if it can, state its dimensions.
(a) BC (b) BD (c) ADC
(d) ADB (e) $DB + 2E$ (f) A^2

In Problems 2 through 9, form the indicated matrix (when possible) if

$$A = \begin{bmatrix} -2 & 3 & 0 \\ -1 & 0 & 2 \\ 5 & -5 & 1 \end{bmatrix} \qquad B = \begin{bmatrix} 3 & 1 \\ -2 & 1 \\ 0 & -1 \end{bmatrix}$$

$$C = \begin{bmatrix} 1 & 0 & 6 & -3 \\ 4 & -2 & -1 & 2 \end{bmatrix} \qquad D = \begin{bmatrix} 1 & 0 \\ 0 & 1 \end{bmatrix} \qquad E = \begin{bmatrix} 2 & 3 & 5 \end{bmatrix}$$

2 AB **3** BD **4** EB **5** CB
6 A^2 **7** ABC **8** EBC **9** A^2BD

10 Let

$$A = \begin{bmatrix} 2 & 3 & 4 \\ -1 & 0 & 7 \\ 1 & -2 & 2 \end{bmatrix} \qquad B = \begin{bmatrix} 1 & 0 & 2 \\ 3 & 1 & 0 \\ 2 & -1 & 0 \end{bmatrix} \qquad C = \begin{bmatrix} 4 & 4 & 1 \\ 3 & 1 & 0 \\ -2 & 1 & -3 \end{bmatrix}$$

(a) Use the distributive property to compute $AB + AC$ efficiently.
(b) Check your answer in part (a) by computing $AB + AC$ directly.

11 (a) For the matrices in Problem 10, use the distributive property to compute $AB + CB$ efficiently.
(b) Check your answer in part (a) by computing $AB + CB$ directly.

12 Using the matrices

$$A = \begin{bmatrix} a_{11} & a_{12} \\ a_{21} & a_{22} \end{bmatrix} \qquad B = \begin{bmatrix} b_{11} & b_{12} \\ b_{21} & b_{22} \end{bmatrix} \qquad \text{and} \qquad C = \begin{bmatrix} c_{11} & c_{12} \\ c_{21} & c_{22} \end{bmatrix}$$

verify the following algebraic properties:
(a) Associative Property: $A(BC) = (AB)C$
(b) Distributive Property: $(A + B)C = AC + BC$
(c) Identity Property: $AI = IA = A$, where I is the 2×2 identity matrix.

In Problems 13 through 16, capital letters stand for matrices, lowercase letters denote scalars, and 0 stands for the zero matrix of appropriate dimensions. In each problem, decide if the given statement is true or false. If you decide the statement is true, explain why. If you decide it is false, justify your answer by giving an example. (Assume that all the indicated operations can be performed.)

13 If $AB = 0$, then $A = 0$ or $B = 0$
14 $A0 = 0A = 0$

15 If $AB = AC$ and $A \neq 0$, then $B = C$
16 $(cA)B = c(AB)$
17 Let

$$A = \begin{bmatrix} -3 & 2 & 1 \\ 5 & -1 & 0 \\ 4 & 1 & 2 \end{bmatrix} \qquad X = \begin{bmatrix} x_1 \\ x_2 \\ x_3 \end{bmatrix} \qquad \text{and} \qquad B = \begin{bmatrix} 1 \\ 0 \\ 3 \end{bmatrix}$$

The matrix equation $AX = B$ is actually a system of three linear equations in three unknowns in disguise. Write this system in its more familiar form. (Do not solve the system.)

18 Write the system

$x_1 - 3x_2 - 5x_3 + x_4 = 9$
$x_1 - x_3 + 2x_4 = 0$
$3x_2 - 2x_3 - 4x_4 = -5$
$-2x_1 + x_2 - 5x_3 - 9x_4 = 6$

as a matrix equation similar to the one in Problem 17. (Do not solve the system.)

19 A manufacturer of redwood furniture has two plants, one in South Carolina and one in New Jersey. Suppose that the matrix

$$\begin{array}{cc} & \text{Skilled} \quad \text{Unskilled} \\ A = \$ & [a_{11} \qquad a_{12}] \end{array}$$

shows the hourly wages of the manufacturer's employees, the matrix

$$B = \begin{array}{l} \text{Skilled} \\ \text{Unskilled} \end{array} \overset{\begin{array}{ccc}\text{Table} & \text{Chair} & \text{Desk}\end{array}}{\begin{bmatrix} b_{11} & b_{12} & b_{13} \\ b_{21} & b_{22} & b_{23} \end{bmatrix}}$$

gives the number of worker-hours of labor required for production, and the matrix

$$C = \begin{array}{r} \text{Table} \\ \text{Chair} \\ \text{Desk} \end{array} \overset{\begin{array}{cc}\text{S.C.} & \text{N.J.}\end{array}}{\begin{bmatrix} c_{11} & c_{12} \\ c_{21} & c_{22} \\ c_{31} & c_{32} \end{bmatrix}}$$

summarizes the orders received at each plant in June.

Describe the information contained in the following matrices:
(a) BC (b) AB (c) ABC

20 An airline has two flights daily from Chicago to Miami, one departing at 9:30 A.M. and the other at 8:35 P.M. The planes used on the morning flight can hold 25 first-class passengers, 60 tourist-class passengers, and 25 economy-class passengers. The planes used on the evening flight are smaller and have only 15 first-class, 40 tourist-class, and 20 economy-class seats. On weekends, the airline charges the full fare of $240 for first class, $200 for tourist class, and $180 for economy class. On weekdays, all fares are reduced by 10 percent.

(a) Display the information about the seating capacities of each flight in a 2×3 matrix. Display the information about fares in a 3×2 matrix. Construct the matrices so that their product will give the airline's receipts from ticket sales for each flight on weekends and weekdays when the planes are filled to capacity.

(b) Compute the product indicated in part (a) and label its rows and columns.

(c) Use the matrix you found in part (b) to decide which generates more revenue for the airline: a Tuesday morning flight booked to capacity or a Saturday evening flight also booked to capacity.

21 The prices (in dollars) of pizza, beer, and salad at three local pizza parlors are summarized in the matrix

$$A = \begin{array}{l} \\ \text{Pizza} \\ \text{Beer} \\ \text{Salad} \end{array} \begin{array}{c} \begin{array}{ccc} \text{Pizza King} & \text{Campus Pizza} & \text{Joe's} \end{array} \\ \begin{bmatrix} 2.65 & 2.80 & 2.75 \\ 0.65 & 0.55 & 0.60 \\ 0.50 & 0.55 & 0.45 \end{bmatrix} \end{array}$$

The orders from three campus groups are summarized in the matrix

$$B = \begin{array}{l} \\ \text{Pizza} \\ \text{Beer} \\ \text{Salad} \end{array} \begin{array}{c} \begin{array}{ccc} \text{Math club} & \text{Student senate} & \text{Rugby team} \end{array} \\ \begin{bmatrix} 2 & 5 & 6 \\ 6 & 20 & 30 \\ 3 & 10 & 1 \end{bmatrix} \end{array}$$

(a) Multiply two matrices (not necessarily A and B) to form a 3×3 matrix that shows the amount each group would be charged for its order at each pizza parlor.

(b) Decide at which pizza parlor each group should place its order.

22 Versatile matrix programs are available on most college and university computer systems. Find out how to use the program that is available at your school and use it to do Problems 2 through 9.

3 • SYSTEMS OF LINEAR EQUATIONS

An equation in n variables $x_1, x_2, \ldots, x_n$ is said to be **linear** if it is of the form

$$a_1x_1 + a_2x_2 + \cdots + a_nx_n = c$$

where $a_1, a_2, \ldots, a_n$, and c are constants. For example, the equation $3x_1 + \frac{2}{3}x_2 - 5x_4 = -2$ is a linear equation, but the equations $2x_3 - x_2^2 + x_1 = 5$, $x_1 - 2x_2x_3 = 6$, and $x_1 + 3/x_2 - 7x_3 = 0$ are not. (Do you see why?) The word linear is used in this context because the graph of a linear equation $a_1x_1 + a_2x_2 = c$ in two variables is a straight line in the x_1x_2 plane.

Two or more linear equations that are to be solved simultaneously are said to form a **system of linear equations** or, simply, a **linear system.** Here is an example to illustrate how a linear system might arise in practice.

EXAMPLE 3.1 ● A nursery stocks three brands of grass seed whose ingredients are summarized in the table

	Ryegrass	Fescue	Bluegrass
Brand P	2	2	6
Brand Q	4	2	4
Brand R	0	6	4

where each entry represents kilograms per bag. The owner of the nursery wishes to determine how many bags of each brand to combine to satisfy a customer who wants a mixture containing 30 kilograms of ryegrass seed, 30 kilograms of fescue seed, and 50 kilograms of bluegrass seed. Formulate this problem in mathematical terms.

SOLUTION We let

$$x_1 = \text{number of bags of brand P in the mixture}$$
$$x_2 = \text{number of bags of brand Q in the mixture}$$
$$x_3 = \text{number of bags of brand R in the mixture}$$

The total number of kilograms of ryegrass seed in the mixture is $2x_1 + 4x_2$. To meet the customer's specifications, this must be equal to 30. Hence the variables must satisfy the linear equation

$$2x_1 + 4x_2 = 30$$

Similarly, the requirements that the mixture should contain 30 kilograms of fescue seed and 50 kilograms of bluegrass seed lead to the two additional equations

$$2x_1 + 2x_2 + 6x_3 = 30$$
$$6x_1 + 4x_2 + 4x_3 = 50$$

To determine the number of bags of each brand that should be used, we must find values of x_1, x_2, and x_3 that satisfy all three of these linear equations. ●

The augmented matrix Linear systems can be written compactly in matrix form. For instance, the system

$$\begin{aligned} 2x_1 + 4x_2 \phantom{{}+6x_3} &= 30 \\ 2x_1 + 2x_2 + 6x_3 &= 30 \\ 6x_1 + 4x_2 + 4x_3 &= 50 \end{aligned}$$

from Example 3.1 can be represented by the 3×4 matrix

$$\left[\begin{array}{ccc|c} 2 & 4 & 0 & 30 \\ 2 & 2 & 6 & 30 \\ 6 & 4 & 4 & 50 \end{array}\right]$$

in which the rows represent the three equations and the first three columns contain the coefficients of the three variables. The last column contains the constants that appear on the right-hand sides of the equations, and a vertical line appears in place of the missing equal signs. This matrix is sometimes called the **augmented matrix** for the linear system because it consists of the matrix of the coefficients augmented by an extra column representing the constants that appear on the right-hand sides of the equations.

Matrix reduction There is a systematic procedure for solving systems of linear equations that are represented in matrix form. The goal (for a system of three equations in three variables) is to reduce the augmented matrix to one of the form

$$\left[\begin{array}{ccc|c} 1 & 0 & 0 & a \\ 0 & 1 & 0 & b \\ 0 & 0 & 1 & c \end{array}\right]$$

from which we can read off the solution

$$x_1 = a \qquad x_2 = b \qquad x_3 = c$$

The operations we use to perform the reduction are the division of each entry in a row by a constant (corresponding to the division of each side of an equation by a constant), the addition of a multiple of one row to another (corresponding to the addition of a multiple of one equation to another), and the interchanging of two rows (corresponding to the interchanging of two equations). In the next example, we use this procedure to solve the linear system from Example 3.1.

EXAMPLE 3.2 ● Solve the following system of linear equations.

$$\begin{aligned} 2x_1 + 4x_2 &= 30 \\ 2x_1 + 2x_2 + 6x_3 &= 30 \\ 6x_1 + 4x_2 + 4x_3 &= 50 \end{aligned}$$

SOLUTION We begin with the augmented matrix

$$\left[\begin{array}{ccc|c} 2 & 4 & 0 & 30 \\ 2 & 2 & 6 & 30 \\ 6 & 4 & 4 & 50 \end{array}\right]$$

The first step is to transform the matrix into one in which the only nonzero entry in the first column is a 1 occurring in the first row. To do this, we divide the first row by 2 and then add -2 times the result to the second row and -6 times the result to the third row. We get

$$\left[\begin{array}{ccc|c} 1 & 2 & 0 & 15 \\ 2 & 2 & 6 & 30 \\ 6 & 4 & 4 & 50 \end{array}\right] \; R_1 \div 2 \qquad \left[\begin{array}{ccc|c} 1 & 2 & 0 & 15 \\ 0 & -2 & 6 & 0 \\ 0 & -8 & 4 & -40 \end{array}\right] \; \begin{array}{c} \\ -2R_1 + R_2 \\ -6R_1 + R_3 \end{array}$$

The next step is to transform the new matrix into one in which the only nonzero entry in the second column is a 1 in the second row. We divide the second row by -2 and add -2 times the result to the first row and 8 times the result to the third to get

$$\left[\begin{array}{ccc|c} 1 & 2 & 0 & 15 \\ 0 & 1 & -3 & 0 \\ 0 & -8 & 4 & -40 \end{array}\right] \; R_2 \div -2 \qquad \left[\begin{array}{ccc|c} 1 & 0 & 6 & 15 \\ 0 & 1 & -3 & 0 \\ 0 & 0 & -20 & -40 \end{array}\right] \; \begin{array}{c} -2R_2 + R_1 \\ \\ 8R_2 + R_3 \end{array}$$

Finally, we divide the third row by -20 and add appropriate multiples of the result to the first and second rows to get the reduced matrix

$$\left[\begin{array}{ccc|c} 1 & 0 & 6 & 15 \\ 0 & 1 & -3 & 0 \\ 0 & 0 & 1 & 2 \end{array}\right] \; R_3 \div -20 \qquad \left[\begin{array}{ccc|c} 1 & 0 & 0 & 3 \\ 0 & 1 & 0 & 6 \\ 0 & 0 & 1 & 2 \end{array}\right] \; \begin{array}{c} -6R_3 + R_1 \\ 3R_3 + R_2 \\ \end{array}$$

From the first row of the reduced matrix we see that $x_1 = 3$, from the second row we see that $x_2 = 6$, and from the third that $x_3 = 2$.

In terms of the practical problem in Example 3.1, we conclude that the nursery should combine 3 bags of brand P, 6 bags of brand Q, and 2 bags of brand R to create the mixture requested by the customer. ●

Systems without solutions Some systems of linear equations have no solutions at all. The next example illustrates what happens when the reduction procedure is applied to such a system.

EXAMPLE 3.3 ● Solve the following system of linear equations.

$$x_1 + 2x_2 = 3$$
$$2x_1 + 6x_2 = 2$$
$$3x_1 + 5x_2 = 0$$

SOLUTION Using the reduction procedure, we obtain

$$\left[\begin{array}{cc|c} 1 & 2 & 3 \\ 2 & 6 & 2 \\ 3 & 5 & 0 \end{array}\right] \qquad \left[\begin{array}{cc|c} 1 & 2 & 3 \\ 0 & 2 & -4 \\ 0 & -1 & -9 \end{array}\right] \quad \begin{array}{l} \\ -2R_1 + R_2 \\ -3R_1 + R_3 \end{array}$$

followed by

$$\left[\begin{array}{cc|c} 1 & 2 & 3 \\ 0 & 1 & -2 \\ 0 & -1 & -9 \end{array}\right] \quad R_2 \div 2 \qquad \left[\begin{array}{cc|c} 1 & 0 & 7 \\ 0 & 1 & -2 \\ 0 & 0 & -11 \end{array}\right] \quad \begin{array}{r} -2R_2 + R_1 \\ \\ R_2 + R_3 \end{array}$$

No further reduction is possible. To interpret the result, we translate the reduced matrix back into algebraic terms. The three rows give us the three equations

$$x_1 = 7 \qquad x_2 = -2 \qquad 0 = -11$$

Since the last of these equations is false, we conclude that the original system has no solution. That is, there are no numbers x_1 and x_2 that satisfy simultaneously all three of the original equations. ●

Systems with infinitely many solutions The linear system in Example 3.2 had a unique solution and the system in Example 3.3 had no solution. A third possibility is illustrated in the next example.

EXAMPLE 3.4 ● Solve the following system of linear equations.

$$x_2 - 2x_3 = 1$$
$$x_1 - x_2 + 2x_3 = 2$$
$$-3x_1 + 5x_2 - 10x_3 = -4$$

SOLUTION The augmented matrix is

$$\left[\begin{array}{rrr|r} 0 & 1 & -2 & 1 \\ 1 & -1 & 2 & 2 \\ -3 & 5 & -10 & -4 \end{array}\right]$$

We begin the reduction by interchanging rows 1 and 2 so that the entry in row 1, column 1 is nonzero. (In algebraic terms, this operation simply interchanges two of the original equations.) We get

$$\left[\begin{array}{rrr|r} 1 & -1 & 2 & 2 \\ 0 & 1 & -2 & 1 \\ -3 & 5 & -10 & -4 \end{array}\right] \begin{array}{l} R_2 \\ R_1 \\ \end{array}$$

Proceeding with the reduction we get

$$\left[\begin{array}{rrr|r} 1 & -1 & 2 & 2 \\ 0 & 1 & -2 & 1 \\ 0 & 2 & -4 & 2 \end{array}\right] \begin{array}{l} \\ \\ 3R_1 + R_3 \end{array} \qquad \left[\begin{array}{rrr|r} 1 & 0 & 0 & 3 \\ 0 & 1 & -2 & 1 \\ 0 & 0 & 0 & 0 \end{array}\right] \begin{array}{l} R_2 + R_1 \\ \\ -2R_2 + R_3 \end{array}$$

Since the entry in row 3, column 3 is zero, no further reduction is possible. Translating the reduced matrix back into algebraic terms we get the reduced linear system

$$x_1 = 3 \qquad x_2 - 2x_3 = 1 \qquad 0 = 0$$

The third equation tells us nothing new. The first tells us that $x_1 = 3$. The only restriction on x_2 and x_3 is contained in the second equation. We conclude that any pair of numbers x_2 and x_3 that satisfy this equation can be used with $x_1 = 3$ to form a solution of the original system.

To state our conclusion in a more convenient way, we solve the second equation for x_2 getting $x_2 = 1 + 2x_3$ and then write the solution as

$$x_1 = 3 \qquad x_3 = \text{anything} \qquad x_2 = 1 + 2x_3$$

That is, the system has infinitely many solutions. To form one, we must take $x_1 = 3$. We can choose any value for x_3, and once this value is chosen, we use the equation $x_2 = 1 + 2x_3$ to find the corresponding value of x_2. ●

Geometric interpretation The preceding examples suggest that a system of linear equations may have a unique solution, no solution at all, or infinitely many solutions. It can be shown that these are the only possibilities. In the case of a system of two linear equations in two variables, this corresponds to the geometric fact (illustrated in

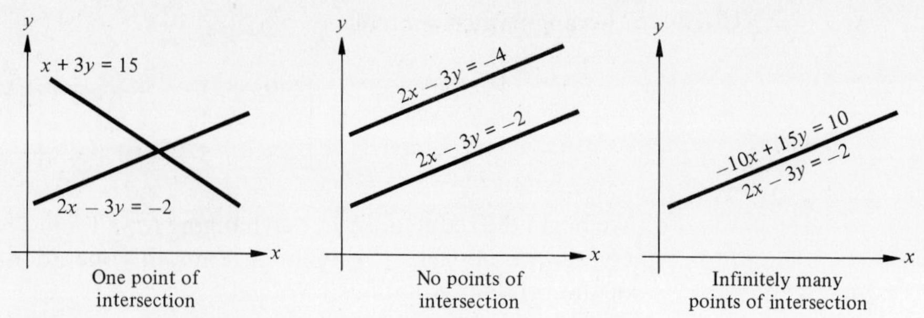

Figure 3.1 Possibilities for the intersection of two lines.

Figure 3.1) that the graphs of two linear equations may intersect at a single point, may be parallel lines and not intersect at all, or may actually be the same line and so intersect at infinitely many points.

Problems

Each of the matrices in Problems 1 through 6 is the reduced augmented matrix for a system of linear equations in the variables $x_1, x_2, x_3 \ldots$. In each case, interpret the information contained in the matrix to obtain the solution (if any) of the linear system.

1 $\left[\begin{array}{ccc|c} 1 & 0 & 0 & 0 \\ 0 & 1 & 0 & -1 \\ 0 & 0 & 1 & 3 \end{array}\right]$

2 $\left[\begin{array}{ccc|c} 1 & 0 & 0 & 2 \\ 0 & 0 & 1 & 5 \\ 0 & 0 & 0 & 4 \end{array}\right]$

3 $\left[\begin{array}{ccc|c} 1 & 2 & 0 & 3 \\ 0 & 0 & 1 & 3 \\ 0 & 0 & 0 & 0 \end{array}\right]$

4 $\left[\begin{array}{ccc|c} 1 & 0 & 1 & -2 \\ 0 & 1 & 0 & 5 \\ 0 & 0 & 0 & 0 \end{array}\right]$

5 $\left[\begin{array}{cccc|c} 0 & 1 & 0 & 0 & 7 \\ 0 & 0 & 1 & 0 & 9 \\ 0 & 0 & 0 & 1 & -8 \\ 0 & 0 & 0 & 0 & 0 \end{array}\right]$

6 $\left[\begin{array}{cccc|c} 1 & 0 & 0 & 0 & 5 \\ 0 & 1 & 4 & 0 & -2 \\ 0 & 0 & 0 & 1 & 0 \\ 0 & 0 & 0 & 0 & 0 \end{array}\right]$

In Problems 7 through 19, solve the given system of linear equations.

7 $x_1 - 2x_2 = -6$
$5x_1 + x_2 = 3$

8 $x_1 - 2x_2 = -3$
$3x_1 + x_2 = 5$

9 $5x_1 - 2x_2 = 1$
$10x_1 - 4x_2 = -1$

10 $x_1 - 3x_2 = 0$
$3x_1 + x_2 = 0$

11 $-2x_1 + 6x_2 = 20$
$3x_1 - 9x_2 = -30$

12 $x_1 + 3x_2 = 5$
$-2x_1 - 4x_2 = 6$
$3x_1 + 9x_2 = 15$

13 $3x_1 - 6x_2 + 9x_3 = 15$
$-2x_1 + 4x_2 - 6x_3 = -10$

14 $x_1 - 3x_2 = 7$
$2x_1 - x_2 = 4$
$-3x_1 + 10x_2 = -5$

15 $-x_1 + 2x_2 + x_3 = 14$
$2x_1 - 3x_2 + 4x_3 = -1$
$-2x_1 + 4x_2 - 5x_3 = 0$

16 $x_1 - 2x_2 + 3x_3 = 2$
$-3x_1 + 8x_2 - 9x_3 = -6$
$-x_1 + 2x_2 - 4x_3 = -2$

17 $x_1 + 2x_2 - 2x_3 = 1$
$2x_1 + 5x_2 - 4x_3 = 2$
$-x_1 + x_2 + 2x_3 = -1$

18 $x_1 - 2x_2 = -1$
$3x_2 - 6x_3 = 2$
$3x_1 - 7x_2 + 2x_3 = 4$

19 $x_1 + 2x_2 - x_3 + 3x_4 = 1$
$-2x_1 - 3x_2 + 4x_3 - 5x_4 = 3$
$4x_1 + 8x_2 - 5x_3 + 9x_4 = 2$
$-x_1 - 4x_2 - 3x_4 = -5$

20 Shipments from one wholesaler contain 10 units of item A and 36 units of item B, while shipments from a second wholesaler contain 30 units of A and 12 units of B. How many shipments should a retailer order from each wholesaler to get exactly 120 units of A and 240 units of B?

21 Ten years ago a father was six times as old as his son. Ten years from now he will be twice as old as his son. How old are they now?

22 A retailer carries two brands of coffee. The first brand is a blend of 20 percent Brazilian coffee and 80 percent Colombian coffee. The second brand is a blend of 60 percent Brazilian and 40 percent Colombian. How many kilograms of each brand should be combined to produce a blend containing 100 kilograms of Brazilian and 100 kilograms of Colombian coffee?

23 A dietician is planning a meal consisting of three foods. A 30-gram serving of the first food contains 5 units of protein, 2 units of carbohydrates, and 3 units of iron. A 30-gram serving of the second food contains 10 units of protein, 3 units of carbohydrates, and 6 units of iron. And a 30-gram serving of the third food contains 15 units of protein, 2 units of carbohydrates, and 1 unit of iron. How many grams of each food should be used to create a meal containing 55 units of protein, 13 units of carbohydrates, and 17 units of iron?

24 Sets of a certain construction toy come in three sizes: A standard set contains 20 rods, 5 wheels, and 30 blocks; a jumbo set contains 40 rods, 15 wheels, and 40 blocks; and a giant set contains 60 rods, 25 wheels, and 60 blocks. At the end of a production run, 240 rods, 85 wheels, and 290 blocks are left over. The manufacturer would like to use these pieces to form complete sets. Is this possible?

25 An investor receives a total of \$5,700 per annum in interest from three stocks yielding 4, 5, and 8 percent per annum, respectively. The amount invested at 4 percent is \$20,000 more than the amount invested at 5 percent, and the interest from the 8 percent investment is eight times the interest from the 5 percent investment. How much money is invested in each stock?

26 A nursery stocks three brands of lawn preparations. The ingredients of each brand are listed in the following table:

	Fertilizer	Weed killer	Insecticide
Brand A	30%	20%	50%
Brand B	60%	30%	10%
Brand C	90%	10%	0%

How many kilograms of each brand should be used to produce a 100-kilogram mixture containing 19 kilograms of weed killer and twice as much fertilizer as insecticide?

27 Versatile programs for solving systems of linear equations are available on most college and university computer systems. Find out how to use the program that is available at your school and use it to solve some of the linear systems in Problems 7 through 19.

4 • MATRIX INVERSION

A square matrix A is said to be **invertible** if there is a matrix B with the property that

$$AB = I \qquad \text{and} \qquad BA = I$$

where I is the identity matrix. A matrix B with this property is said to be an **inverse** of A.

EXAMPLE 4.1 ● If

$$A = \begin{bmatrix} 1 & 2 \\ 3 & 4 \end{bmatrix} \qquad \text{and} \qquad B = \begin{bmatrix} -2 & 1 \\ \frac{3}{2} & -\frac{1}{2} \end{bmatrix}$$

show that B is an inverse of A.

SOLUTION We compute the products AB and BA and find that

$$AB = \begin{bmatrix} 1 & 2 \\ 3 & 4 \end{bmatrix} \begin{bmatrix} -2 & 1 \\ \frac{3}{2} & -\frac{1}{2} \end{bmatrix} = \begin{bmatrix} 1 & 0 \\ 0 & 1 \end{bmatrix}$$

and

$$BA = \begin{bmatrix} -2 & 1 \\ \frac{3}{2} & -\frac{1}{2} \end{bmatrix} \begin{bmatrix} 1 & 2 \\ 3 & 4 \end{bmatrix} = \begin{bmatrix} 1 & 0 \\ 0 & 1 \end{bmatrix}$$

as required. ●

Actually, we did not have to verify both of the equations $AB = I$ and $BA = I$ in Example 4.1. It can be shown in general that if one holds, the other holds automatically.

It is not hard to show that each invertible matrix has exactly one inverse. If A is invertible, it is customary to use the symbol A^{-1} (read "A inverse") to denote the inverse of A.

A procedure for inverting matrices

Here is a systematic procedure for inverting matrices that is closely related to the procedure for solving systems of linear equations.

How to find the inverse of a matrix
To find the inverse of an invertible $n \times n$ matrix A, form the augmented matrix $[A \mid I]$ by adjoining the $n \times n$ identity matrix to A. Then apply the reduction procedure from Section 3. When the matrix to the left of the vertical line has been reduced to the identity I, you will have a matrix $[I \mid A^{-1}]$ in which the $n \times n$ matrix to the right of the vertical line is A^{-1}.

Let us apply this procedure to the matrix we considered in Example 4.1.

EXAMPLE 4.2 ● Find A^{-1} if

$$A = \begin{bmatrix} 1 & 2 \\ 3 & 4 \end{bmatrix}$$

SOLUTION Applying the reduction procedure to the augmented matrix $[A \mid I]$ we get

$$\left[\begin{array}{cc|cc} 1 & 2 & 1 & 0 \\ 3 & 4 & 0 & 1 \end{array}\right] \qquad \left[\begin{array}{cc|cc} 1 & 2 & 1 & 0 \\ 0 & -2 & -3 & 1 \end{array}\right] \quad -3R_1 + R_2$$

followed by

$$\left[\begin{array}{cc|cc} 1 & 2 & 1 & 0 \\ 0 & 1 & \frac{3}{2} & -\frac{1}{2} \end{array}\right] \quad R_2 \div -2 \qquad \left[\begin{array}{cc|cc} 1 & 0 & -2 & 1 \\ 0 & 1 & \frac{3}{2} & -\frac{1}{2} \end{array}\right] \quad -2R_2 + R_1$$

and conclude, as expected, that

$$A^{-1} = \begin{bmatrix} -2 & 1 \\ \frac{3}{2} & -\frac{1}{2} \end{bmatrix}$$

●

It is not hard to see why the procedure illustrated in Example 4.2 works. To invert the matrix A, we had to find a 2×2 matrix B for which $AB = I$. That is, we had to find numbers b_{11}, b_{12}, b_{21}, and b_{22} for which

$$\begin{bmatrix} 1 & 2 \\ 3 & 4 \end{bmatrix} \begin{bmatrix} b_{11} & b_{12} \\ b_{21} & b_{22} \end{bmatrix} = \begin{bmatrix} 1 & 0 \\ 0 & 1 \end{bmatrix}$$

When the product on the left-hand side is multiplied out, this matrix equation becomes

$$\begin{bmatrix} b_{11} + 2b_{21} & b_{12} + 2b_{22} \\ 3b_{11} + 4b_{21} & 3b_{12} + 4b_{22} \end{bmatrix} = \begin{bmatrix} 1 & 0 \\ 0 & 1 \end{bmatrix}$$

which is equivalent to the pair of linear systems

$$\begin{aligned} b_{11} + 2b_{21} &= 1 \\ 3b_{11} + 4b_{21} &= 0 \end{aligned} \qquad \text{and} \qquad \begin{aligned} b_{12} + 2b_{22} &= 0 \\ 3b_{12} + 4b_{22} &= 1 \end{aligned}$$

Reducing the augmented matrix

$$\left[\begin{array}{cc|cc} 1 & 2 & 1 & 0 \\ 3 & 4 & 0 & 1 \end{array}\right]$$

is just a compact way of solving these two linear systems at the same time.

Noninvertible matrices Not every square matrix is invertible. If the procedure for matrix inversion is applied to a matrix that is not invertible, the reduced matrix to the left of the vertical line will not be equal to the identity matrix but instead will contain at least one row of zeros. (Think of the connection between matrix inversion and linear systems and see if you can explain why this happens.) In the next example, we use this fact to determine whether a given matrix has an inverse.

EXAMPLE 4.3 ● Decide if the matrix

$$A = \begin{bmatrix} 1 & 2 \\ 3 & 6 \end{bmatrix}$$

is invertible and if so, find A^{-1}.

SOLUTION We reduce the augmented matrix

$$\left[\begin{array}{cc|cc} 1 & 2 & 1 & 0 \\ 3 & 6 & 0 & 1 \end{array}\right]$$

and get

$$\left[\begin{array}{cc|cc} 1 & 2 & 1 & 0 \\ 0 & 0 & -3 & 1 \end{array}\right]$$

Since the reduced matrix to the left of the vertical line contains a row of zeros, we conclude that A is not invertible. ●

The solution of linear systems Matrix inversion can be used to solve certain systems of linear equations in which the number of variables is equal to the number of equations. For example, a system of three linear equations in three variables

$$\begin{aligned} a_{11}x_1 + a_{12}x_2 + a_{13}x_3 &= c_1 \\ a_{21}x_1 + a_{22}x_2 + a_{23}x_3 &= c_2 \\ a_{31}x_1 + a_{32}x_2 + a_{33}x_3 &= c_3 \end{aligned}$$

can be rewritten as the matrix equation

$$AX = C$$

where

$$A = \begin{bmatrix} a_{11} & a_{12} & a_{13} \\ a_{21} & a_{22} & a_{23} \\ a_{31} & a_{32} & a_{33} \end{bmatrix} \qquad X = \begin{bmatrix} x_1 \\ x_2 \\ x_3 \end{bmatrix} \qquad \text{and} \qquad C = \begin{bmatrix} c_1 \\ c_2 \\ c_3 \end{bmatrix}$$

(To see this, compute the product AX and set it equal to C. The result will be a matrix equation that is clearly equivalent to the original linear system.)

If A is invertible, we can solve the equation $AX = C$ by multiplying both sides by A^{-1} to get

$$A^{-1}AX = A^{-1}C \qquad \text{or} \qquad X = A^{-1}C$$

This technique for solving linear systems has obvious disadvantages. It applies only to linear systems in which the number of variables is equal to the number of equations. It will fail if the coefficient matrix is not invertible. And, the work required to invert the coefficient matrix is even more time-consuming than the work required to solve the system directly. There are, however, some situations in which this method offers more versatility than the direct approach. One such situation is illustrated in the next example.

EXAMPLE 4.4 ● A nursery stocks three brands of grass seed whose ingredients are summarized in the following table:

	Ryegrass	Fescue	Bluegrass
Brand P	2	2	6
Brand Q	4	2	4
Brand R	0	6	4

where each entry represents kilograms per bag. From time to time, a customer will ask the owner of the nursery to mix these brands to create a blend containing specified amounts of each type of seed. Develop a simple procedure that the owner can use to determine the number of bags of each brand to combine to satisfy such a customer.

SOLUTION Suppose a customer requests a mixture containing c_1 kilograms of ryegrass seed, c_2 kilograms of fescue seed, and c_3 kilograms of bluegrass seed. If we let x_1, x_2, and x_3 denote the number of bags of brands P, Q, and R, respectively, in the mixture, the customer's specifications for ryegrass, fescue, and bluegrass, respectively, lead to the three equations

$$2x_1 + 4x_2 = c_1$$
$$2x_1 + 2x_2 + 6x_3 = c_2$$
$$6x_1 + 4x_2 + 4x_3 = c_3$$

The corresponding matrix equation is $AX = C$ where

$$A = \begin{bmatrix} 2 & 4 & 0 \\ 2 & 2 & 6 \\ 6 & 4 & 4 \end{bmatrix} \qquad X = \begin{bmatrix} x_1 \\ x_2 \\ x_3 \end{bmatrix} \qquad \text{and} \qquad C = \begin{bmatrix} c_1 \\ c_2 \\ c_3 \end{bmatrix}$$

Its solution, if A is invertible, is $X = A^{-1}C$.

Applying the procedure for matrix inversion to A, we find that

$$A^{-1} = \begin{bmatrix} -\frac{1}{5} & -\frac{1}{5} & \frac{3}{10} \\ \frac{7}{20} & \frac{1}{10} & -\frac{3}{20} \\ -\frac{1}{20} & \frac{1}{5} & -\frac{1}{20} \end{bmatrix}$$

Hence, we can rewrite the matrix equation $X = A^{-1}C$ as

$$\begin{bmatrix} x_1 \\ x_2 \\ x_3 \end{bmatrix} = \begin{bmatrix} -\frac{1}{5} & -\frac{1}{5} & \frac{3}{10} \\ \frac{7}{20} & \frac{1}{10} & -\frac{3}{20} \\ -\frac{1}{20} & \frac{1}{5} & -\frac{1}{20} \end{bmatrix} \begin{bmatrix} c_1 \\ c_2 \\ c_3 \end{bmatrix}$$

This describes a simple procedure involving nothing more than matrix multiplication by which the owner of the nursery can now determine the number of bags of each brand to use to satisfy any customer's request for c_1 kilograms of ryegrass, c_2 kilograms of fescue, and c_3 kilograms of bluegrass.

For example, for a mixture containing 30 kilograms of ryegrass, 30 of fescue, and 50 of bluegrass, the owner would compute the product

$$\begin{bmatrix} -\frac{1}{5} & -\frac{1}{5} & \frac{3}{10} \\ \frac{7}{20} & \frac{1}{10} & -\frac{3}{20} \\ -\frac{1}{20} & \frac{1}{5} & -\frac{1}{20} \end{bmatrix} \begin{bmatrix} 30 \\ 30 \\ 50 \end{bmatrix} = \begin{bmatrix} 3 \\ 6 \\ 2 \end{bmatrix}$$

and would conclude that 3 bags of brand P, 6 of brand Q, and 2 of brand R are needed. ●

Problems In Problems 1 through 4, decide if the given matrices are inverses of one another.

1 $\begin{bmatrix} 1 & 3 \\ 2 & 5 \end{bmatrix}$ and $\begin{bmatrix} -5 & 3 \\ 2 & -1 \end{bmatrix}$

2 $\begin{bmatrix} 1 & -1 \\ 2 & 0 \end{bmatrix}$ and $\begin{bmatrix} 0 & -\frac{1}{2} \\ 1 & \frac{1}{2} \end{bmatrix}$

3 $\begin{bmatrix} -1 & 2 & 3 \\ 2 & 0 & -5 \\ 1 & -3 & 4 \end{bmatrix}$ and $\begin{bmatrix} 5 & 2 & 3 \\ 0 & 1 & 0 \\ 2 & 0 & 1 \end{bmatrix}$

4 $\begin{bmatrix} 1 & 0 & 2 \\ 0 & 1 & 3 \\ -2 & 2 & 0 \end{bmatrix}$ and $\begin{bmatrix} 3 & -2 & 1 \\ 3 & -2 & \frac{3}{2} \\ -1 & 1 & -\frac{1}{2} \end{bmatrix}$

In Problems 5 through 12, decide if the given matrix is invertible, and if it is, find its inverse.

5 $\begin{bmatrix} -1 & 2 \\ 5 & -9 \end{bmatrix}$

6 $\begin{bmatrix} 2 & 6 \\ -1 & 0 \end{bmatrix}$

7 $\begin{bmatrix} -2 & 6 \\ 3 & -9 \end{bmatrix}$

8 $\begin{bmatrix} 2 & 0 \\ 0 & 2 \end{bmatrix}$

9 $\begin{bmatrix} 1 & 3 & -1 \\ 2 & 8 & -6 \\ -3 & -9 & 1 \end{bmatrix}$

10 $\begin{bmatrix} 1 & -2 & 0 \\ 0 & 3 & -6 \\ 3 & -7 & 2 \end{bmatrix}$

11 $\begin{bmatrix} 1 & -2 & 3 \\ -3 & 8 & -9 \\ -1 & 2 & -4 \end{bmatrix}$

12 $\begin{bmatrix} 1 & 2 & -1 & 3 \\ -2 & -3 & 4 & -5 \\ 4 & 8 & -5 & 9 \\ -1 & -4 & 0 & 5 \end{bmatrix}$

In Problems 13 through 18, use matrix inversion to solve the given linear system.

13 $\begin{aligned} x_1 + 5x_2 &= 2 \\ -3x_1 - 14x_2 &= -1 \end{aligned}$

14 $\begin{aligned} 2x_1 + 4x_2 &= 1 \\ -2x_1 - 3x_2 &= -5 \end{aligned}$

15 $\begin{aligned} x_1 - x_2 + 2x_3 &= 2 \\ -4x_1 + 5x_2 - 5x_3 &= 3 \\ 2x_1 + 3x_2 - 6x_3 &= 9 \end{aligned}$

16 $\begin{aligned} x_1 - 3x_2 + 2x_3 &= 0 \\ -2x_1 + 6x_2 - 5x_3 &= -4 \\ -x_1 + 4x_2 + 4x_3 &= 1 \end{aligned}$

17 $\begin{aligned} x_1 - 2x_2 + 2x_3 &= 4 \\ -3x_1 + 4x_2 - 4x_3 &= -8 \\ 5x_1 - 7x_2 + 9x_3 &= 6 \end{aligned}$

18 $\begin{aligned} x_1 + x_4 &= 3 \\ x_3 + x_4 &= 0 \\ x_2 + x_4 &= 5 \\ x_1 + x_2 &= 4 \end{aligned}$

In Problems 19 through 22, find the matrix X for which $AX = B$.

19 $A = \begin{bmatrix} 1 & 2 \\ 0 & 7 \end{bmatrix}$ and $B = \begin{bmatrix} -2 & 1 \\ 0 & -4 \end{bmatrix}$

20 $A = \begin{bmatrix} 2 & 0 \\ 0 & 2 \end{bmatrix}$ and $B = \begin{bmatrix} -8 & 10 \\ 0 & 6 \end{bmatrix}$

21 $A = \begin{bmatrix} 1 & 0 & 6 \\ 2 & 1 & 10 \\ -1 & 3 & -11 \end{bmatrix}$ and $B = \begin{bmatrix} 1 & 1 & 1 \\ 1 & 1 & 1 \\ 1 & 1 & 1 \end{bmatrix}$

22 $A = \begin{bmatrix} 1 & 0 & -2 \\ 0 & 1 & 3 \\ 0 & 2 & 8 \end{bmatrix}$ and $B = \begin{bmatrix} 4 & 1 & 8 \\ 5 & 9 & 1 \\ 7 & 1 & 1 \end{bmatrix}$

23 A nursery stocks three brands of lawn preparations whose ingredients are summarized in the following table:

	Fertilizer	Weed killer	Insecticide
Brand I	3	2	5
Brand II	6	3	1
Brand III	9	1	0

where each entry represents kilograms per bag. From time to time, a customer will ask for a mixture containing specified amounts of each ingredient.

(a) Develop a simple procedure involving nothing more than matrix multiplication that the owner of the nursery can use to determine the number of bags of each brand to combine to satisfy such a customer.

(b) Suppose a customer requests a mixture containing 36 kilograms of fertilizer, 17 of weed killer, and 27 of insecticide. Use the procedure developed in part (a) to determine how many bags of each brand the owner should combine to satisfy the customer.

24 A retailer carries two brands of coffee. The first brand is a blend of 20 percent Brazilian coffee and 80 percent Colombian coffee. The second brand is a blend of 60 percent Brazilian and 40 percent Colombian. From time to time, a customer will ask the retailer to mix these brands to create a blend containing specified amounts of Brazilian and Colombian coffees.

(a) Develop a simple procedure involving nothing more than matrix multiplication that the retailer can use to determine the number of kilograms of each brand to combine to satisfy such a customer.

(b) Suppose a customer requests a mixture containing 8 kilograms of Brazilian and 12 of Colombian coffee. How many kilograms of each brand should the retailer combine to satisfy the customer?

25 A furniture manufacturer has two plants, one in North Carolina and one in South Dakota. Suppose that the matrix

$$A = \begin{array}{l} \text{Skilled} \\ \text{Unskilled} \end{array} \overset{\begin{array}{cc}\text{Table} & \text{Chair}\end{array}}{\begin{bmatrix} a_{11} & a_{12} \\ a_{21} & a_{22} \end{bmatrix}}$$

gives the number of worker-hours of labor required for production, and the matrix

$$B = \begin{array}{l} \text{Skilled} \\ \text{Unskilled} \end{array} \overset{\begin{array}{cc}\text{N.C.} & \text{S.D.}\end{array}}{\begin{bmatrix} b_{11} & b_{12} \\ b_{21} & b_{22} \end{bmatrix}}$$

summarizes the total number of worker-hours of labor used during the month of July. Describe the information contained in the matrix $A^{-1}B$ and label the rows and columns of this matrix.

26 Sets of a certain construction toy contain rods, wheels, and blocks and come in three sizes, standard, jumbo, and giant. The sets are manufactured at plants in three cities, Los Angeles, Kansas City, and New York. Suppose the ingredients of the sets are summarized in a matrix of the form

$$A = \begin{array}{l} \text{Standard} \\ \text{Jumbo} \\ \text{Giant} \end{array} \overset{\begin{array}{ccc}\text{Rods} & \text{Wheels} & \text{Blocks}\end{array}}{\begin{bmatrix} a_{11} & a_{12} & a_{13} \\ a_{21} & a_{22} & a_{23} \\ a_{31} & a_{32} & a_{33} \end{bmatrix}}$$

and suppose the parts that were manufactured and packaged in each city during the month of November are summarized in a matrix of the form

$$B = \begin{array}{l} \text{L.A.} \\ \text{K.C.} \\ \text{N.Y.} \end{array} \overset{\begin{array}{ccc}\text{Rods} & \text{Wheels} & \text{Blocks}\end{array}}{\begin{bmatrix} b_{11} & b_{12} & b_{13} \\ b_{21} & b_{22} & b_{23} \\ b_{31} & b_{32} & b_{33} \end{bmatrix}}$$

What practical information is contained in the matrix BA^{-1}?

27 Versatile matrix programs are available on most college and university computer systems. Find out how to use the matrix inversion program that is available at your school and use it to do Problems 5 through 12.

5 • MATRIX MODELS

In this section, you will see two typical applications of matrix algebra. The first involves communications networks and the second comes from economics.

Communications matrices

Matrices can be used to represent communications networks. Suppose, for example, that a community consists of four individuals, p_1, p_2, p_3, and p_4, and

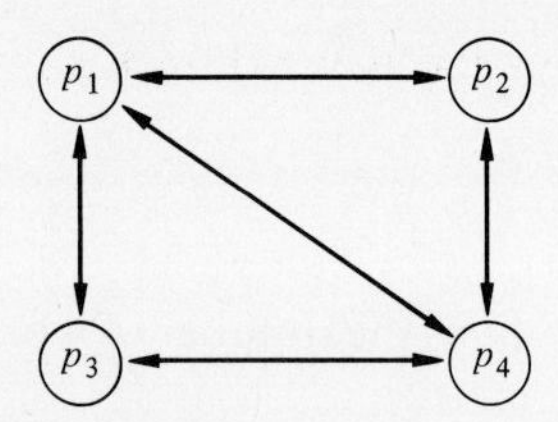

Figure 5.1 A communications network.

that certain pairs of these individuals are in direct communication as indicated in Figure 5.1.

To represent this communications network in matrix form, we construct a 4×4 matrix A with one row and one column for each of the four individuals. To indicate that two individuals p_i and p_j are in direct communication, we place the number 1 in the matrix as the entry in row i, column j. We also put 1 in row j, column i because we assume that if p_i communicates with p_j, then p_j also communicates with p_i. To indicate that two individuals are not in direct communication, we make the corresponding entries equal to zero. (For technical reasons, we assume that an individual does not communicate with himself.)

The procedure for constructing a communications matrix A can be summarized compactly as follows:

$$a_{ij} = \begin{cases} 1 & \text{if } p_i \text{ communicates directly with } p_j \\ 0 & \text{if } p_i \text{ does not communicate directly with } p_j \end{cases}$$

The communications matrix for the network in Figure 5.1 is represented as follows:

$$A = \begin{matrix} p_1 \\ p_2 \\ p_3 \\ p_4 \end{matrix} \overset{\begin{matrix} p_1 & p_2 & p_3 & p_4 \end{matrix}}{\begin{bmatrix} 0 & 1 & 1 & 1 \\ 1 & 0 & 0 & 1 \\ 1 & 0 & 0 & 1 \\ 1 & 1 & 1 & 0 \end{bmatrix}}$$

If a communications matrix A is multiplied by itself, the result is a matrix A^2 that summarizes the number of ways each pair of individuals can communicate through an intermediary. In particular, if $C = A^2$, then

$$c_{ij} = \text{total number of ways that } p_i \text{ can communicate with } p_j \text{ through one intermediary}$$

For the network in Figure 5.1, for example,

$$A^2 = \begin{matrix} & p_1 & p_2 & p_3 & p_4 \\ p_1 & 3 & 1 & 1 & 2 \\ p_2 & 1 & 2 & 2 & 1 \\ p_3 & 1 & 2 & 2 & 1 \\ p_4 & 2 & 1 & 1 & 3 \end{matrix}$$

According to this matrix, there should be two ways that p_3 can communicate with p_2 through an intermediary. If we refer back to Figure 5.1, we see that there are indeed two paths of communication between p_3 and p_2 that go through an intermediary: $p_3 \leftrightarrow p_4 \leftrightarrow p_2$, and $p_3 \leftrightarrow p_1 \leftrightarrow p_2$.

To see why A^2 gives the number of ways each pair of individuals can communicate through one intermediary, observe that if $C = A^2$, the definition of matrix multiplication tells us that

$$c_{ij} = a_{i1}a_{1j} + a_{i2}a_{2j} + \cdots + a_{in}a_{nj}$$

The first term $a_{i1}a_{1j}$ in this sum can be interpreted as follows:

$$\begin{aligned} a_{i1}a_{1j} &= \begin{bmatrix}\text{number of ways (either 0 or 1) that } p_i \\ \text{can communicate directly with } p_1\end{bmatrix} \\ &\qquad \cdot \begin{bmatrix}\text{number of ways that } p_1 \text{ can} \\ \text{communicate directly with } p_j\end{bmatrix} \\ &= \text{number of ways that } p_i \text{ can communicate} \\ &\quad\ \text{with } p_j \text{ using } p_1 \text{ as an intermediary} \end{aligned}$$

Similarly,

$$\begin{aligned} a_{i2}a_{2j} &= \text{number of ways that } p_i \text{ can communicate} \\ &\quad\ \text{with } p_j \text{ using } p_2 \text{ as an intermediary} \end{aligned}$$

The remaining terms in the sum have analogous interpretations, and it follows that

$$\begin{aligned} c_{ij} &= \text{total number of ways } p_i \text{ can communicate with} \\ &\quad\ p_j \text{ through one intermediary} \end{aligned}$$

In general, the matrix A^k summarizes the number of ways each pair of individuals can communicate through exactly $k - 1$ intermediaries.

The Leontief input-output model

In a well-known mathematical model developed by the Nobel Prize-winning economist Wassily Leontief, matrices are used to analyze the interaction among the various sectors of an economic society. Suppose that an economic society consists of n sectors, $S_1, S_2, \ldots, S_n$. Typical sectors in this context include agriculture, manufacturing, transportation, and law enforcement. Portions of the output of each sector are used as input by the others. For example, manufacturers use food from agriculture to feed their workers, require transportation to distribute their products, and rely on law enforcement personnel for security at their plants. A sector may also use some of its own output as input for further production. For example, farmers grow alfalfa to feed their livestock.

To represent these interactions in matrix form, we let

$$a_{ij} = \text{number of units from sector } S_i \text{ consumed in the production of one unit of output by sector } S_j$$

and we form the **input-output matrix**

$$A = \underset{\text{providing input}}{\text{Sector}}\begin{matrix} S_1 \\ S_2 \\ \vdots \\ S_n \end{matrix}\overset{\displaystyle\begin{matrix}\text{Sector producing output}\\ S_1 \quad\; S_2 \quad\; \cdots \quad\; S_n\end{matrix}}{\begin{bmatrix} a_{11} & a_{12} & \cdots & a_{1n} \\ a_{21} & a_{22} & \cdots & a_{2n} \\ \vdots & \vdots & & \vdots \\ a_{n1} & a_{n2} & \cdots & a_{nn} \end{bmatrix}}$$

If X is the **output vector**

$$X = \begin{bmatrix} x_1 \\ x_2 \\ \vdots \\ x_n \end{bmatrix} \qquad \text{where} \quad x_j = \text{total number of units produced by sector } S_j$$

then the product AX is an $n \times 1$ matrix that gives the total output from each sector that is used by the society as input for production. To see this, observe that if

$$AX = C = \begin{bmatrix} c_1 \\ c_2 \\ \vdots \\ c_n \end{bmatrix}$$

then the definition of matrix multiplication tells us that

$$\begin{aligned} c_i &= a_{i1}x_1 + a_{i2}x_2 + \cdots + a_{in}x_n \\ &= \begin{bmatrix} \text{number of units} \\ \text{from } S_i \text{ used to produce} \\ \text{one unit in } S_1 \end{bmatrix} \cdot \begin{bmatrix} \text{number of units} \\ \text{produced in } S_1 \end{bmatrix} + \cdots \end{aligned}$$

$$+ \begin{bmatrix} \text{number of units} \\ \text{from } S_i \text{ used to produce} \\ \text{one unit in } S_n \end{bmatrix} \cdot \begin{bmatrix} \text{number of units} \\ \text{produced in } S_n \end{bmatrix}$$

$$= \text{total number of units from } S_i \text{ used by the society for production}$$

In a typical application of this model, consumer demand is known and the goal is to determine how many units each sector should produce to meet this demand with no surpluses. That is, we know the demand vector

$$D = \begin{bmatrix} d_1 \\ d_2 \\ \vdots \\ d_n \end{bmatrix} \quad \text{where } d_j = \text{total number of units from } S_j \text{ demanded by consumers}$$

and we wish to determine the output vector

$$X = \begin{bmatrix} x_1 \\ x_2 \\ \vdots \\ x_n \end{bmatrix} \quad \text{where } x_j = \text{total number of units produced by } S_j$$

for which consumer demand will be satisfied. (Convince yourself that the solution is not simply $x_1 = d_1, x_2 = d_2, \ldots, x_n = d_n$.)

To solve this problem, we observe that

$$\text{Total output} = \text{consumer demand} + \text{input for production}$$

The total output is given by the vector X, the consumer demand by the vector D, and (as we just saw) the number of units from each sector used as input for production is given by the product AX. Hence,

$$X = D + AX \qquad \text{or} \qquad X - AX = D$$

We can rewrite this equation using the identity matrix I as

$$IX - AX = D$$

and then, using the distributive property as

$$(I - A)X = D$$

If $I - A$ is invertible, we solve for X by multiplying both sides of the equation by $(I - A)^{-1}$ and conclude that

$$X = (I - A)^{-1}D$$

Problems **1** A communications network is represented by the following graph.

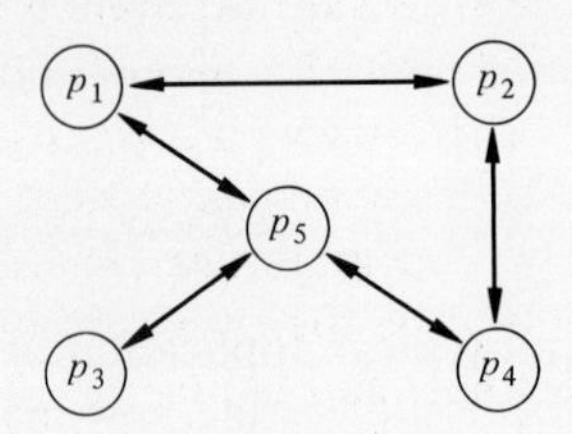

(a) Construct the communications matrix for this network.
(b) Find the matrix that gives the number of ways each pair of individuals can communicate through one intermediary.
(c) Use the matrix in part (b) to determine the number of ways p_4 and p_1 can communicate through one intermediary. Then, referring to the graph, list the corresponding paths of communication.

2 A communications network is represented by the following graph.

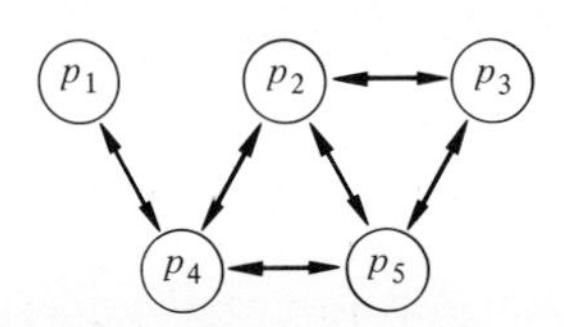

(a) Construct the communications matrix for this network.
(b) Find the matrix that gives the number of ways each pair of individuals can communicate through one intermediary. Then, referring to the graph, list the corresponding paths of communication.
(c) Add the matrix in part (b) to the matrix in part (a) and use the result to decide if there are any two individuals in the community who cannot communicate with each other either directly or through one intermediary. Check your answer by referring to the graph.

3 Draw a graph of the communications network summarized in the following matrix.

$$\begin{array}{c} \\ p_1 \\ p_2 \\ p_3 \\ p_4 \\ p_5 \end{array}
\begin{array}{c} \begin{array}{ccccc} p_1 & p_2 & p_3 & p_4 & p_5 \end{array} \\
\begin{bmatrix} 0 & 0 & 0 & 1 & 0 \\ 0 & 0 & 1 & 1 & 1 \\ 0 & 1 & 0 & 0 & 1 \\ 0 & 1 & 0 & 0 & 1 \\ 0 & 1 & 1 & 1 & 0 \end{bmatrix} \end{array}$$

4 Using an argument similar to the one in the text justifying the interpretation of A^2, show that the cube A^3 of a communications matrix A gives the number of ways that each pair of individuals can communicate through two intermediaries. (*Hint:* Think of A^3 as the product A^2A and use the interpretations of A and A^2.)

5 Suppose A is the communications matrix for a community of n individuals.

(a) What information is contained in the matrix $A + A^2 + A^3 + \cdots + A^{n-1}$?

(b) How would you use the matrix in part (a) to determine if the community contains two individuals who have neither direct nor indirect communication with each other?

6 Let A be the communications matrix for a community containing five individuals and suppose

$$B = A + A^2 + A^3 + A^4 = \begin{array}{c} \\ p_1 \\ p_2 \\ p_3 \\ p_4 \\ p_5 \end{array} \begin{array}{c} \begin{array}{ccccc} p_1 & p_2 & p_3 & p_4 & p_5 \end{array} \\ \begin{bmatrix} 2 & 0 & 2 & 0 & 0 \\ 0 & 3 & 0 & 3 & 3 \\ 2 & 0 & 2 & 0 & 0 \\ 0 & 3 & 0 & 3 & 3 \\ 0 & 3 & 0 & 3 & 3 \end{bmatrix} \end{array}$$

(a) The contents of some secret tape recordings have become known to each member of the community. An investigator theorizes that at least two individuals had direct knowledge of the contents of the tapes. Use the information in matrix B to verify the investigator's theory.

(b) Graph the communications network represented by matrix A. (*Hint:* You can do this without first calculating A.)

7 A simple economic society consists of three sectors, agriculture (A), manufacturing (M), and transportation (T). Suppose that the input-output matrix for this society is

$$\begin{array}{c} \\ \text{A} \\ \text{M} \\ \text{T} \end{array} \begin{array}{c} \begin{array}{ccc} \text{A} & \text{M} & \text{T} \end{array} \\ \begin{bmatrix} 0.4 & 0.1 & 0.2 \\ 0.2 & 0.4 & 0.4 \\ 0.2 & 0.3 & 0.1 \end{bmatrix} \end{array}$$

(a) How many agriculture units are consumed in the production of one manufacturing unit?

(b) Which sector uses the greatest number of transportation units in the production of one of its own units?

(c) How many transportation units are needed to produce a total of 5 agriculture units, 8 manufacturing units, and 2 transportation units?

8 The interaction among the sectors S_1, S_2, and S_3 of a simple economic society is described by the Leontief input-output matrix

$$A = \begin{array}{c} \\ S_1 \\ S_2 \\ S_3 \end{array} \begin{array}{c} \begin{array}{ccc} S_1 & S_2 & S_3 \end{array} \\ \begin{bmatrix} 0.1 & 0.4 & 0.2 \\ 0{,}2 & 0.3 & 0.4 \\ 0.5 & 0.2 & 0.1 \end{bmatrix} \end{array}$$

and the number of units produced by each sector is summarized in the output vector

$$X = \begin{matrix} S_1 \\ S_2 \\ S_3 \end{matrix} \begin{bmatrix} 400 \\ 600 \\ 500 \end{bmatrix}$$

(a) Calculate the number of units of output from each sector that are used by the society as input for production and summarize your results in a 3×1 matrix.

(b) Combine A and X in an appropriate way to form a 3×1 matrix giving the number of units of output from each sector that are *not* consumed by the society as input for production.

9 The interaction among the sectors S_1 and S_2 of a simple economic society is described by the Leontief input-output matrix

$$A = \begin{matrix} \\ S_1 \\ S_2 \end{matrix} \begin{matrix} \begin{matrix} S_1 & S_2 \end{matrix} \\ \begin{bmatrix} 0.8 & 0.2 \\ 0.1 & 0.4 \end{bmatrix} \end{matrix}$$

and the number of units from each sector demanded by consumers is given by the demand vector

$$D = \begin{matrix} S_1 \\ S_2 \end{matrix} \begin{bmatrix} 50 \\ 36 \end{bmatrix}$$

How many units must each sector produce to meet the consumer demand?

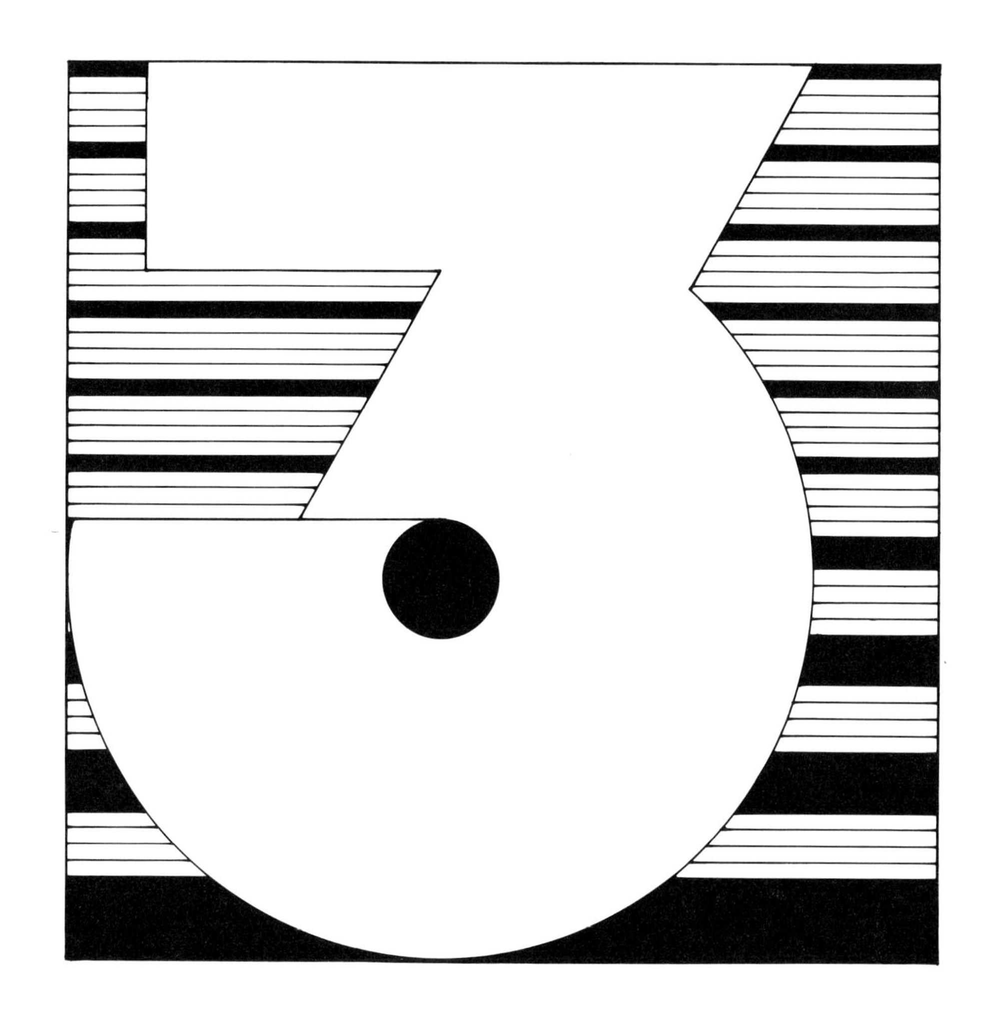

LINEAR PROGRAMMING

1 • LINEAR MODELS

A manufacturer may wish to determine how to use a limited supply of raw materials to generate the largest possible profit. A personnel manager may wish to determine how to allocate work among various employees to accomplish certain required tasks in the shortest possible time. A farmer may wish to determine how to combine several grains to produce a feed mixture that meets certain minimal nutritional requirements at the smallest possible cost.

In each of these situations, the goal is to maximize or minimize a particular quantity without violating certain restrictions or constraints. For example, in the first situation, profit is to be maximized subject to the constraint that only a limited supply of raw materials is available. In the second situation, employees' hours are to be minimized subject to the constraint that certain tasks must be accomplished. And in the third situation, cost is to be minimized subject to the constraint that certain nutritional requirements have to be met.

Linear programming is a branch of mathematics that deals with constrained optimization problems of this sort. In this section, you will see some practical problems that can be formulated mathematically as linear-programming problems. Techniques for solving these will be developed in Sections 2 and 3. Here is an example to illustrate basic features of a linear-programming problem.

A maximization problem

EXAMPLE 1.1 ● A manufacturer makes two grades of concrete. Each bag of the high-grade concrete contains 10 kilograms of gravel and 5 of cement, while each bag of the low-grade concrete contains 12 kilograms of gravel and 3 of cement. There are 1,920 kilograms of gravel and 780 of cement currently available. The manufacturer can make a profit of $1.20 on each bag of the high-grade concrete and $1.00 on each bag of the low-grade concrete, and wishes to determine how many bags of each grade to make from the available supplies to generate the largest possible profit. Formulate this problem in mathematical terms.

SOLUTION For convenience, we begin by constructing a table that shows the number of kilograms of gravel and cement in a bag of each grade of concrete, the available supply of each of these ingredients, and the profit the manufacturer can make on each bag.

	High-grade	Low-grade	Supply
Gravel	10	12	1,920
Cement	5	3	780
Profit	1.20	1.00	

If we let

$$x_1 = \text{number of bags of high-grade concrete produced}$$

$$x_2 = \text{number of bags of low-grade concrete produced}$$

we can represent the manufacturer's profit, f, by the equation

$$f = 1.2x_1 + x_2$$

The constraints that no more than 1,920 kilograms of gravel and 780 kilograms of cement are available can be represented mathematically by the pair of inequalities

$$10x_1 + 12x_2 \leq 1{,}920 \qquad \text{and} \qquad 5x_1 + 3x_2 \leq 780$$

Since negative values of x_1 and x_2 make no sense in this problem, the variables must also satisfy the inequalities

$$x_1 \geq 0 \qquad \text{and} \qquad x_2 \geq 0$$

Our goal, then, is to maximize the function

$$f = 1.2x_1 + x_2$$

subject to the constraints

$$10x_1 + 12x_2 \leq 1{,}920 \qquad 5x_1 + 3x_2 \leq 780 \qquad x_1 \geq 0 \qquad x_2 \geq 0$$ ●

Linear functions and inequalities

The function we wished to maximize in Example 1.1 was a **linear function,** that is, a function of the form

$$f = a_1x_1 + a_2x_2$$

where a_1 and a_2 are constants. In addition, each of the inequalities representing the constraints was a **linear inequality,** that is, one of the form

$$b_1x_1 + b_2x_2 \leq c$$

where b_1, b_2, and c are constants.

The linearity of the function and the constraints are the features that characterize linear-programming problems in general.

Linear programming

A *linear-programming problem* is one in which a linear function of several variables is to be maximized or minimized subject to a set of constraints, each of which can be represented as a linear inequality (or linear equation) and subject to the further restriction that all the variables be nonnegative.

The function to be maximized or minimized is sometimes called the **objective function.** In the next example, we consider a practical situation that leads to a linear-programming problem in which the goal is to *minimize* the objective function.

A minimization problem

EXAMPLE 1.2 ● A farmer prepares feed for livestock by combining two grains. Each unit of the first grain costs 20 cents and contains 2 units of protein and 5 units of iron, while each unit of the second grain costs 30 cents and contains 4 units of protein and 1 unit of iron. Each animal must receive at least 10 units of protein and 16 units of iron each day, and the farmer wishes to determine the number of units of each grain to feed daily to each animal to satisfy these nutritional requirements at the smallest possible cost. Formulate the farmer's problem mathematically as a linear-programming problem.

SOLUTION For convenience, we construct a table showing the number of units of protein and iron in one unit of each grain, the minimal daily requirements for each of these nutrients, and the cost per unit of each grain.

	Grain I	Grain II	Minimal requirement
Protein	2	4	10
Iron	5	1	16
Cost	20	30	

If we let

$$x_1 = \text{number of units of grain I fed daily to each animal}$$

$$x_2 = \text{number of units of grain II fed daily to each animal}$$

we can represent the farmer's cost by the linear function $f = 20x_1 + 30x_2$ and the nutritional requirements by the linear inequalities $2x_1 + 4x_2 \geq 10$ and $5x_1 + x_2 \geq 16$. The resulting linear-programming problem is to minimize the objective function

$$f = 20x_1 + 30x_2$$

subject to the constraints

$$2x_1 + 4x_2 \geq 10 \qquad 5x_1 + x_2 \geq 16 \qquad x_1 \geq 0 \qquad x_2 \geq 0$$

●

A transportation problem

The situation described in the next example is typical of a class of linear-programming problems known as **transportation problems.**

EXAMPLE 1.3 ● A manufacturer of copying machines has two plants, P_1 and P_2, that can produce up to 200 units and 140 units per month, respectively. The units are shipped to two warehouses, W_1 and W_2, that require at least 180 units and 150 units per month, respectively. Shipping costs from P_1 to W_1 are \$75 per unit and from P_1 to W_2 \$40 per unit, while shipping costs from P_2 to W_1 are \$80 per unit and from P_2 to W_2 \$30 per unit. The manufacturer wishes to determine the number of units from each plant that should be shipped monthly to each warehouse to satisfy the requirements of the warehouses while keeping shipping costs as small as possible. Formulate the manufacturer's problem mathematically as a linear-programming problem.

SOLUTION For convenience, we construct a table summarizing the relevant information. Since the goal is to determine how many units from each of the two plants P_1 and P_2 should be shipped to each of the two warehouses W_1 and W_2, it is natural to introduce four variables, x_{11}, x_{12}, x_{21}, and x_{22} where

$$x_{ij} = \text{number of units shipped monthly from } P_i \text{ to } W_j$$

	W_1	W_2	Maximum supply
P_1	\$75/unit; x_{11} units	\$40/unit; x_{12} units	200 units
P_2	\$80/unit; x_{21} units	\$30/unit; x_{22} units	140 units
Minimum demand	180 units	150 units	

The total monthly shipping cost, f, is given by the linear function

$$f = 75x_{11} + 40x_{12} + 80x_{21} + 30x_{22}$$

The production capacities of the two plants can be represented by the linear inequalities

$$x_{11} + x_{12} \leq 200 \quad \text{and} \quad x_{21} + x_{22} \leq 140$$

and the minimal requirements of the two warehouses can be represented by the inequalities

$$x_{11} + x_{21} \geq 180 \quad \text{and} \quad x_{12} + x_{22} \geq 150$$

The resulting linear-programming problem is to minimize the objective function

$$f = 75x_{11} + 40x_{12} + 80x_{21} + 30x_{22}$$

subject to the constraints

$$x_{11} + x_{12} \leq 200 \qquad x_{21} + x_{22} \leq 140$$

$$x_{11} + x_{21} \geq 180 \qquad x_{12} + x_{22} \geq 150$$

$$x_{11} \geq 0 \qquad x_{12} \geq 0 \qquad x_{21} \geq 0 \qquad x_{22} \geq 0$$

●

Problems For each of the following problems, construct an appropriate linear-programming model. (Do not try to solve the resulting linear-programming problem at this time.)

1 At a local leather shop 1 hour of skilled labor and 1 hour of unskilled labor are required to produce a briefcase, while 1 hour of skilled labor and 2 hours of unskilled labor are required to produce a suitcase. The owner of the shop can make a profit of $15 on each briefcase and $20 on each suitcase. On a particular day, only 7 hours of skilled labor and 11 hours of unskilled labor are available, and the owner wishes to determine how many briefcases and how many suitcases to make that day to generate the largest profit possible.

2 Work Problem 1 again, this time taking into account the additional information that 0.2 square meters of leather are used in the production of a briefcase, 0.6 square meters are used in the production of a suitcase, and, on the particular day in question, only 1.8 square meters of leather are available.

3 Work Problem 2 again, this time taking into account the further information that on the morning of the day in question, a customer ordered 2 suitcases which the owner promised to have ready by the end of the day.

4 A manufacturer makes two grades of concrete. Each bag of the high-grade concrete contains 15 kilograms of sand, 10 of gravel, and 5 of cement, while each bag of the low-grade concrete contains 15 kilograms of sand, 12 of gravel, and 3 of cement. There are 2,400 kilograms of sand, 1,920 of gravel, and 780 of cement currently available. The manufacturer can make a profit of $1.20 on each bag of the high-grade concrete and $1.00 on each bag of the low-grade concrete, and wishes to determine how many bags of each grade to make from the available supplies to generate the largest possible profit.

5 A distributor packages three types of coffee: Type I which is 100 percent Colombian, type II which is 100 percent Brazilian, and type III which is a mixture containing 40 percent Colombian and 60 percent Brazilian coffee. The distributor can make a profit of $1.20 per kilogram from the sale of type I, $1.00 per kilogram from the sale of type II, and $1.20 per kilogram from the sale of type III. There are 400 kilograms of Colombian and 480 kilograms of Brazilian coffee currently on hand, and the distributor wishes to determine how many kilograms of each of the three types of coffee to package to generate the largest possible profit.

6 A manufacturer makes three commodities, C_1, C_2, and C_3, from three raw materials, M_1, M_2, and M_3. The number of units of each raw material consumed in the production of one unit of each commodity, and the number of units of each of the raw materials currently available are shown in the following table:

	C_1	C_2	C_3	Supply
M_1	a_{11}	a_{12}	a_{13}	s_1
M_2	a_{21}	a_{22}	a_{23}	s_2
M_3	a_{31}	a_{32}	a_{33}	s_3

The manufacturer can make a profit of p_1, p_2, and p_3 dollars per unit, respectively, from the sale of the commodities C_1, C_2, and C_3, and wishes to determine how many units of each commodity to make from the available supply of raw materials to generate the largest possible profit.

7 A dietician is planning a meal consisting of three foods whose ingredients are summarized in the following table:

	One unit of		
	Food I	Food II	Food III
Units of protein	5	10	15
Units of carbohydrates	2	3	2
Units of iron	3	6	1
Calories	60	140	120

The dietician wishes to determine the number of units of each food to use to create a meal containing at least 30 units of protein, 8 units of carbohydrates, and 10 units of iron, and as few calories as possible.

8 A group of students has gathered for dinner at a small Chinese restaurant near campus that has the following menu:

Menu

Dinner #1: Wonton soup, egg roll, chow mein $2.40
Dinner #2: Egg roll, sweet-and-sour pork $3.20
Dinner #3: Egg roll, chow mein, sweet-and-sour pork $3.50

The students want a total of 4 bowls of wonton soup, 9 egg rolls, 6 orders of chow mein, and 5 orders of sweet-and-sour pork, and, naturally, wish to keep the total cost as small as possible.

9 Shipments from one wholesaler contain 3 units of item A, 6 units of item B, 4 units of item C, and cost $20. Shipments from a second wholesaler contain

12 units of A, 3 units of B, 3 units of C, and cost \$26. A retailer wishes to determine how many shipments to order from each wholesaler to get at least 396 units of A, 288 units of B, and 255 units of C at the smallest possible cost.

10 A manufacturing firm with two plants, P_1 and P_2, ships its product to two markets, M_1 and M_2. The consumer demand for the product at M_1 is 80 units per month and at M_2 is 125 units per month. P_1 can produce up to 100 units per month and P_2 can produce up to 120 units per month. The per-unit shipping costs between the plants and markets are summarized in the following table:

From \ To	M_1	M_2
P_1	\$50	\$15
P_2	\$25	\$45

The manufacturer wishes to determine the number of units that should be shipped monthly from each plant to each market to meet the consumer demand while keeping the total shipping cost as small as possible.

11 Work Problem 10 again, this time taking into account the additional assumptions that the firm is required to ship at least 40 units each month from P_1 to M_1 and is allowed to ship no more than 50 units each month from P_1 to M_2.

12 At the beginning of the evening shift, the dispatcher for a city police department finds that three outlying precincts, A_1, A_2, and A_3, have a surplus of patrol cars in service, while two central precincts, B_1 and B_2, are short several cars. In particular, precincts A_1, A_2, and A_3 have 5, 8, and 6 extra cars, respectively, while precincts B_1 and B_2 each need 9 additional cars. The distances (in miles) between the outlying and central precincts are shown in the following table:

From \ To	B_1	B_2
A_1	6	3
A_2	4	5
A_3	8	6

To bring the number of patrol cars in each of the central precincts up to the required level, surplus cars from the outlying precincts are to be assigned to duty in the central precincts. The dispatcher would like to determine how this should be done to minimize the total distance that the surplus cars must travel to reach their destinations.

13 A firm owns two manufacturing plants, P_1 and P_2, that can produce up to s_1 and s_2 units of a certain commodity per month, respectively, and two warehouses, W_1 and W_2, that require at least d_1 and d_2 units of the commodity per month, respectively. The per-unit shipping costs between the plants and warehouses are shown in the following table:

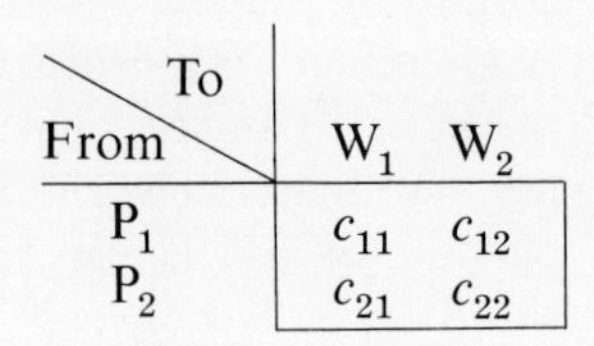

From \ To	W_1	W_2
P_1	c_{11}	c_{12}
P_2	c_{21}	c_{22}

The manufacturer wishes to determine how many units to ship monthly from each plant to each warehouse in order to satisfy the requirements of the warehouses while keeping the total shipping cost as small as possible.

2 • THE GEOMETRIC METHOD

In a linear-programming problem, the goal is to maximize (or minimize) a linear objective function subject to constraints that form a system of linear inequalities. When only two variables are involved, this system of inequalities can be represented as a set of points in a plane and the problem can be solved geometrically. The procedure is based on the following geometric properties of linear inequalities.

The graphs of linear inequalities

The graph of the linear inequality $a_1x_1 + a_2x_2 \leq c$ (or $a_1x_1 + a_2x_2 \geq c$) consists of the straight line $a_1x_1 + a_2x_2 = c$ together with one of the **half-planes** on either side of this line. To determine which, choose any convenient point that is not on the line and see if its coordinates satisfy the given inequality. If they do, the appropriate half-plane is the one containing the chosen point; and if they do not, the appropriate half-plane is the one on the other side of the line. Here is an example.

EXAMPLE 2.1 ● Graph the linear inequality $2x_1 + x_2 \geq 4$.

SOLUTION We begin by graphing the straight line $2x_1 + x_2 = 4$. (An easy way to do this is to plot the x_1 and x_2 intercepts.) Next we choose any point (x_1, x_2) that is not on the line and see if its coordinates satisfy the given inequality. In this case, (0, 0) is a convenient choice. Since $2(0) + 0 = 0 < 4$, the coordinates of (0, 0) do not satisfy the inequality $2x_1 + x_2 \geq 4$ and we conclude that the appropriate half-plane is the one that does not contain this point. The situation is illustrated in Figure 2.1. ●

The graph of a system of linear inequalities is the set of points that satisfy all of the inequalities simultaneously. Here is an example.

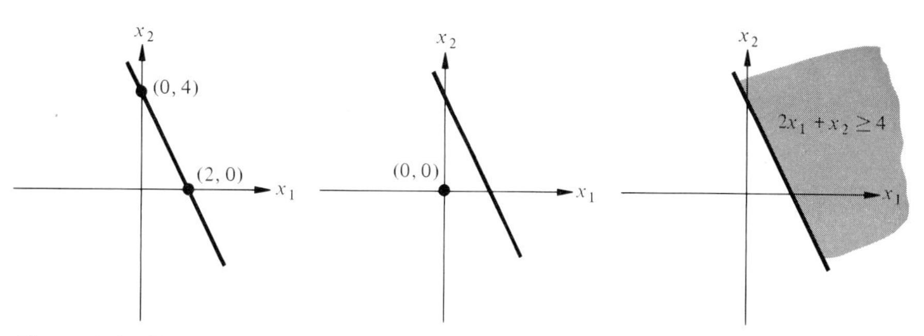

Figure 2.1 Steps leading to the graph of the inequality $2x_1 + x_2 \geq 4$.

EXAMPLE 2.2 ● Graph the following system of linear inequalities.

$$x_1 + x_2 \leq 4 \qquad 2x_1 + 3x_2 \leq 9 \qquad x_2 \leq 2 \qquad x_1 \geq 0 \qquad x_2 \geq 0$$

SOLUTION The last two inequalities tell us that the region in question lies in the first quadrant. We begin by drawing in this quadrant the three lines $x_1 + x_2 = 4$, $2x_1 + 3x_2 = 9$, and $x_2 = 2$. The points at which the pairs of lines intersect are found by solving simultaneously the corresponding pairs of linear equations.

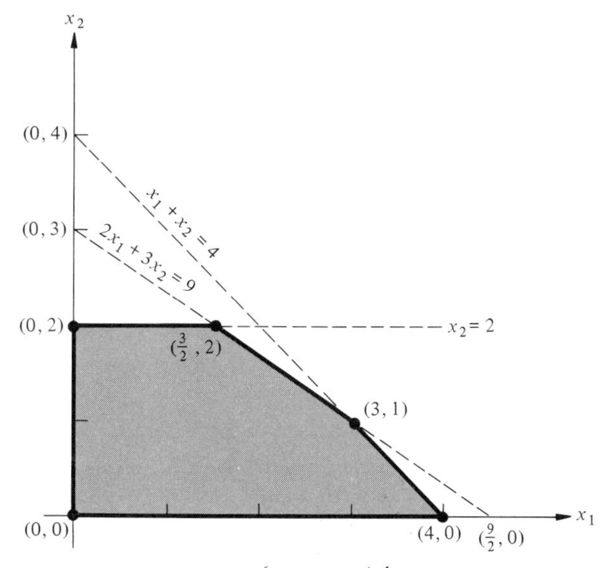

Figure 2.2 The graph of the system $\begin{cases} x_1 + x_2 \leq 4 \\ 2x_1 + 3x_2 \leq 9 \\ x_2 \leq 2 \quad x_1 \geq 0 \quad x_2 \geq 0. \end{cases}$

Next, we identify the half-planes defined by each of the three inequalities, $x_1 + x_2 \leq 4$, $2x_1 + 3x_2 \leq 9$, and $x_2 \leq 2$. We find in each case that the half-plane in question is the one that lies below the corresponding straight line. The resulting graph is shown in Figure 2.2. ●

The solution of linear-programming problems

The set of points that satisfy the inequalities in a linear-programming problem is called the **set of feasible solutions.** To solve the linear-programming problem, we must select from the set of feasible solutions an **optimal solution** that maximizes (or minimizes) the objective function. For problems involving only two variables, we can graph the set of feasible solutions. It can be shown that an optimal solution (if one exists) can always be found at a corner or **vertex** of this set. Hence, to solve a two-variable linear-programming problem, we simply graph the set of feasible solutions and select the vertex at which the value of the objective function is maximal (or minimal). We illustrate the procedure in the next example by solving the linear-programming problem from Example 1.1.

EXAMPLE 2.3 ● Maximize the function $f = 1.2x_1 + x_2$ subject to the constraints

$$10x_1 + 12x_2 \leq 1{,}920 \qquad 5x_1 + 3x_2 \leq 780 \qquad x_1 \geq 0 \qquad x_2 \geq 0$$

SOLUTION First we graph the corresponding set of feasible solutions and label its vertices as shown in Figure 2.3. Then we compute the value of the objective function at

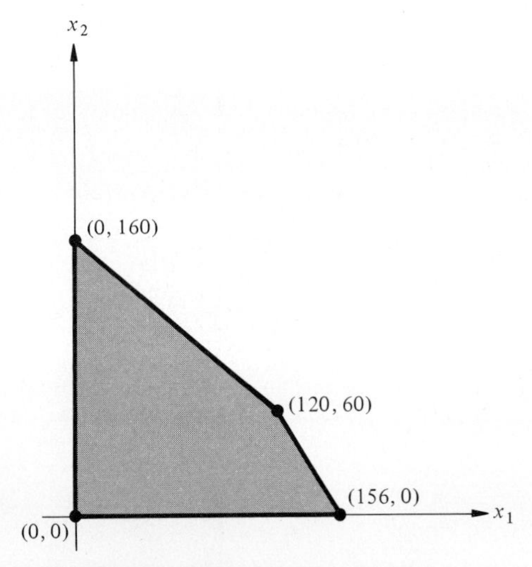

Figure 2.3 The set of feasible solutions for Example 2.3.

each of the four vertices. Using the functional notation $f(x_1, x_2)$ to denote the value of f at the point (x_1, x_2), we get

$$f(0, 0) = 0 \qquad f(0, 160) = 160$$
$$f(120, 60) = 204 \qquad f(156, 0) = 187.2$$

Since 204 is the largest of these values, the corresponding vertex (120, 60) must be an optimal solution to the linear-programming problem. That is, subject to the given constraints, the maximum value of f is 204, which occurs when $x_1 = 120$ and $x_2 = 60$. In terms of the practical problem in Example 1.1, we conclude that to generate the largest possible profit, the manufacturer should produce 120 bags of high-grade concrete and 60 bags of low-grade concrete. If this is done, the resulting profit will be $204. ●

A rigorous explanation of why optimal solutions can be found among the vertices of the sets of feasible solutions is beyond the scope of this text. However, there is a simple geometric argument that you should find convincing. To illustrate this argument, let us reconsider the problem we solved in Example 2.3.

Our goal in Example 2.3 was to find the point in the set of feasible solutions at which the objective function $f = 1.2x_1 + x_2$ had the largest value. To visualize the relationship between the set of feasible solutions and the values of the objective function, let us choose some arbitrary values c and graph the sets of points (x_1, x_2) for which $f(x_1, x_2) = c$. That is, for each c we graph the points (x_1, x_2) that satisfy the linear equation

$$1.2x_1 + x_2 = c$$

Putting this equation in slope-intercept form we get

$$x_2 = -1.2x_1 + c$$

from which we see that the graph in question is the straight line in the x_1x_2 plane whose slope is -1.2 and whose x_2 intercept is $(0, c)$. In Figure 2.4 we draw several of these lines (corresponding to several values of c), together with the set of feasible solutions from Figure 2.3.

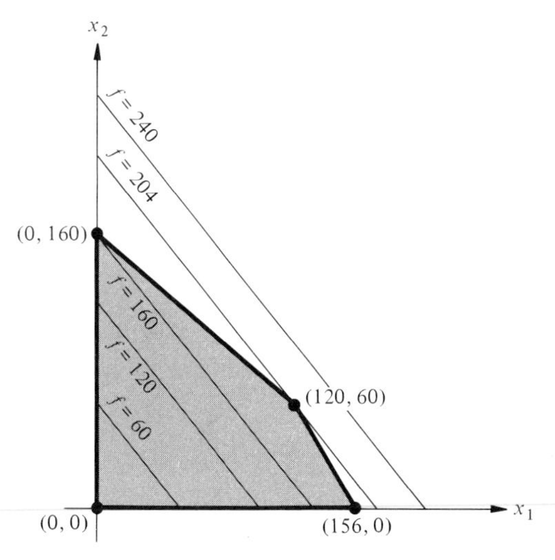

Figure 2.4 The set of feasible solutions and lines $f = c$.

Notice that the lines $f = c$ are parallel (since the slope of each is -1.2) and that the lines corresponding to the larger values of c are further from the origin than those corresponding to the smaller values of c.

To maximize the objective function subject to the given constraints, we must find the largest value of c for which the line $f = c$ intersects the set of feasible solutions. From Figure 2.4 we see that this value is $c = 204$ and that the line $f = 204$ intersects the set of feasible solutions at a vertex (120, 60), as predicted. Moreover, since the line $f = 204$ intersects the set of feasible solutions exactly once, the point (120, 60) is the only optimal solution to this problem.

In the next example, we solve the feed-mix problem from Example 1.2.

EXAMPLE 2.4 ● Minimize the function $f = 20x_1 + 30x_2$ subject to the constraints

$$2x_1 + 4x_2 \geq 10 \qquad 5x_1 + x_2 \geq 16 \qquad x_1 \geq 0 \qquad x_2 \geq 0$$

SOLUTION The set of feasible solutions is shown in Figure 2.5. At the three vertices, the values of the objective function are

$$f(0, 16) = 480 \qquad f(3, 1) = 90 \qquad f(5, 0) = 100$$

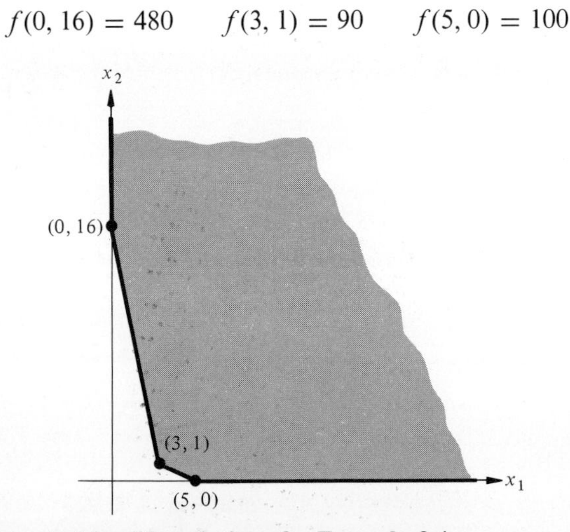

Figure 2.5 The set of feasible solutions for Example 2.4.

Since 90 is the smallest of these values, the corresponding vertex (3, 1) must be an optimal solution. We conclude that to satisfy the nutritional requirements at the smallest possible cost, the farmer in Example 1.2 should feed each animal 3 units of grain I and 1 unit of grain II each day.

To see why the optimal solution occurs at a vertex, draw some of the lines $f = c$ and determine graphically the feasible solution at which the objective function is minimized. In addition, convince yourself that there is only one optimal solution to this problem. ●

Problems In Problems 1 through 14, assume that the given linear-programming problem has an optimal solution. In each case, solve the problem by evaluating the objective function at the vertices of the set of feasible solutions. Verify your answer by graphing some of the lines $f = c$.

1 Maximize the function $f = 5x_1 + x_2$ subject to the constraints

$$4x_1 + 3x_2 \leq 24 \qquad x_1 + 3x_2 \leq 15 \qquad x_1 \geq 0 \qquad x_2 \geq 0$$

2 Maximize the function $f = x_1 + 3x_2$ subject to the constraints in Problem 1.

3 Minimize the function $f = 10x_1 + 6x_2$ subject to the constraints

$$5x_1 + x_2 \geq 6 \qquad x_1 + 3x_2 \geq 4 \qquad x_1 \geq 0 \qquad x_2 \geq 0$$

4 Minimize the function $f = 20x_1 + 80x_2$ subject to the constraints

$$\begin{array}{lll} 4x_1 + x_2 \geq 8 & 2x_1 + 5x_2 \geq 18 & \\ 2x_1 + 3x_2 \geq 14 & x_1 \geq 0 & x_2 \geq 0 \end{array}$$

5 Minimize the function $f = 20x_1 + 30x_2$ subject to the constraints in Problem 4.

6 Maximize the function $f = x_1 + 2x_2$ subject to the constraints

$$\begin{array}{lll} 2x_1 + 3x_2 \leq 23 & x_1 + 4x_2 \leq 24 & \\ 3x_1 + x_2 \leq 24 & x_1 \geq 0 & x_2 \geq 0 \end{array}$$

7 Maximize the function $f = 2x_1 + x_2$ subject to the constraints in Problem 6.

8 Minimize the function $f = 2x_1 + x_2$ subject to the constraints in Problem 6.

9 Minimize the function $f = 2x_1 + 3x_2$ subject to the constraints

$$\begin{array}{lll} x_1 + 3x_2 \geq 19 & 3x_1 + 2x_2 \geq 22 & \\ 4x_1 + 5x_2 \geq 34 & x_1 \geq 0 & x_2 \geq 0 \end{array}$$

10 Maximize the function $f = 20x_1 + 30x_2$ subject to the constraints

$$\begin{array}{lll} x_1 + x_2 \leq 8 & x_1 + 2x_2 \leq 11 & \\ x_1 + 3x_2 \leq 12 & x_1 \geq 0 & x_2 \geq 0 \end{array}$$

11 Minimize the function $f = 3x_1 + x_2$ subject to the constraints

$$3x_1 - 2x_2 \geq 0 \qquad 3x_1 + x_2 \leq 18 \qquad x_2 \geq 3 \qquad x_1 \geq 0 \qquad x_2 \geq 0$$

12 Maximize the function $f = 3x_1 + x_2$ subject to the constraints in Problem 11.

13 Minimize the function $f = 3x_1 + x_2$ subject to the constraints

$$3x_1 + 5x_2 \leq 38 \qquad 2x_1 + x_2 \geq 9$$
$$x_1 - 3x_2 \leq -6 \qquad x_1 \geq 0 \qquad x_2 \geq 0$$

14 Maximize the function $f = 5x_1 + 4x_2$ subject to the constraints

$$3x_1 + 2x_2 \leq 16 \qquad 2x_1 + x_2 \leq 10$$
$$x_1 + x_2 \geq 5 \qquad x_2 \leq 5 \qquad x_1 \leq 5 \qquad x_1 \geq 0 \qquad x_2 \geq 0$$

In Problems 15 through 19, solve the indicated practical problem from Section 1.

15 Problem 1 **16** Problem 2 **17** Problem 3
18 Problem 4 **19** Problem 9

20 Consider the problem of maximizing the function $f = 10x_1 + 6x_2$ subject to the constraints

$$5x_1 + x_2 \geq 6 \qquad x_1 + 3x_2 \geq 4 \qquad x_1 \geq 0 \qquad x_2 \geq 0$$

Graph the set of feasible solutions and some of the lines $f = c$. What conclusion can you draw about the problem's solution?

21 Consider the problem of minimizing the function $f = x_1 - x_2$ subject to the constraints

$$2x_1 - x_2 \leq -2 \qquad 2x_1 - 3x_2 \leq 0$$
$$x_1 + x_2 \geq 5 \qquad x_1 \geq 0 \qquad x_2 \geq 0$$

Graph the set of feasible solutions and some of the lines $f = c$. What conclusion can you draw about the problem's solution?

3 • THE SIMPLEX METHOD

There is an algebraic procedure for solving linear-programming problems that is more versatile than the geometric method introduced in Section 2. It is called the **simplex method** and can be applied to problems involving any number of variables and any number of constraints. Because it has led to the development of efficient computer programs for solving linear-programming problems, the simplex method has had tremendous impact since its introduction in the 1940s.

Most interesting linear-programming problems that arise in practical work involve so many variables and constraints that they must be solved on a computer. For this reason, it is not the purpose of this section to make you proficient in solving linear-programming problems by hand using the simplex method. Instead, the purpose is to give you a feel for the method so that you will have some idea what lies behind the computer programs you will encounter.

Essentially, the simplex method is a systematic procedure for evaluating the objective function at a specially selected sequence of vertices of the set of feasible solutions. The efficiency of the method comes from the fact that each vertex is chosen so that the corresponding value of the objective function is closer to the optimal value than it was at the preceding vertex. Here is an example to illustrate how the simplex method works. So that you will be able to compare the simplex method with the geometric method of Section 2, the set of feasible solutions and the values of the objective function at all its vertices are shown in Figure 3.1.

EXAMPLE 3.1 ● Maximize the function $f = 4x_1 + 5x_2$ subject to the constraints

$$2x_1 + 3x_2 \leq 9 \qquad x_1 + x_2 \leq 4 \qquad x_1 \geq 0 \qquad x_2 \geq 0$$

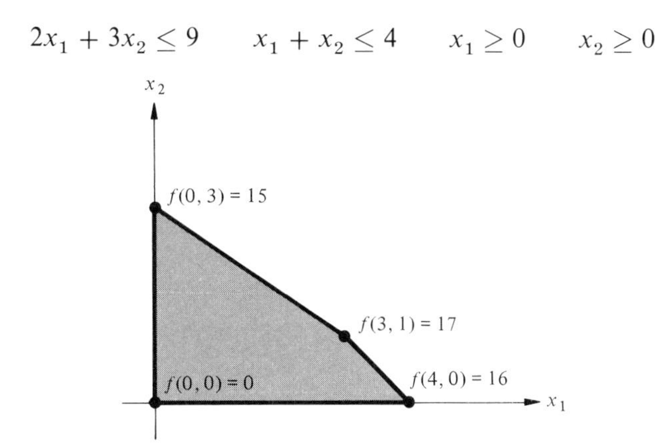

Figure 3.1 The set of feasible solutions for Example 3.1.

SOLUTION Because equations are easier to work with than inequalities, we begin by rewriting the two constraints

$$2x_1 + 3x_2 \leq 9 \qquad \text{and} \qquad x_1 + x_2 \leq 4$$

as linear equations by adding an appropriate nonnegative quantity to the left-hand side of each. That is, we take up the slack between the left- and right-hand side of each inequality by introducing nonnegative variables y_1 and y_2 such that

$$2x_1 + 3x_2 + y_1 = 9 \qquad \text{and} \qquad x_1 + x_2 + y_2 = 4$$

For obvious reasons, the variables y_1 and y_2 are called **slack variables.**

We also rewrite the equation defining the objective function as

$$-4x_1 - 5x_2 + f = 0$$

This gives us the following system of linear equations:

$$2x_1 + 3x_2 + y_1 = 9$$
$$x_1 + x_2 + y_2 = 4$$
$$-4x_1 - 5x_2 + f = 0$$

which can be represented in matrix form by the **simplex tableau**

$$\begin{array}{ccccc|c} x_1 & x_2 & y_1 & y_2 & f & \\ \hline 2 & 3 & 1 & 0 & 0 & 9 \\ 1 & 1 & 0 & 1 & 0 & 4 \\ \hline -4 & -5 & 0 & 0 & 1 & 0 \end{array}$$

Our goal is to find the solution of this linear system for which x_1, x_2, y_1, and y_2 are nonnegative and f is as large as possible. Our strategy will be to find a sequence of nonnegative solutions of the system with the property that at each stage, the corresponding value of f will be closer to the optimal value than the preceding one was.

For simplicity, we start by taking $x_1 = 0$ and $x_2 = 0$. From the first row of the simplex tableau we conclude that $y_1 = 9$ when $x_1 = 0$ and $x_2 = 0$, and from the second row we conclude that $y_2 = 4$. The third tells us that $f = 0$. The resulting solution is

$$x_1 = 0 \qquad x_2 = 0 \qquad y_1 = 9 \qquad y_2 = 4 \qquad f = 0$$

(If you look back at Figure 3.1 you will see that this solution corresponds to the fact that $f(0, 0) = 0$.)

To decide if this solution is optimal, we look at the entries in the last row of the simplex tableau. The negative entries -4 and -5 become positive coefficients when the objective function is written algebraically as

$$f = 4x_1 + 5x_2$$

and indicate that the solution is *not* optimal because we can increase the value of f by increasing x_1 or x_2. Since things become too complicated if we try to increase both variables at once, we get the next solution by increasing x_1 while keeping x_2 equal to zero. (We could just as well have kept x_1 equal to zero and increased x_2.)

The amount by which we can increase x_1 is limited by the constraints represented by the first two rows of the simplex tableau. The first row says that

$$2x_1 + 3x_2 + y_1 = 9$$

As long as x_2 remains equal to zero, this equation reduces to

$$2x_1 + y_1 = 9$$

which is equivalent to the inequality

$$2x_1 \leq 9$$

since y_1 is nonnegative. Dividing both sides of this inequality by 2, we conclude that

$$x_1 \leq \frac{9}{2} = 4.5$$

Notice that the value 4.5 in the inequality $x_1 \leq 4.5$ is simply the quotient obtained from the first row of the simplex tableau by dividing the entry in the last column by the entry in the column corresponding to x_1. The corresponding quotient from the second row tells us that

$$x_1 \leq \frac{4}{1} = 4$$

Comparing the inequalities $x_1 \leq 4.5$ and $x_1 \leq 4$, we see that 4 is the largest we can make x_1 if we are to satisfy both constraints.

The column in the simplex tableau corresponding to the variable that we decide to increase is sometimes called the **pivot column** and the row from which we obtain the upper bound on the size of this variable is known as the **pivot row.** The entry in the pivot row and pivot column is called the **pivot term.** To find the new solution, we use the operations introduced in Chapter 2, Section 3, to transform the matrix to one in which the pivot term is equal to 1 and all the other entries in the pivot column are equal to zero.

In this case, column 1 is the pivot column and row 2 is the pivot row. To transform the current matrix

$$\begin{array}{c} \text{Pivot} \\ \text{column} \\ \downarrow \end{array}$$

$$\begin{array}{c} \begin{matrix} x_1 & x_2 & y_1 & y_2 & f & \end{matrix} \\ \left[\begin{array}{rrrrr|r} 2 & 3 & 1 & 0 & 0 & 9 \\ 1 & 1 & 0 & 1 & 0 & 4 \\ \hline -4 & -5 & 0 & 0 & 1 & 0 \end{array}\right] \begin{array}{l} \\ \leftarrow \text{Pivot row} \\ \\ \end{array} \end{array}$$

we add -2 times the second row to the first row and 4 times the second row to the third row to get

$$\begin{array}{c} \begin{matrix} x_1 & x_2 & y_1 & y_2 & f & \end{matrix} \\ \left[\begin{array}{rrrrr|r} 0 & 1 & 1 & -2 & 0 & 1 \\ 1 & 1 & 0 & 1 & 0 & 4 \\ \hline 0 & -1 & 0 & 4 & 1 & 16 \end{array}\right] \begin{array}{l} -2R_2 + R_1 \\ \\ 4R_2 + R_3 \end{array} \end{array}$$

From the second row we see that $x_2 = 0$ and $y_2 = 0$ if we let $x_1 = 4$ as planned. (This is because x_2 and y_2 cannot be negative.) From the first row we see that $y_1 = 1$ when $x_2 = 0$ and $y_2 = 0$. The third row tells us that $f = 16$. We conclude that the new solution is

$$x_1 = 4 \qquad x_2 = 0 \qquad y_1 = 1 \qquad y_2 = 0 \qquad f = 16$$

(corresponding to $f(4, 0) = 16$ in Figure 3.1).

The presence of the negative entry -1 in the last row indicates that the solution is still not optimal and that we can increase the value of f by increasing x_2 from its present value of zero. In each of the first and second rows, we divide the entry in the last column by the entry in the column corresponding to x_2 to get the inequalities

$$x_2 \le \frac{1}{1} = 1 \qquad \text{and} \qquad x_2 \le \frac{4}{1} = 4$$

from which we conclude that 1 is the largest we can make x_2 if we are to satisfy both constraints.

To find the next solution we transform the matrix again, this time using column 2 and row 1 as the pivot column and pivot row, respectively. We get

$$\begin{array}{c} \begin{matrix} x_1 & x_2 & y_1 & y_2 & f & \end{matrix} \\ \left[\begin{array}{ccccc|c} 0 & 1 & 1 & -2 & 0 & 1 \\ 1 & 0 & -1 & 3 & 0 & 3 \\ \hline 0 & 0 & 1 & 2 & 1 & 17 \end{array}\right] \begin{array}{l} \\ -R_1 + R_2 \\ R_1 + R_3 \end{array} \end{array}$$

Since y_2 is still equal to zero and x_2 has been increased to 1, we see that the new solution is

$$x_1 = 3 \qquad x_2 = 1 \qquad y_1 = 0 \qquad y_2 = 0 \qquad f = 17$$

The last row of the matrix no longer contains any negative entries. This indicates that the current solution is optimal. To see this, rewrite the equation represented by the last row as

$$f = 17 - y_1 - 2y_2$$

Because the coefficients of y_1 and y_2 are negative, the only way we can increase f is to decrease y_1 or y_2. But this is impossible, since y_1 and y_2 are currently equal to zero and have to remain nonnegative.

We conclude that subject to the given constraints, the maximum value of the objective function is $f(3, 1) = 17$. Notice that this is precisely the conclusion we would draw from the geometric information summarized in Figure 3.1. ●

In the preceding example the objective function was to be maximized and all of the constraints were of the form

$$b_1x_1 + b_2x_2 + \cdots + b_nx_n \leq c$$

where $c \geq 0$. For problems in which the objective function is to be minimized, or in which some of the constraints are of the form

$$b_1x_1 + b_2x_2 + \cdots + b_nx_n \geq c$$

modifications must be made in the construction of the initial simplex tableau and in the process of finding an initial solution. These modifications are discussed in more advanced texts.

Problems

In Problems 1 through 6, use the simplex method to solve the given linear-programming problem. For the problems involving only two variables, check your answer geometrically.

1 Maximize the function $f = x_1 - 2x_2$ subject to the constraints

$$x_1 + x_2 \leq 6 \qquad 2x_1 + 3x_2 \leq 15 \qquad x_1 \geq 0 \qquad x_2 \geq 0$$

2 Maximize the function $f = 5x_1 + x_2$ subject to the constraints

$$4x_1 + 3x_2 \leq 24 \qquad x_1 + 3x_2 \leq 15 \qquad x_1 \geq 0 \qquad x_2 \geq 0$$

3 Maximize the function $f = x_1 + 3x_2$ subject to the constraints in Problem 2.

4 Maximize the function $f = 20x_1 + 30x_2$ subject to the constraints

$$x_1 + x_2 \leq 8 \qquad x_1 + 2x_2 \leq 11$$
$$x_1 + 3x_2 \leq 12 \qquad x_1 \geq 0 \qquad x_2 \geq 0$$

5 Minimize the function $f = x_1 - 3x_2$ subject to the constraints

$$3x_1 + 2x_2 \leq 18 \qquad -x_1 - 4x_2 \geq -16 \qquad x_1 \geq 0 \qquad x_2 \geq 0$$

(*Hint:* Put this problem in the standard form by multiplying the objective function and the second constraint by -1.)

6 Minimize the function $f = -2x_1 - x_2$ subject to the constraints

$$x_1 + x_2 \leq 6 \qquad -3x_1 - 2x_2 \geq -13$$
$$2x_1 + x_2 \leq 8 \qquad x_1 \geq 0 \qquad x_2 \geq 0$$

(*Hint:* Put this problem in the standard form by multiplying the objective function and the second constraint by -1.)

Versatile programs for solving linear-programming problems are available on most college and university computer systems. Use the program that is available at your school to solve the following practical problems from Section 1.

7 Example 1.1 **8** Example 1.2 **9** Example 1.3
10 Problem 5 **11** Problem 7 **12** Problem 8
13 Problem 9 **14** Problem 10 **15** Problem 11
16 Problem 12

4 • NETWORK MODELS

A **network** is a collection of points or small circles called **nodes,** pairs of which are connected by lines or curves called **arcs.** Arrows at the ends of the arcs indicate the direction of flow between the corresponding nodes. Figure 4.1 shows a network with five nodes and eight arcs. Notice that the flow between nodes 2 and 5 may be in either direction and that no direct flow is possible between nodes 1 and 5 and between nodes 1 and 3.

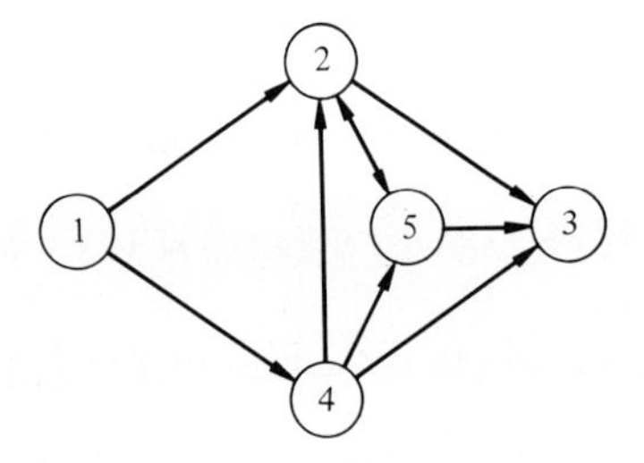

Figure 4.1 A network.

Many practical situations can be represented mathematically by networks. In Chapter 2, Section 5, for example, you saw some communications networks in which the nodes represented individuals and the arcs lines of communication. In a model of traffic flow, the arcs might represent streets and the nodes intersections. In marketing models, the nodes often represent factories and warehouses and the arcs shipping routes.

Many network problems can be formulated as linear-programming problems and solved on a computer. In this section, you will see a sampling of such problems.

A transshipment problem

EXAMPLE 4.1 ● A manufacturing firm has two factories, F_1 and F_2, that ship goods to two warehouses, W_1 and W_2, for subsequent delivery to three markets, M_1, M_2, and M_3. Each factory has a certain supply of a particular commodity and each market a certain demand. The supplies, demands, and per-unit shipping costs are summarized in the network in Figure 4.2. The manufacturer wishes to determine how many units to ship from each factory to each warehouse and how many from each warehouse to each market to meet the demand while keeping the total shipping cost as small as possible. Formulate this as a linear-programming problem.

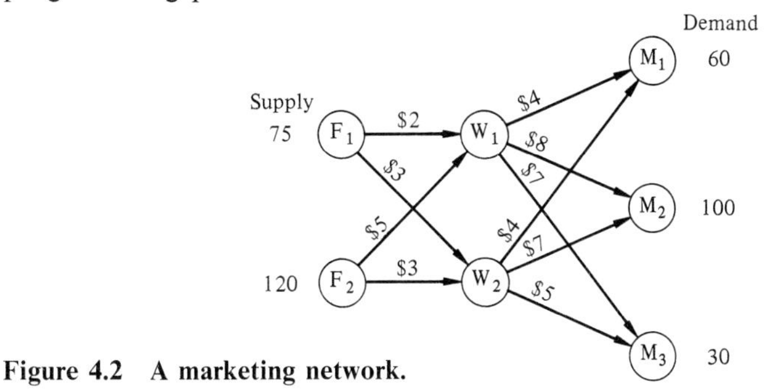

Figure 4.2 A marketing network.

SOLUTION We let

$$x_{ij} = \text{number of units shipped from } F_i \text{ to } W_j$$

and

$$y_{jk} = \text{number of units shipped from } W_j \text{ to } M_k$$

The total shipping cost is then

$$f = 2x_{11} + 3x_{12} + 5x_{21} + 3x_{22} + 4y_{11} + 8y_{12} + 7y_{13} + 4y_{21} + 7y_{22} + 5y_{23}$$

Our goal is to minimize this function subject to the following constraints: The **supply constraints**

$$x_{11} + x_{12} \leq 75 \qquad x_{21} + x_{22} \leq 120$$

stating that the number of units leaving each factory cannot exceed the existing supply; the **demand constraints**

$$y_{11} + y_{21} = 60 \qquad y_{12} + y_{22} = 100 \qquad y_{13} + y_{23} = 30$$

stating that the number of units arriving at each market must equal the corresponding demand; the **conservation of flow constraints**

$$x_{11} + x_{21} = y_{11} + y_{12} + y_{13}$$
$$x_{12} + x_{22} = y_{21} + y_{22} + y_{23}$$

stating that the number of units leaving each warehouse is the same as the number that entered; and the nonnegativity constraints

$$x_{11} \geq 0 \qquad x_{12} \geq 0 \qquad x_{21} \geq 0 \qquad x_{22} \geq 0 \qquad y_{11} \geq 0$$
$$y_{12} \geq 0 \qquad y_{13} \geq 0 \qquad y_{21} \geq 0 \qquad y_{22} \geq 0 \qquad y_{23} \geq 0$$

This linear-programming problem is not easy to solve by hand. We solved it using a computer and found that

$$x_{11} = 60 \qquad x_{12} = 10 \qquad x_{21} = 0 \qquad x_{22} = 120$$
$$y_{11} = 60 \qquad y_{12} = 0 \qquad y_{13} = 0 \qquad y_{21} = 0 \qquad y_{22} = 100 \qquad y_{23} = 30$$

resulting in a minimal shipping cost of $1,600. ●

A shortest-path problem

The next problem involves the selection of the shortest route between two nodes in a network. The choice is obvious in this simple illustrative example, but mathematical formulations and computer solutions are required in more realistic situations.

EXAMPLE 4.2 ● A motorist wishes to drive from C_1 to C_6 in the shortest possible time. The available routes and estimated driving times (in hours) are shown in the network in Figure 4.3. Formulate this as a linear-programming problem.

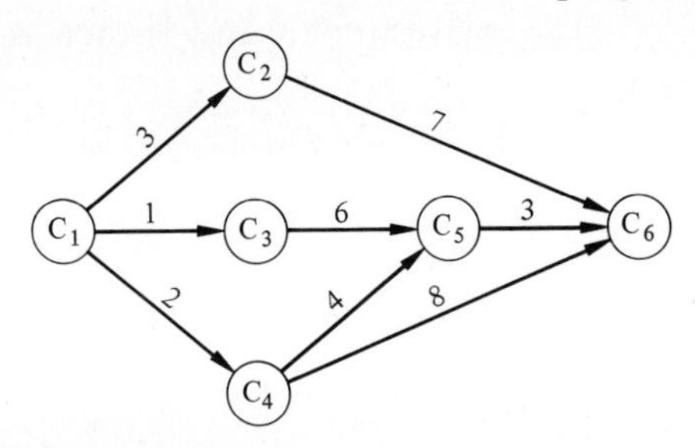

Figure 4.3 A network of highways.

SOLUTION We let

$$x_{ij} = \begin{cases} 1 & \text{if the road from } C_i \text{ to } C_j \text{ is part of the chosen route} \\ 0 & \text{if the road from } C_i \text{ to } C_j \text{ is not part of the chosen route} \end{cases}$$

Each possible route can be described by assigning appropriate values to the variables x_{ij}. For example, the route $C_1 \rightarrow C_2 \rightarrow C_6$ can be represented by the values

$$x_{12} = x_{26} = 1 \qquad \text{and} \qquad x_{13} = x_{35} = x_{56} = x_{14} = x_{45} = x_{46} = 0$$

Our goal is to find the route for which the total driving time is minimal. The driving time is given by the function

$$f = 3x_{12} + 7x_{26} + x_{13} + 6x_{35} + 3x_{56} + 2x_{14} + 4x_{45} + 8x_{46}$$

Observe that for any particular route, the variables corresponding to roads on that route will be 1 and all the others will be zero. Therefore, the only nonzero terms in the sum f will be the driving times for the roads that make up the route in question.

The restrictions imposed by the network are represented by the following algebraic constraints: The constraint at the origin

$$x_{12} + x_{13} + x_{14} = 1$$

stating that exactly one road leaving C_1 is chosen; the constraint at the destination

$$x_{26} + x_{56} + x_{46} = 1$$

stating that exactly one road entering C_6 is chosen; the conservation of flow constraints

$$x_{12} = x_{26} \text{ (for } C_2\text{)} \qquad x_{13} = x_{35} \text{ (for } C_3\text{)}$$
$$x_{14} = x_{45} + x_{46} \text{ (for } C_4\text{)} \qquad x_{35} + x_{45} = x_{56} \text{ (for } C_5\text{)}$$

stating that the number of roads in the chosen route leaving each of C_2, C_3, C_4, and C_5 is equal to the number entering; and the nonnegativity constraints

$$x_{12} \geq 0 \quad x_{13} \geq 0 \quad x_{14} \geq 0 \quad x_{26} \geq 0 \quad x_{35} \geq 0$$
$$x_{45} \geq 0 \quad x_{46} \geq 0 \quad x_{56} \geq 0$$

(In addition, we require that the value of each variable is either 0 or 1. We do not list these additional constraints explicitly because it turns out that the solutions to shortest-path problems of this sort always satisfy these conditions anyway.)

Our goal, then, is to minimize the objective function f subject to the constraint at the origin, the constraint at the destination, the conservation of flow constraints, and the nonnegativity constraints. The solution is

$$x_{14} = x_{45} = x_{56} = 1 \quad \text{and} \quad x_{12} = x_{13} = x_{26} = x_{35} = x_{46} = 0$$

indicating that the shortest route is $C_1 \rightarrow C_4 \rightarrow C_5 \rightarrow C_6$. ●

The critical-path method

Certain problems involving the scheduling of interrelated tasks or activities can be formulated as network problems in which the goal is to find the longest path between two nodes. Here is an example.

EXAMPLE 4.3 ● A traveler planning a European trip must complete certain activities before departing. Some of these activities cannot be started until others have been completed. The activities, their predecessors, and their durations are listed in the following table.

Description of activity	Duration (in days)	Immediate predecessors
1. Plan itinerary	5	None
2. Obtain passport picture	4	None
3. Make plane reservations	1	Itinerary
4. Obtain international driver's license	1	Passport picture
5. Obtain passport	28	Passport picture
6. Make hotel reservations	45	Plane reservations
7. Arrange automobile rental	3	Plane reservations and international driver's license
8. Purchase Eurailpass	7	Passport

The traveler wishes to determine how quickly all these activities can be completed. Formulate this as a linear-programming problem.

SOLUTION To represent the situation as a network, we construct the **project graph** in Figure 4.4 in which the nodes correspond to activities, the arcs join activities and their immediate predecessors, and the number next to each arc is the duration of the activity from which the arc is drawn.

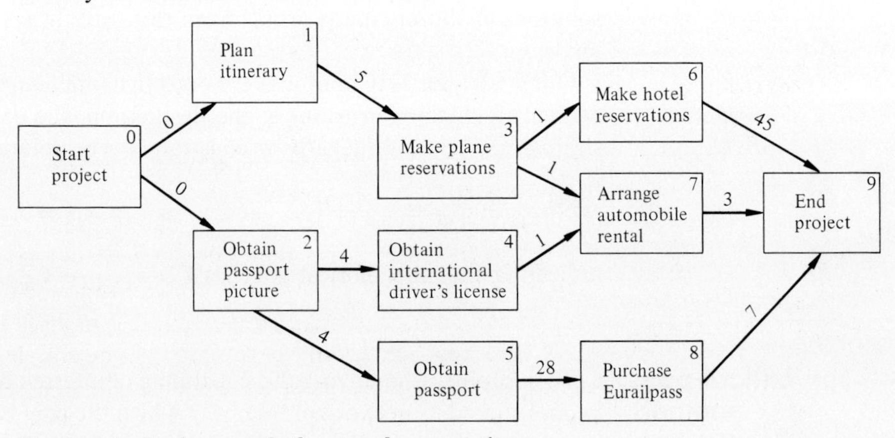

Figure 4.4 Project graph for travel preparations.

Since all the activities must be performed, the minimum time required for the completion of the project is the length (in days) of the *longest* path from the node representing the start of the project to the node representing its end. To construct the linear-programming model for this longest-path problem, we let

$$x_{ij} = \begin{cases} 1 & \text{if the arc from node } i \text{ to node } j \text{ is on the path} \\ 0 & \text{if the arc from node } i \text{ to node } j \text{ is not on the path} \end{cases}$$

For any path through the network, the function

$$f = 5x_{13} + 4x_{24} + 4x_{25} + x_{36} + x_{37} + x_{47} + 28x_{58} + 45x_{69} + 3x_{79} + 7x_{89}$$

gives the number of days needed to complete all the activities on the path. Our goal is to maximize this function subject to the constraints

$x_{01} + x_{02} = 1$ (for node 0) $x_{01} = x_{13}$ (for node 1)

$x_{02} = x_{24} + x_{25}$ (for node 2) $x_{13} = x_{36} + x_{37}$ (for node 3)

$x_{24} = x_{47}$ (for node 4) $x_{25} = x_{58}$ (for node 5)

$x_{36} = x_{69}$ (for node 6) $x_{37} + x_{47} = x_{79}$ (for node 7)

$x_{58} = x_{89}$ (for node 8) $x_{69} + x_{79} + x_{89} = 1$ (for node 9)

and $x_{ij} \geq 0$ for all relevant i and j. (As in Example 4.2, the unstated condition that the value of each variable should be 0 or 1 will be satisfied anyway.)

In this simple illustrative example, the solution can be read off from the network itself. It is

$$x_{01} = x_{13} = x_{36} = x_{69} = 1$$
$$x_{02} = x_{24} = x_{47} = x_{37} = x_{79} = x_{25} = x_{58} = x_{89} = 0$$

indicating that the longest path is $0 \to 1 \to 3 \to 6 \to 9$. Since the length of this path is $0 + 5 + 1 + 45 = 51$, we conclude that at least 51 days are required for the completion of the project. ●

The longest path through a project graph is sometimes referred to as a **critical path** because the activities on such a path cannot be delayed without delaying the completion of the entire project. There is more flexibility in the scheduling of activities that are not on a critical path. For example, the traveler in Example 4.3 need not send for a passport as soon as the passport pictures are ready but may do so anytime up to $28 + 7 = 35$ days before departure. (Do you see why?)

Problems Formulate each of the following as a linear-programming problem. Solve the problem, using a computer if necessary.

1 A manufacturing firm has three factories, F_1, F_2, and F_3, that ship goods to two warehouses, W_1 and W_2, for subsequent delivery to two markets, M_1 and M_2. Factories F_1, F_2, and F_3 have supplies of 75, 125, and 150 units, respectively, and markets M_1 and M_2 demand 150 and 200 units, respectively. The per-unit shipping costs are summarized in the following tables.

From \ To	W_1	W_2
F_1	\$3	\$5
F_2	\$2	\$4
F_3	\$5	\$1

From \ To	M_1	M_2
W_1	\$1	\$5
W_2	\$3	\$6

How many units should the manufacturer ship from each factory to each warehouse and how many from each warehouse to each market to meet the demand while keeping the total shipping cost as small as possible?

2 An oil pipeline is to be built between a newly discovered oil field and a major industrial area. The possible routes for the pipeline and the installation cost of each segment are shown in the following diagram.

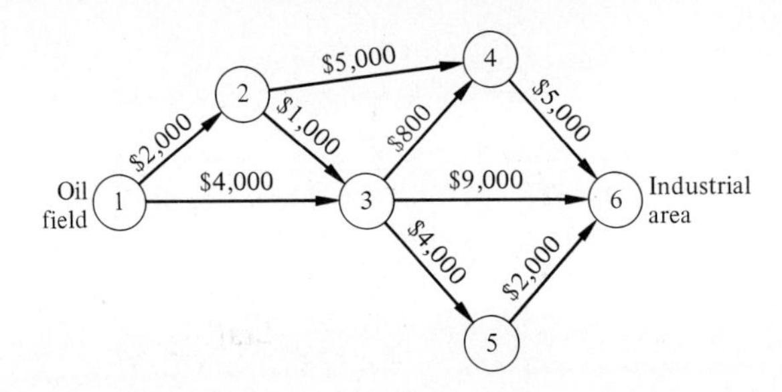

For what route is the total installation cost the smallest?

3 The following diagram shows the times (in hours) required for a message to travel between pairs of individuals in a certain communications network.

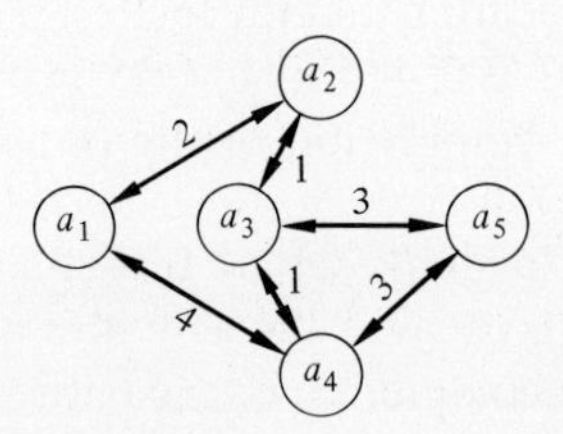

Through which intermediaries should a_1 send a message so that it will reach a_5 in the shortest possible time?

4 The following table shows the activities involved in building a house, their durations, and their predecessors.

Description of activity	Duration (in days)	Immediate predecessors
1. Prepare site	4	None
2. Lay foundation	3	Site
3. Build frame	15	Foundation
4. Install wiring	5	Frame
5. Install plumbing	4	Frame
6. Finish interior	8	Wiring and plumbing
7. Finish exterior	10	Frame
8. Install carpeting	2	Interior and exterior
9. Do landscaping	3	Exterior

How quickly can the house be built?

5 Some of the psychology courses at a certain college are listed below with their prerequisites.

Course number	Title	Prerequisites
1	Introduction to psychology	None
2	Intermediate psychology	Psych. 1
28	Child development	Psych. 1
30	Social psychology	Psych. 1
55	Statistics	None
124	Experimental design	Psych. 2 and 55
286	Psychometrics	Psych. 30 and 124
341	Seminar in testing and evaluation	Psych. 28 and 286

Students at the college may take up to three courses per semester in any one department. For at least how many semesters must a student be enrolled at the college before being eligible to take the seminar on testing and evaluation?

6 To eliminate overcrowding in some of its schools, a school board has decided to bus some students to nearby schools in the district. The following diagram shows the seven schools in the district, the possible bus routes, and the driving times (in minutes) between pairs of schools.

A surplus of students at a school is indicated by a positive number next to the corresponding node and a shortage by a negative number. The zero next

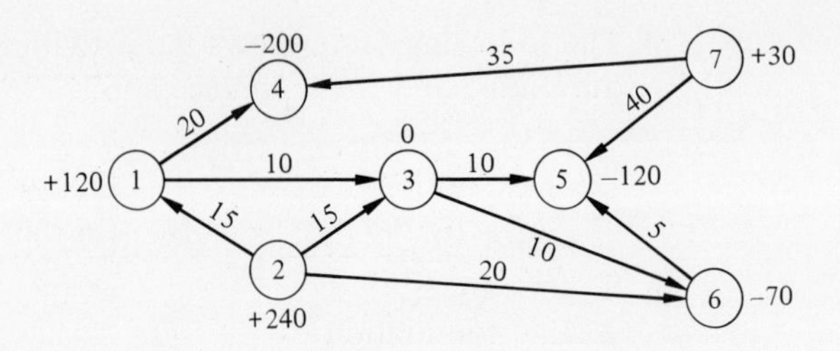

to node 3 indicates that this school is filled to capacity with neither a surplus nor a shortage. However, it is possible for students from schools 1 or 2 to be bused to school 3 while students currently at school 3 are bused to schools 5 or 6. How should the school board redistribute the students to eliminate all overcrowding in the district while keeping the total number of hours students spend on buses at a minimum?

7 A car rental company has offices in five cities. To eliminate shortages of cars at some of the offices, the company is planning to hire drivers to take some of the cars from one office to another. The following diagram shows the five offices, the possible routes, the mileage between pairs of offices, and the surplus or shortage of cars at each office. How should the company redistribute the cars to eliminate all shortages while minimizing the total distance driven?

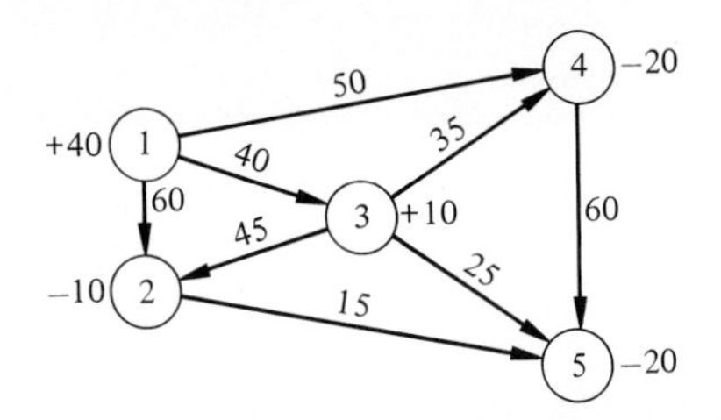

8 The following diagram shows the pipeline network through which water flows from A to B. The number next to each arc is the capacity in units of a thousand gallons per second of the corresponding section of pipe. What is the maximum rate at which water can flow into B?

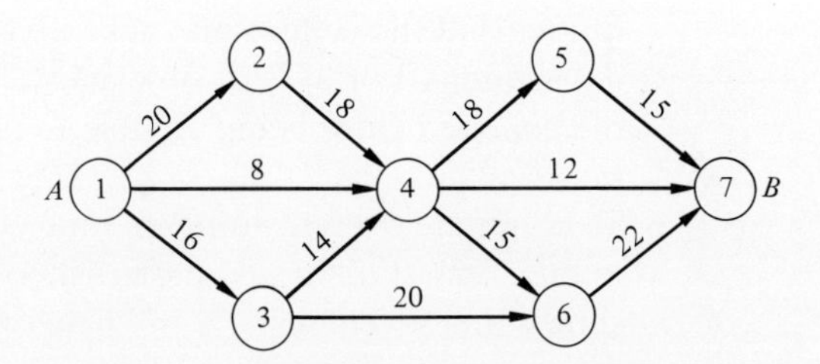

9 The highways connecting the downtown business district of a certain city with one of its residential suburbs are shown in the following diagram. The

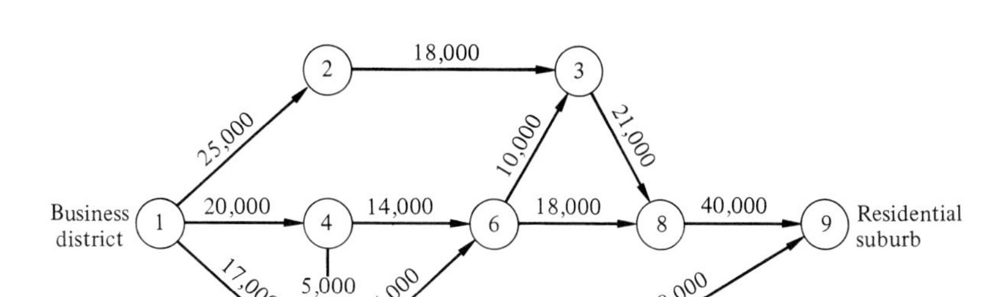

number next to each segment of highway is the capacity of the segment, that is, the maximum number of vehicles per hour that can flow past a point on the segment. What is the maximum number of vehicles from the business district that can arrive at the suburb each hour?

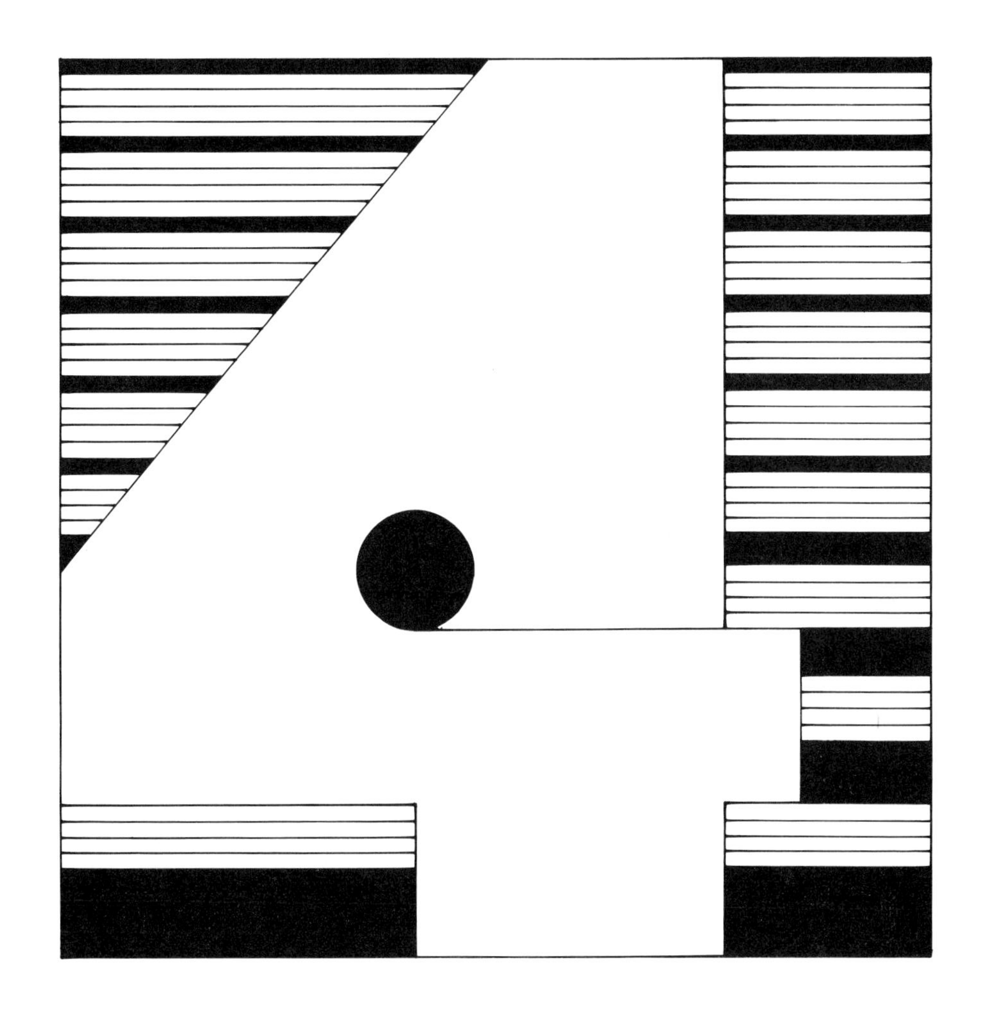

SETS AND COUNTING

1 • SETS

A **set** is any collection of objects. The objects belonging to a set are called **elements.** If the number of elements in a set is small, it is traditional to describe the set by listing its elements and enclosing the list in braces. For example, $\{2, 4, 6, 8\}$ is the set whose elements are the first four positive even numbers, and {Cleveland, Muskogee, Weed} is a set whose elements are American cities.

If there are too many elements in a set to list conveniently, we often use what is known as **set builder notation** to describe the set. For example, $\{x \mid x$ is an integer$\}$ describes the set of integers and is read "the set of all x such that x is an integer." With this notation, x represents an arbitrary element of the set, the slash means "such that" and a description of a typical element follows the slash.

It is customary to use the symbol $\in$ to denote the phrase "is an element of" and $\notin$ for the phrase "is not an element of." For example, Weed $\in \{x \mid x$ is a city in California$\}$ and Cleveland $\notin \{x \mid x$ is a city with clean air$\}$.

A set can have a finite or an infinite number of elements. The set containing no elements is known as the **empty set** and is denoted by the symbol $\varnothing$. As we shall see, the empty set appears naturally in many situations.

Equality of sets

Two sets are said to be **equal** if and only if they have the same elements. It makes no difference how the elements are described or the order in which they are listed. For example,

$$\begin{aligned}\{1, 3, 5, 7\} &= \{7, 3, 5, 1\}\\ &= \{x \mid x \text{ is an odd number between 1 and 7, inclusive}\}\end{aligned}$$

Equality of sets
Two sets are *equal* if and only if they have the same elements.

Subsets

A set A whose elements are all elements of a set B is said to be a **subset** of B. To denote this, we write $A \subset B$. For example, $\{2, 4, 6\} \subset \{2, 4, 6, 8, 10\}$ and $\{x \mid x$ is a young turtle on the Los Angeles freeway$\} \subset \{x \mid x$ is a turtle with a short life expectancy$\}$. Every set is a subset of itself, and the empty set is a subset of every set. (Do you see why?)

The notion of subset is illustrated in Figure 1.1. Diagrams like the one in Figure 1.1 are known as **Venn diagrams** and are named after the British logician John Venn (1834–1923), who used them to illustrate logical relations.

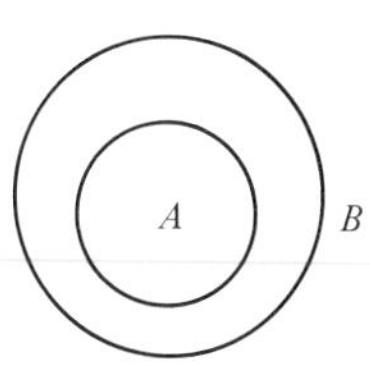

Figure 1.1 ***A* is a subset of *B*: $A \subset B$.**

Subsets
A is a *subset* of B if and only if every element of A is also an element of B. If A is a subset of B we write $A \subset B$.

The following example further illustrates the notion of a subset.

EXAMPLE 1.1 ● Let $B = \{x, y, z\}$. List all the subsets of B.

SOLUTION The subsets of B are listed as follows: $\{x\}$, $\{y\}$, $\{z\}$, $\{x, y\}$, $\{x, z\}$, $\{y, z\}$, $\{x, y, z\}$, $\varnothing$. ●

Union and intersection

Two important ways of combining sets are by **union** and **intersection.**

The union of sets
The *union* $A \cup B$ of two sets A and B is the set of all elements that are elements of A or B or both. In general, the union of any collection of sets is the set of all elements that are in one or more of the sets in the collection.

In Figure 1.2, the shaded regions represent the indicated unions.

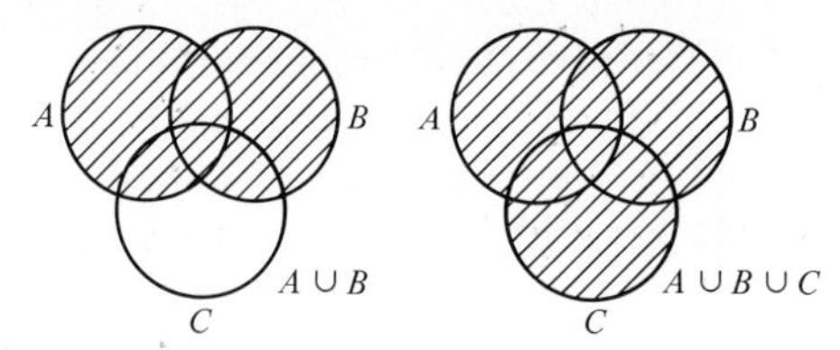

Figure 1.2 The union of sets.

The intersection of sets
The *intersection* $A \cap B$ of two sets A and B is the set of all elements that are elements of both A and B. The intersection of any collection of sets is the set of all elements that are in every set in the collection.

In Figure 1.3 the shaded regions represent the indicated intersections.

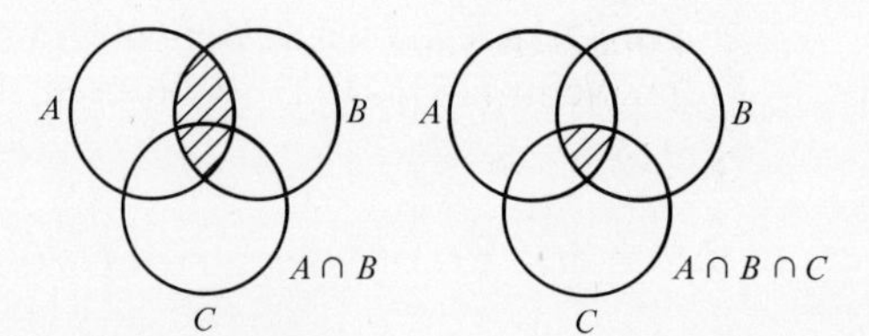

Figure 1.3 The intersection of sets.

The next example further illustrates the concepts of union and intersection.

EXAMPLE 1.2 ● Let $A = \{1, 3, 5, 7\}$, $B = \{1, 3, 9, 0\}$, $C = \{1, 3, 5\}$, $D = \{2, 8, 0\}$. Find:

(a) $A \cup B$ (b) $A \cup C$ (c) $A \cap C$
(d) $A \cap D$ (e) $A \cap B \cap C$ (f) $B \cap D$

SOLUTION (a) $A \cup B = \{1, 3, 5, 7, 9, 0\}$. Notice that each element is listed only once.
(b) $A \cup C = \{1, 3, 5, 7\}$.
(c) $A \cap C = \{1, 3, 5\}$.
(d) $A \cap D = \varnothing$, the empty set.
(e) $A \cap B \cap C = \{1, 3\}$.
(f) $B \cap D = \{0\}$. Note: $\{0\}$ is *not* the empty set. ●

Disjoint sets If two sets have no common elements, as in part (d) of Example 1.2, they are said to be **disjoint** or **mutually exclusive.** That is, A and B are disjoint if and only if $A \cap B = \varnothing$. The Venn diagram in Figure 1.4 shows two disjoint sets.

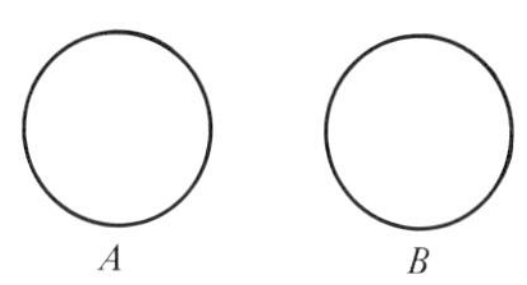

Figure 1.4 A **and** B **are disjoint:** $A \cap B = \varnothing$.

Complements and universal sets In the context of a given problem, there is usually a natural set that contains all objects under study. This set is called the **universal set** and will be denoted by Ω, the Greek letter omega. The **complement** of a given set is the set of all elements of the universal set that are not elements of the given set. We shall use A' to denote the complement of A. In Figure 1.5 the shaded region represents the complement of A in the universal set Ω.

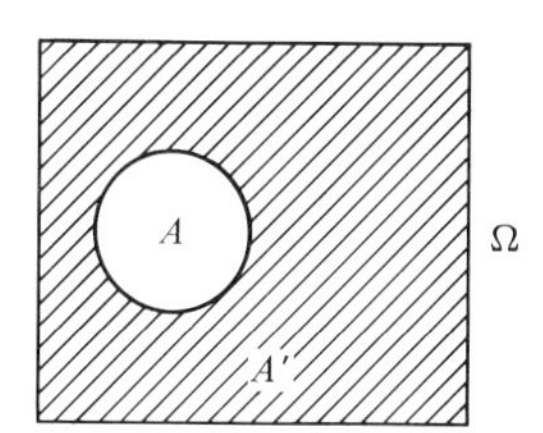

Figure 1.5 The complement of a set A**, relative to the universal set** Ω.

> **Complement of a set**
> The *complement* A' of a set A is the set of all elements in the universal set that are not in A.

The next example illustrates these ideas.

EXAMPLE 1.3 ● Let $\Omega = \{x \mid x \text{ is a nonnegative integer}\}$, $A = \{x \mid x \text{ is even}\}$, $B = \{x \mid x \geq 4\}$. Find:

(a) A' (b) B' (c) Ω' (d) $\varnothing'$

SOLUTION (a) $A' = \{x \mid x \text{ is odd}\}$.
(b) $B' = \{0, 1, 2, 3\}$.
(c) Since there are no elements in the universal set which are not in the universal set, $\Omega' = \varnothing$.
(d) $\varnothing' = \Omega$. ●

It is always true that $A \cup A' = \Omega$, $A \cap A' = \varnothing$, and $(A')' = A$. (Why?)

Distributive laws for union and intersection

Some problems involving both union and intersection can be simplified by use of **distributive laws** that are similar to those used for addition and multiplication of numbers.

Distributive laws for union and intersection
For any sets A, B, and C,

1 $A \cup (B \cap C) = (A \cup B) \cap (A \cup C)$
2 $A \cap (B \cup C) = (A \cap B) \cup (A \cap C)$

The next example illustrates how Venn diagrams can be used to establish the first distributive law.

EXAMPLE 1.4 ● Use Venn diagrams to establish the first distributive law $A \cup (B \cap C) = (A \cup B) \cap (A \cup C)$.

SOLUTION Starting with the set $B \cap C$ in Figure 1.6*a*, we get the set $A \cup (B \cap C)$ in Figure 1.6*b*. Starting with $A \cup B$ in Figure 1.6*c* and $A \cup C$ in Figure 1.6*d*, we get $(A \cup B) \cap (A \cup C)$ in Figure 1.6*e*. Comparing Figures 1.6*b* and 1.6*e* we conclude that the first distributive law holds. ●

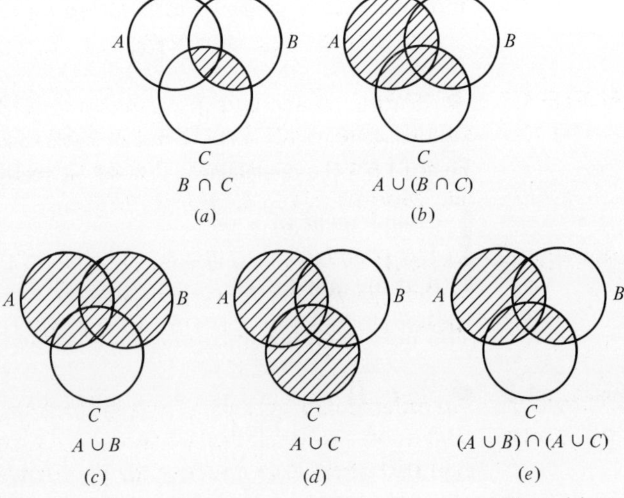

Figure 1.6 The first distributive law: $A \cup (B \cap C) = (A \cup B) \cap (A \cup C)$.

The next example illustrates the use of the second distributive law.

EXAMPLE 1.5 ● Let $A = \{x \mid x \text{ is a positive integer less than } 14\}$, $B = \{x \mid x \text{ is odd}\}$, $C = \{x \mid x \text{ is a positive multiple of } 3\}$. Use the second distributive law to obtain $A \cap (B \cup C)$.

SOLUTION Since $A \cap B = \{1, 3, 5, 7, 9, 11, 13\}$ and $A \cap C = \{3, 6, 9, 12\}$, we obtain $A \cap (B \cup C) = (A \cap B) \cup (A \cap C) = \{1, 3, 5, 6, 7, 9, 11, 12, 13\}$. If you try to solve this problem without using the second distributive law, you will see that the sets you have to work with are more complicated. ●

De Morgan's laws

De Morgan's laws, named after the British mathematician Augustus De Morgan (1806–1871), provide useful formulas for the complements of unions and intersections.

De Morgan's laws
For any sets A and B

1 $(A \cup B)' = A' \cap B'$
2 $(A \cap B)' = A' \cup B'$

Similar formulas hold for more than two sets.

The next two examples illustrate the use of De Morgan's laws.

EXAMPLE 1.6 ● Let $\Omega = \{x \mid x \text{ is a positive integer}\}$, $A = \{x \mid x \text{ is greater than } 20\}$, and $B = \{x \mid x \text{ is odd}\}$. Use De Morgan's laws to find $(A \cup B)'$ and $(A \cap B)'$.

SOLUTION Since $A' = \{x \mid x \text{ is less than or equal to } 20\}$, and $B' = \{x \mid x \text{ is even}\}$, it follows from De Morgan's laws that $(A \cup B)' = A' \cap B' = \{x \mid x \text{ is an even number that is less than or equal to } 20\}$, and $(A \cap B)' = A' \cup B' = \{x \mid x \text{ is even or } x \text{ is less than or equal to } 20\}$. ●

EXAMPLE 1.7 ● Farmer Jones' chickens are either speckled or brown. Some are big and the others are small. Let $\Omega = \{x \mid x \text{ is a chicken owned by Farmer Jones}\}$, $A = \{x \mid x \text{ is a speckled chicken}\}$, and $B = \{x \mid x \text{ is a big chicken}\}$. Use De Morgan's laws to find $(A \cup B)'$.

SOLUTION Since $A' = \{x \mid x \text{ is a brown chicken}\}$ and $B' = \{x \mid x \text{ is a small chicken}\}$, it follows from De Morgan's laws that $(A \cup B)' = A' \cap B' = \{x \mid x \text{ is a small brown chicken}\}$. Try to do this directly, without using De Morgan's laws. ●

Cartesian products

An **ordered pair** of objects is a pair (a, b) whose members are listed in a specific order. For example, the coordinates of a point in the xy plane form an ordered pair with the first member of the pair denoting the x coordinate and the second member denoting the y coordinate. Two ordered pairs (a, b) and (c, d) are equal if and only if $a = c$ and $b = d$. An **ordered n-tuple** is a sequence of n objects listed in a specific order. For example, in baseball it is traditional to use an ordered triple of numbers to specify runs, hits, and errors. Sets of ordered n-tuples are important in the social and managerial sciences, particularly when processes that occur in a specific order or over a period of time are being studied. Certain sets whose elements are ordered n-tuples are called **cartesian products,** named after René Descartes, the famous seventeenth century French mathematician and philosopher. Here is the formal definition.

Cartesian product
The *cartesian product* $A \times B$ of two sets A and B, is the set of all ordered pairs (a, b), for which $a \in A$ and $b \in B$. In general, $A_1 \times A_2 \times \cdots \times A_n$ is the set of all ordered n-tuples $(a_1, a_2, \ldots, a_n)$, where $a_i \in A_i$ for $i = 1, 2, \ldots, n$.

The next two examples illustrate this concept.

EXAMPLE 1.8 ● List the elements in the cartesian product $A \times B$ if $A = \{a, b\}$ and $B = \{c, d, e\}$.

SOLUTION $A \times B = \{(a, c), (a, d), (a, e), (b, c), (b, d), (b, e)\}$ ●

EXAMPLE 1.9 ● Show how the set of three-letter "words," in which the first and last letters can be any letters of the alphabet and the second letter must be a vowel, can be expressed as a cartesian product.

SOLUTION If we let $A = C = \{x \mid x \text{ is a letter of the alphabet}\}$ and $B = \{\text{a, e, i, o, u}\}$, then the set of three-letter words (ordered triples) just described is the cartesian product $A \times B \times C$. ●

Problems

1 Let $A = \{x \mid x \text{ is an integer between 1 and 15, inclusive}\}$. Decide whether the following statements are true or false.

(a) $1 \in A$ (b) $16 \in A$ (c) $15 \notin A$
(d) $15 \subset A$ (e) $\{15\} \subset A$ (f) $\{1, 2, 11\} \subset A$
(g) $A \subset A$ (h) $\{A\} \subset A$ (i) $A \in A$
(j) $\varnothing \in A$ (k) $\{\varnothing\} \subset A$ (l) $\varnothing \subset A$

2 Let $A = \{1, 2, 3, 4, 5, 6\}$. Decide which of the following sets are equal to A.

(a) $\{x \mid x \text{ is an integer between 1 and 6, inclusive}\}$
(b) $\{6, 5, 3, 2, 4, 1\}$
(c) $\{1, 2, 3, 4, 5, 6, 7\}$
(d) $\{1, 2, 3\}$
(e) $\{x \mid x \text{ is a positive integer whose square is less than 40}\}$

3 Denote the following sets using set-builder notation.
(a) The set of stars in the Milky Way.
(b) The set of U.S. cities west of the Mississippi River.
(c) The set of positive integers that are divisible by 11.

4 Let $A = \{1, 3, 5, 7, 8\}$. Decide whether the following statements are true or false.
(a) $A \subset \{x \mid x \text{ is an odd integer}\}$
(b) $A \subset \{x \mid x \text{ is an integer less than } 10\}$
(c) $A \subset \{1, 3, 5, 7\}$
(d) $A \subset \{3, 1, 5, 7, 8, 9\}$
(e) $A \subset A$

5 List all the subsets of the set $\{w, x, y, z\}$.

6 Let $A = \{0, 1, 2\}$, $B = \{1, 2, 3\}$, and $C = \{3, 4\}$. Find:
(a) $A \cup B$ (b) $A \cap B$ (c) $B \cap C$
(d) $A \cap C$ (e) $A \cup B \cup C$ (f) $A \cap B \cap C$

7 Using Venn diagrams, prove the second distributive law: $A \cap (B \cup C) = (A \cap B) \cup (A \cap C)$.

8 Use Venn diagrams to prove De Morgan's laws.

9 Use Venn diagrams to justify the following versions of De Morgan's laws for three sets:
(a) $(A \cup B \cup C)' = A' \cap B' \cap C'$
(b) $(A \cap B \cap C)' = A' \cup B' \cup C'$

10 Show, using Venn diagrams, that if $A \subset B$, then $B' \subset A'$.

11 Shade the indicated regions in the following Venn diagram.

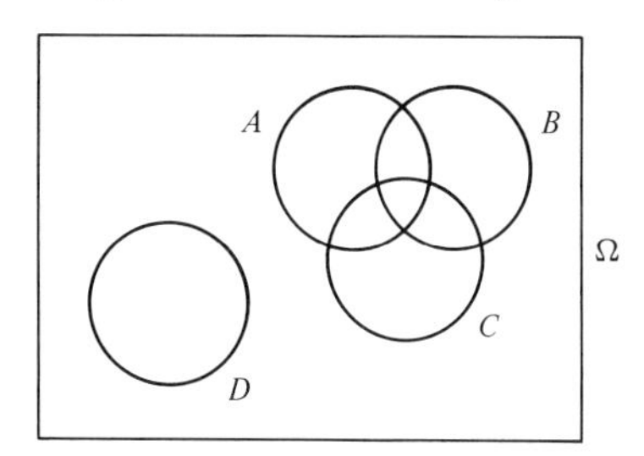

(a) $A \cap B$ (b) $A \cap B \cap C$ (c) $A \cup D$
(d) $(A \cup D)'$ (e) $A \cap D$ (f) $A \cap (B \cup C)$
(g) $A \cap (B \cup D)$

12 Let $A = \{x \mid x \text{ is an odd prime number}\}$, $B = \{x \mid x \text{ is a positive integer divisible by } 6\}$, and $C = \{x \mid x \text{ is an odd number}\}$.
(a) Define an appropriate universal set Ω (b) Find $A \cap B$
(c) Find $A \cap C$ (d) Find $A \cup C$
(e) Find $(A \cap C)'$ (f) Find $(B \cup C)'$

13 Let $A = \{1\}$ and $B = \{\{1\}\}$. Decide whether the following statements are true or false.
(a) $A \subset B$ (b) $A \in B$ (c) $A = B$
(d) $A \cup B = \{1, \{1\}\}$ (e) $A \cap B = \varnothing$

14 All the apples in a certain bag are either red or green. Let $\Omega = \{x \mid x \text{ is an apple in the bag}\}$, $A = \{x \mid x \text{ is a red apple}\}$, $B = \{x \mid x \text{ is a green apple}\}$, and $C = \{x \mid x \text{ is a large apple}\}$. Find the following sets:
(a) $A \cup B$ (b) $A \cap C$ (c) $B \cup C$
(d) $(B \cup C)'$ (e) $B' \cap C'$ (f) $B \cap C'$

In Problems 15 and 16, $\#(A)$ denotes the number of elements in the set A.

15 Suppose there are a total of 20 apples in the bag in Problem 14, and that 12 are red and 11 are large, with 7 of the red apples being large. Find:
(a) $\#(A)$ (b) $\#(B \cap C)$ (c) $\#(A \cup C)$
(d) $\#(A \cup C)'$ (e) $\#(B \cap C')$

16 Explain why the following formulas are true:
(a) $\#(A) + \#(A') = \#(\Omega)$
(b) $\#(A \cup B) = \#(A) + \#(B) - \#(A \cap B)$
(c) $\#(A \cap B') + \#(B \cap A') = \#(A \cup B) - \#(A \cap B)$

17 The **difference,** $A - B$, of two sets A and B is the set of all elements in A that are not in B. Use Venn diagrams to illustrate the following sets.
(a) $A - B$ (b) $(A - B) \cup (B - A)$
(c) $(A \cup B) - (A \cap B)$ (d) $\Omega - A$

18 Using Venn diagrams (or some other method) show that the following equations are true in general.
(a) $A - B = A \cap B'$
(b) $\Omega - A = A'$
(c) $(A - B) \cup (B - A) = (A \cap B) - (A \cap B)$. This set is known as the **symmetric difference** of A and B.
(d) $A - A' = A$

19 One way to describe an infinite set whose elements form a recognizable sequence is to write down the first few elements and then three dots to indicate continuation of the sequence. For example, we could write $\{1, 2, 3, \ldots\}$ to denote the set of positive integers. The following sets are denoted in this way. Describe them in words.
(a) $\{2, 4, 6, 8, \ldots\}$ (b) $\{1, 4, 9, 16, 25, \ldots\}$
(c) $\{8, 16, 24, 32, \ldots\}$ (d) $\{19, 20, 21, 22, \ldots\}$

20 Let $A_1 = \{1, 2, 3, \ldots\}$, $A_2 = \{2, 3, 4, \ldots\}$, $A_3 = \{3, 4, 5, \ldots\}$, $A_4 = \{4, 5, 6, \ldots\}$, ... and so on, so that for any positive integer n, $A_n = \{n, n + 1, n + 2, \ldots\}$. Find the intersection of all the sets A_n, that is, find $A_1 \cap A_2 \cap A_3 \cap \cdots$.

21 List the elements in the cartesian product $A \times B$ for the following pairs of sets.
(a) $A = \{0, 1\}$, $B = \{0, 1\}$
(b) $A = \{x, y, z\}$, $B = \{5, 6\}$
(c) $A = \{r, s, t\}$, $B = \{r, s, t\}$

22 List the elements in the cartesian product $A \times B \times C$, where $A = \{0, 1\}$, $B = \{0, 1\}$, and $C = \{0, 1\}$.

23 You bowl four games with a friend. Show how the record of your wins and losses can be expressed as an element of a cartesian product.

24 Find the number of elements in each of the cartesian products $A \times B$ in Problem 21. Derive a general formula for the number of elements in $A \times B$ in terms of the number of elements in A and the number of elements in B.

25 Find the number of elements in the cartesian product $A \times B \times C$ in Problem 22. Derive a general formula for the number of elements in a cartesian product $A_1 \times A_2 \times \cdots \times A_n$ in terms of the number of elements in each set A_i.

2 • COUNTING

In this section, we shall develop some important techniques for counting various collections and arrangements of objects. In the next chapter, we shall use some of these methods to solve probability problems. All the sets we work with in this section are assumed to contain a finite number of elements. We shall use the symbol $\#(A)$ to denote the number of elements in a particular set A. For example, we shall write $\#(A) = 2$ to indicate that A contains 2 elements.

Counting the elements of a cartesian product

The first counting formula we shall derive is for the number of elements in the cartesian product of two sets. The formula is easy to discover.

EXAMPLE 2.1 ● Find $\#(A \times B)$ if $A = \{p, q\}$ and $B = \{r, s, t\}$.

SOLUTION We list the elements of $A \times B$ in the following 2×3 rectangular array.

	r	s	t
p	(p, r)	(p, s)	(p, t)
q	(q, r)	(q, s)	(q, t)

The array has two rows, one for each element of A, and three columns, one for each element of B. It follows that $\#(A \times B) = 2(3) = 6$. ●

The reasoning in Example 2.1 can be used for any pair of sets to derive the formula $\#(A \times B) = \#(A)\#(B)$. This formula can be generalized to any finite number of sets.

Number of elements in a cartesian product
The number of elements in the cartesian product of a finite number of sets is the product of the numbers of elements in the individual sets. That is,

$$\#(A_1 \times A_2 \times \cdots \times A_n) = \#(A_1)\#(A_2) \cdots \#(A_n)$$

The next example shows how this formula is derived for a cartesian product of three sets.

EXAMPLE 2.2 ● If $\#(A) = 2$, $\#(B) = 2$, and $\#(C) = 3$, show that $\#(A \times B \times C) = \#(A)\#(B)\#(C)$.

SOLUTION We begin by choosing three arbitrary sets with the appropriate numbers of elements: $A = \{a, b\}$, $B = \{p, q\}$, and $C = \{x, y, z\}$. We can list the elements of $A \times B \times C$ in a rectangular array with $\#(A)\#(B) = 4$ rows, one for each

	x	y	z
(a, p)	(a, p, x)	(a, p, y)	(a, p, z)
(a, q)	(a, q, x)	(a, q, y)	(a, q, z)
(b, p)	(b, p, x)	(b, p, y)	(b, p, z)
(b, q)	(b, q, x)	(b, q, y)	(b, q, z)

element of $A \times B$, and $\#(C) = 3$ columns, one for each element of C. We conclude that $\#(A \times B \times C) = \#(A)\#(B)\#(C)$. ●

Here are two applications of the formula for the number of elements in a cartesian product.

EXAMPLE 2.3 ● (a) How many two-letter "words" can be made using the letters of the alphabet?
(b) How many five-letter words can be made?
(c) How many three-letter words can be made in which the second letter is a vowel?

SOLUTION (a) A word is an ordered list of letters. The two-letter words correspond to the elements of the cartesian product $A \times A$, where $A = \{x \mid x \text{ is a letter of the alphabet}\}$. Thus, the number of two-letter words is $\#(A \times A) = \#(A)\#(A) = 26(26) = 676$.
(b) Let $A = \{x \mid x \text{ is a letter of the alphabet}\}$. Then the five-letter words correspond to the elements of the cartesian product $A \times A \times A \times A \times A$. The number of five-letter words is therefore $\#(A)\#(A)\#(A)\#(A)\#(A) = (26)^5 = 11{,}881{,}376$.
(c) Let $A = \{x \mid x \text{ is a letter of the alphabet}\}$, and let $B = \{\text{a, e, i, o, u}\}$. Then the three-letter words in which the second letter is a vowel correspond to the elements of the cartesian product $A \times B \times A$. The number of such words is $\#(A)\#(B)\#(A) = 3{,}380$. ●

EXAMPLE 2.4 ● (a) How many three-digit sequences can be made from the digits 0, 1, 2, 3, 4, 5, 6, 7, 8, 9?
(b) How many three-digit numbers (in which the first digit is not 0) are there?
(c) How many three-digit numbers (in which the first digit is not 0) are there with no 6's?

SOLUTION (a) Let $A = \{0, 1, 2, 3, 4, 5, 6, 7, 8, 9\}$. The three-digit sequences correspond to the elements of the set $A \times A \times A$ and so there are $\#(A \times A \times A) = 10(10)(10) = 1{,}000$ three-digit sequences.
(b) The three-digit numbers in which the first digit is not 0 correspond to the element of the set $A \times B \times B$, where $A = \{1, 2, 3, 4, 5, 6, 7, 8, 9\}$, and

$B = \{0, 1, 2, 3, 4, 5, 6, 7, 8, 9\}$. Hence there are $\#(A)\#(B)\#(B) = 900$ such numbers.

(c) The three-digit numbers with no 6's and in which the first digit is not 0 correspond to the elements of the set $A \times B \times B$, where $A = \{1, 2, 3, 4, 5, 7, 8, 9\}$, $B = \{, 1, 2, 3, 4, 5, 7, 8, 9\}$. There are $\#(A)\#(B)\#(B) = 648$ such numbers. ●

Permutations

To **permute** is to rearrange. If you have a collection of different symbols or objects, the number of possible orderings or **permutations** of the collection can be calculated with a simple formula. We shall derive this formula after the next example.

EXAMPLE 2.5 ● How many permutations are there of the letters A, B, and C?

SOLUTION There are three possible letters to start with: A, B, or C. For each starting letter, there are two letters that can follow it, and when the first two letters are determined, there is only one choice left. Thus, there are $3(2)(1) = 6$ permutations of the letters A, B, C as illustrated in Figure 2.1. ●

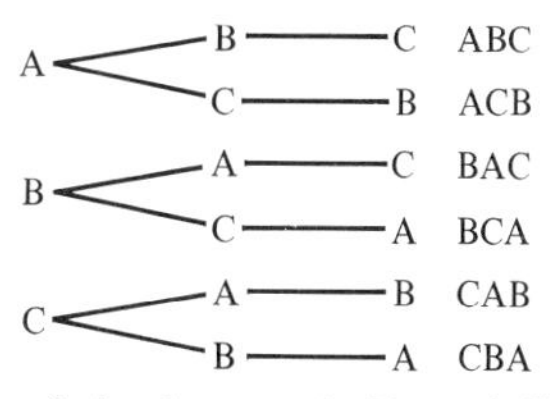

Figure 2.1 The permutations of the letters A, B, and C.

In Example 2.5, all the permutations are easily listed; however, the reasoning we used can be applied in general. For a collection of n objects, there are n objects that can be first in an ordering. Once the first object is selected there are $n - 1$ objects that can go second; once the first and second objects are selected there are $n - 2$ objects that can go third, and so on, yielding a total of $n(n - 1)(n - 2) \cdots (1)$ permutations altogether.

Permutations formula

The number of permutations or orderings of a collection of n distinct objects is denoted by $n!$ (read "n factorial") and is given by the formula

$$n! = n(n - 1) \ldots (1) \qquad \text{(By convention, } 0! = 1\text{)}$$

The next two examples illustrate the use of the permutations formula.

EXAMPLE 2.6 ● A bookshelf contains 15 different books. How many different book arrangements are there?

SOLUTION The number of permutations of a collection of size 15 is $15! = 1{,}307{,}674{,}368{,}000$. That is, there are one trillion, three hundred seven billion, six hundred seventy-four million, three hundred sixty-eight thousand ways to arrange the books! ●

EXAMPLE 2.7 ● How many different eight-digit numbers in which no digit is repeated can be made with the digits 1 through 8?

SOLUTION There are $8! = (8)(7)(6)(5)(4)(3)(2)(1) = 40{,}320$ such numbers. (Use a cartesian product to show that if repeated digits are allowed, there are $(8)^8 = 16{,}777{,}216$ such numbers.) ●

Counting ordered subcollections

So far we have been dealing with orderings that use all the objects in the collection. It turns out that a modification of the permutations formula can be used to calculate the number of **ordered subcollections** of a fixed size that can be made from a given collection. The next example illustrates this technique.

EXAMPLE 2.8 ● How many three-digit numbers with no repeated digits can be made using the digits 1 through 9?

SOLUTION There are 9 digits that can go first. For each of these, there are 8 digits that can follow, and for each of these two-digit sequences, there are 7 digits that can go third. It follows that there are $(9)(8)(7) = 504$ three-digit numbers with no repeated digits. We shall not attempt to list them. ●

It is customary to call the ordered subcollections of size k that can be made from a set with n elements the **permutations of n objects taken k at a time,** and to denote the number of such permutations by $P_{n,k}$.

Ordered subcollections formula
The number of ordered subcollections of size k that can be made from a set containing n elements is denoted by $P_{n,k}$ and is given by the formula

$$P_{n,k} = n(n-1)\cdots(n-k+1)$$

The next example illustrates the use of this formula.

EXAMPLE 2.9 ● How many four-letter words in which no letter is repeated can be made with the 26 letters of the alphabet?

SOLUTION The number of permutations of 26 objects taken 4 at a time is given by

$$P_{26,4} = 26(25)(24)(23) = 358{,}800$$

That is, there are 358,800 four-letter words with no repeated letters. ●

Combinations

In many applications, it is necessary to know the number of **unordered subcollections** of a fixed size that can be formed from the elements of a set. That is, we wish to know the number of subsets with a fixed number of elements that can be made from a given set. This number can be computed using the formula $P_{n,k}$ for *ordered* subcollections. In particular, the number of unordered subcollections is equal to the number of ordered subcollections divided by the number of permutations of each subcollection. In the next example, we shall develop this idea.

EXAMPLE 2.10 ● There are 5 people in a room; Mary, Keri, Harry, Larry, and Sherry. How many different committees of 3 people can be made from these 5 people?

SOLUTION From the previous discussion, we know that there are $P_{5,3} = 60$ *ordered* collections with 3 elements that can be made from a set of 5 elements. In this case, we are interested in the number of subsets or *unordered* collections with 3 elements. There are fewer of these because each unordered collection can be rearranged to give many ordered collections. For example the different ordered lists (Mary, Keri, Harry) and (Mary, Harry, Keri) are rearrangements of the same subset.

From the permutations formula we see that each subset with 3 elements can be arranged in $3! = 6$ different ways. This means that the 60 ordered collections consist of a certain number of 3-element subsets, each of which is ordered in 6 different ways. It follows that there are $60/6 = 10$ different subsets with 3 elements:

$$\begin{array}{ccccc} \{M, L, H\} & \{M, L, S\} & \{M, H, S\} & \{L, H, K\} & \{L, K, S\} \\ \{M, L, K\} & \{M, H, K\} & \{M, K, S\} & \{L, H, S\} & \{H, K, S\} \end{array}$$ ●

In the preceding example we found the number of subsets of a fixed size that could be made from a given set by dividing the total number of ordered subsets by the number of permutations of each. In general, to determine the number of ways to select k objects from a set of n objects, we first calculate the number of ordered subsets of size k, namely $P_{n,k} = n(n-1) \cdots (n-k+1)$, and then divide by $k!$, the number of permutations of each subset, to get

$$\frac{n(n-1) \cdots (n-k+1)}{k!}$$

Some people can remember this expression more easily in the form

$$\frac{n!}{k!\,(n-k)!}$$

which is obtained by multiplying both numerator and denominator by $(n - k)!$.

Selection formula
The number of ways to select k objects from a collection of n objects, or equivalently, the number of subsets with k elements that can be made from a set with n elements, is denoted by $\binom{n}{k}$ (read "n choose k") and is given by the formula

$$\binom{n}{k} = \frac{n(n-1)\cdots(n-k+1)}{k!} = \frac{n!}{k!(n-k)!} \qquad \text{for } k = 0, 1, \ldots, n$$

Because $0! = 1$, $\binom{n}{n} = \binom{n}{0} = 1$. Unordered subsets are sometimes called **combinations,** and the expression $\binom{n}{k}$ is sometimes called **the number of combinations of n objects taken k at a time.**

The selection formula is one of the most important combinatorial formulas in mathematics and has applications in many areas. Its use is illustrated in the next example.

EXAMPLE 2.11 ● Find:

(a) The number of ways to form a committee of size 4 from a group of size 10.
(b) The number of subsets with 3 elements that can be made from a set with 8 elements.
(c) The number of subsets with 3 elements that can be made from a set with 9 elements.
(d) The number of 5-card combinations that can be dealt from a deck of 52 cards.

SOLUTION In each case we apply the selection formula.

(a) There are $\binom{10}{4} = \frac{10!}{4!\,6!} = 210$ ways to form a committee of size 4 from a group of size 10.
(b) There are $\binom{8}{3} = 56$ subsets with 3 elements that can be made from a set with 8 elements.
(c) There are $\binom{9}{3} = 84$ subsets with 3 elements that can be made from a set with 9 elements.
(d) There are $\binom{52}{5} = 2{,}598{,}960$ different 5-card combinations that can be dealt from a deck of 52 cards. ●

Joint selections In the next example, we shall use the selection formula when selections are made from more than one set.

EXAMPLE 2.12 ● In a third-grade class there are 10 boys and 12 girls. How many ways can you select 6 children so that 3 are boys and 3 are girls?

SOLUTION There are $\binom{10}{3} = 120$ ways to select 3 boys from 10 and $\binom{12}{3} = 220$ ways to select 3 girls from 12. It follows that there are $\binom{12}{3}$ sets of 3 girls for each set of 3 boys and hence a total of $\binom{10}{3}\binom{12}{3} = 120(220) = 26{,}400$ groups of 6 containing 3 boys and 3 girls. ●

The reasoning used in Example 2.12 can be used to show that there are $\binom{n}{j}\binom{m}{k}$ ways to select j elements from a set containing n elements and k elements from a set containing m elements. Here is the general formula.

> **Joint selections formula**
> There are $\binom{n_1}{j_1}\binom{n_2}{j_2}\cdots\binom{n_m}{j_m}$ ways to simultaneously select j_1 elements from a set containing n_1 elements, j_2 elements from a set containing n_2 elements, . . . , and j_m elements from a set containing n_m elements.

The following example illustrates the use of the joint selections formula.

EXAMPLE 2.13 ● An interior decorator has 8 shades of paint from which to choose 2 shades, 5 rugs from which to choose 3, and 7 light fixtures, from which to choose 4. How many different combinations are there altogether?

SOLUTION By the joint selections formula, there are $\binom{8}{2}\binom{5}{3}\binom{7}{4} = 9{,}800$ combinations altogether. ●

Sequences with repeated symbols In the next two examples the selection formula is used to calculate the number of distinguishable orderings of a sequence with repeated symbols.

EXAMPLE 2.14 ● A coin is tossed 10 times. How many different ways are there to get 6 heads and 4 tails?

SOLUTION Each sequence of 10 tosses with 6 heads and 4 tails can be represented by a sequence of 10 symbols, 6 of which are H's and 4 of which are T's. For example, HHHHHHTTTT indicates that the first 6 tosses are heads and the last 4 are tails. The number of ways to get 6 heads and 4 tails is the same as the number of different sequences of 10 symbols, of which 6 are H's and 4 are T's. Such a sequence is determined when the 6 places for the H's are specified, the remaining 4 places being allocated to the T's. But this is simply the number of ways to choose 6 places from 10, that is, $\binom{10}{6}$. Thus, there are $\binom{10}{6} = 210$ ways to get 6 heads and 4 tails in 10 tosses. ●

EXAMPLE 2.15 ● How many distinguishable orderings are there of the seven symbols $++++---$?

SOLUTION The number of ways to arrange 4 +'s and 3 −'s is the same as the number of different ways to choose 4 objects from 7, where each object stands for a place in the sequence. Once 4 places are chosen for the +'s, the −'s go in the other 3 places. Thus, there are $\binom{7}{4} = 35$ orderings of $++++---$. ●

Here is a generalization of the technique illustrated in the preceding example.

EXAMPLE 2.16 ● How many distinguishable sequences can be made from the letters AAAABBBCC?

SOLUTION There are $\binom{9}{4}$ ways to choose 4 places out of 9 to put the A's. Once the A's are placed, there are 5 places left, and so there are $\binom{5}{3}$ ways to choose 3 places out of 5 for the B's. Finally, there are 2 places left and this is where the C's must go. Thus, there are $\binom{9}{4}\binom{5}{3} = 1{,}260$ distinguishable sequences. ●

EXAMPLE 2.17 ● A class of 25 children is to be split into 4 groups: group A is to have 8 children, group B is to have 9, group C is to have 5, and group D is to have 3. How many ways are there to do this?

SOLUTION There are $\binom{25}{8}$ ways to choose 8 children for group A. Once these children have been selected, there are 17 children left and so there are $\binom{17}{9}$ ways to select 9 children for group B. Once these children have been selected, 8 children are left, and so there are $\binom{8}{5}$ ways to choose 5 children for group C. The other 3 children must then be in group D. Thus, there are $\binom{25}{8}\binom{17}{9}\binom{8}{5} =$ 1,472,412,942,000 ways to split the class into the 4 prescribed groups. ●

Problems

1 Let $A = \{x \mid x \text{ is an integer between 1 and 4, inclusive}\}$ and $B = \{a, b, c\}$. Find $\#(A \times B)$ and list the elements of this cartesian product.

2 Let $A = \{x \mid x \text{ is a letter of the alphabet}\}$, $B = \{x \mid x \text{ is a number between 1 and 100, inclusive}\}$, and $C = \{+, -\}$. Find $\#(A \times B \times C)$.

3 Let $A = \{\text{H, T}\}$.
(a) Find $\#(A \times A \times A)$.
(b) List the elements of $A \times A \times A$.
(c) Explain how $A \times A \times A$ can be used to represent the possible outcomes of three tosses of a coin.
(d) Find the number of possible outcomes of 10 tosses of a coin.

4 Let $A = \{1, 2, 3, 4, 5, 6\}$.
(a) Find $\#(A \times A)$.
(b) List the elements of $\#(A \times A)$.
(c) Explain how $A \times A$ can be used to represent the possible outcomes when a pair of dice is rolled.

5 Each day the price of a certain stock goes up, goes down, or remains the same as the previous day. Explain how a record of these fluctuations over a 20-day period can be expressed as an element of an appropriate cartesian product. How many different 20-day records of this type are possible?

6 (a) How many four-letter words can be made using only vowels?
(b) How many six-digit numbers can be made using only the digits 2, 4, 6, and 8?
(c) How many three-digit numbers are there in which the first digit must be 4, 5, or 6, the second digit must be 7, 8, or 9, and the third digit must be 0 or 1?

7 (a) How many subsets can be made from the set $\Omega = \{x, y, z\}$? (Remember that $\varnothing \subset \Omega$, and $\Omega \subset \Omega$.)
(b) How many subsets can be made from a set with 4 elements?
(c) How many subsets can be made from a set with 6 elements? (*Hint:* Suppose the set is $\Omega = \{1, 2, 3, 4, 5, 6\}$. Represent each subset by a sequence of length 6 of 1's and 0's, where a 1 in the ith place means that i is in the subset, and a 0 in the ith place means that i is not in the subset. For example 0, 0, 0, 1, 1, 0 represents the subset $\{4, 5\}$; 1, 0, 1, 0, 1, 0 represents the subset $\{1, 3, 5\}$; 0, 0, 0, 0, 0, 0 represents $\varnothing$; and so on. Notice that sequences of length 6 of 1's and 0's are the elements of the cartesian product of 6 copies of the set $\{0, 1\}$, so you can easily calculate how many there are.)
(d) Using the reasoning of part (c) find the number of subsets that can be made from a set with 10 elements.
(e) How many subsets can be made from a set containing n elements?

8 You are one of 8 people in line at a teller's window in a bank, when the teller suddenly closes the window to take a coffee break, and a new window opens up a few yards away. As the other seven people scramble to line up at the new window, you pause to calculate the number of possible ways there are for the 8 people to line up at the new window. In the process, you end up last in line.

(a) How many ways can the 8 people line up?

(b) How many ways can the 8 people line up so that you're last in line?

9 A bookshelf has room for 15 books, and you have 20. How many ways can you choose 15 of your 20 books and arrange them on the shelf?

10 (a) How many seven-letter words, in which no letter is used more than once, can you make from the letters of the alphabet?

(b) How many seven-letter words can you make if repeated letters are allowed?

11 (a) How many four-digit numbers without repeated digits can you make with the digits 1, 3, 5, 7, 9?

(b) How many four-digit numbers can you make with the digits 1, 3, 5, 7, 9 if repeated digits are allowed?

12 How many combinations of 5 players can you make from a team consisting of 12 players?

13 How many juries of size 12 can be formed from a jury pool of 50 people?

14 (a) How many subsets of size 4 can be made from a set of 7 elements?

(b) How many subsets of size 3 can be made from a set of 7 elements?

15 Show that for $k = 0, 1, \ldots, n$, $\binom{n}{k} = \binom{n}{n-k}$.

16 Derive the **binomial expansion:** For any numbers a and b and any positive integer n, $(a+b)^n = \binom{n}{0}a^n + \binom{n}{1}a^{n-1}b + \binom{n}{2}a^{n-2}b^2 + \binom{n}{3}a^{n-3}b^3 + \cdots + \binom{n}{n-1}ab^{n-1} + \binom{n}{n}b^n$. (*Hint:* Notice that the term $(a+b)^n = (a+b)(a+b)\ldots(a+b)$, where there are n factors in all. Think of what happens when you expand this and find the coefficient of each term a^ib^j.)

17 The number of subsets of a set with n elements can be found by adding up the number of subsets with k elements, for $k = 0, 1, 2, \ldots, n$. In other words, the number of subsets of a set with n elements is $\binom{n}{0} + \binom{n}{1} + \binom{n}{2} + \cdots + \binom{n}{n}$. Using the binomial expansion (Problem 16) with $a = 1$ and $b = 1$, evaluate this sum and compare it with the answer you obtained in Problem 7(e).

18 In the grids below find the number of paths from A to B in which each step has to be either to the north or east.

(a) 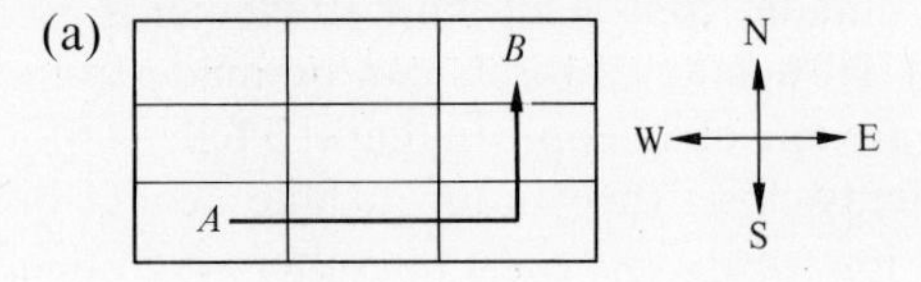

(*Hint:* The path indicated by the arrow can be represented by the sequence *EENN*, where *E* means east and *N* means north. Every path from *A* to *B* can then be represented by a sequence of 2 *E*'s and 2 N's.)

(b)

					B
A					

(c)

				B
A				

19 Using a technique similar to the one you used in Problem 18, find the number of paths from *A* to *B* in the following grids that pass through all the *X*'s. (Only north and east steps are allowed.)

(a)

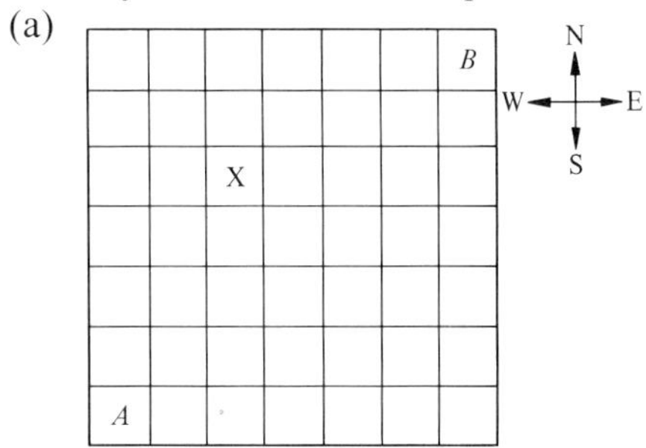

(b)

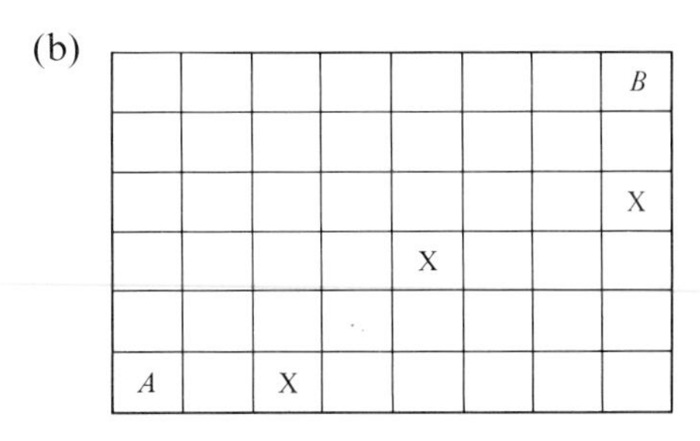

20 How many different ways are there to get exactly 9 heads in 15 tosses of a coin?

21 Find the number of arrangements for the following lists of letters:
(a) ABCD (b) ABCDA
(c) ABCBA (d) HHTTHHTTTTTT

22 A toy store has 8 types of Hot Wheels cars, 11 types of Matchbox cars, and 9 types of Dinky cars. An indulgent parent lets a child buy 3 Hot Wheels cars, 4 Matchbox cars, and 2 Dinky cars. How many possible combinations are there?

23 How many different five-card hands selected from an ordinary deck of 52 playing cards contain:
(a) 3 queens and 2 eights
(b) 4 sevens and 1 three
(c) 5 hearts
(d) Exactly 3 eights

24 There are 5 boxes, A, B, C, D, and E, and 12 *indistinguishable* balls. The balls are put in the boxes.
(a) How many different arrangements are there of balls in boxes? (*Hint:* Each arrangement can be represented by a sequence of 12 zeros and 4 slashes, with each zero representing a ball and each slash representing a division between two boxes. For example, the arrangment consisting of 5 balls in box A, 1 ball in box B, 0 balls in box C, 4 balls in box D, and 2 balls in box E would be represented by the sequence 00000|0||0000|00. The number of arrangements of balls in boxes is the same as the number of arrangements of this sequence.)
(b) How many different arrangements are there in which no box remains empty? (*Hint:* Since every such arrangement must have at least one ball in each box, you can start with one ball in each of the 5 boxes and arrange the remaining 7 balls in any way.)

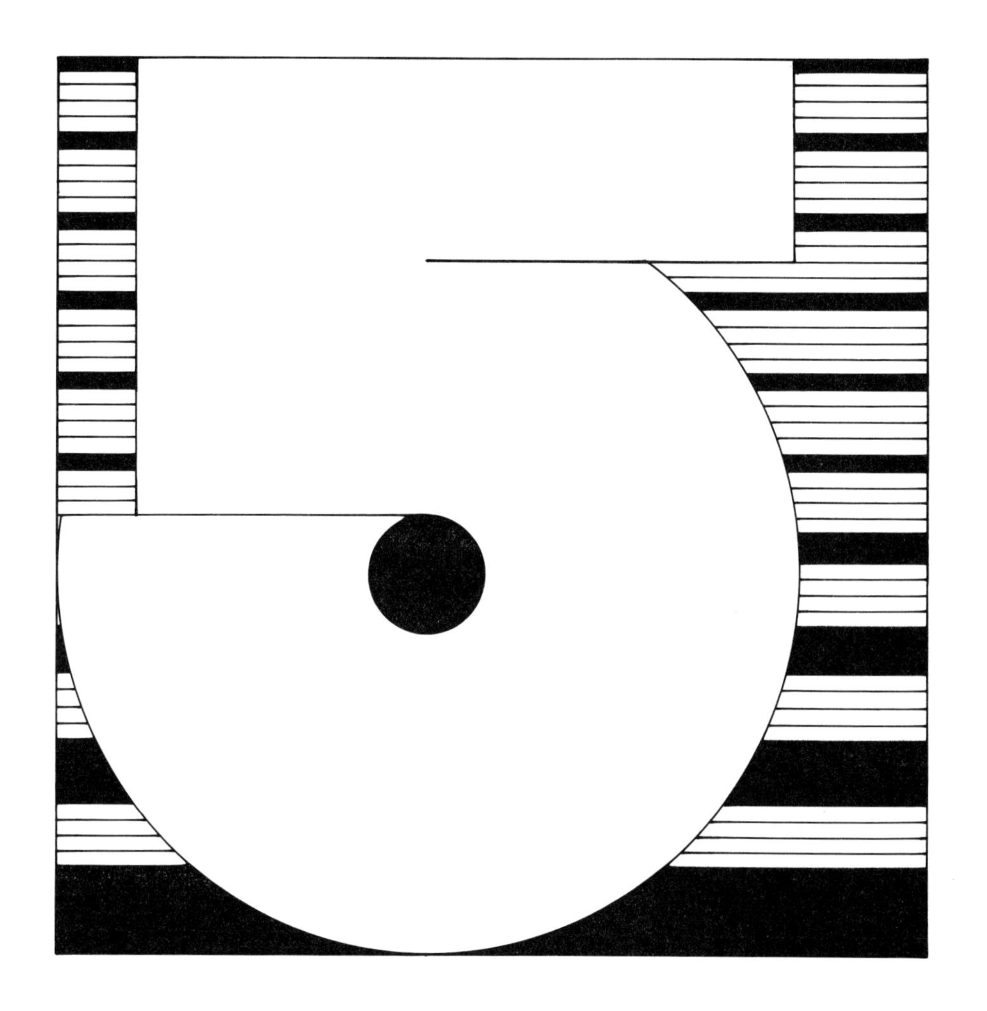

PROBABILITY

1 • BASIC CONCEPTS

The formal study of **probability** began in the seventeenth century when European aristocrats hired mathematicians to develop good strategies for gambling games. Since then, probability has become essential to many disciplines, including statistics, economics, genetics, and physics. Since statistics plays a central role in the analysis of research data, probability has had a strong influence in the social and managerial sciences.

In this chapter you will learn the basic concepts and laws of probability and will see some of the reasons why the theory of games of chance can be applied to so many practical problems. In addition, you will learn how to use the methods of counting developed in Chapter 4 to solve a variety of probability problems.

Outcomes and events

A chance experiment is one in which the results cannot be predicted with certainty. We shall call a result of a chance experiment an **outcome.** Any collection of outcomes is called an **event.** The event consisting of all the possible outcomes for a given experiment is called the **sample space.** For example, suppose you write the integers from 1 through 10 on separate slips of paper and place them in a hat and then select one without looking. An outcome of this experiment is the selection of one of the numbered slips of paper. There are 10 outcomes. The event that an even-numbered ticket is selected occurs if any of the tickets marked 2, 4, 6, 8, or 10 is selected. The event that an odd-numbered ticket is selected occurs if any of the tickets marked 1, 3, 5, 7, or 9 is selected.

> **Outcomes and events**
> An *outcome* is a result of a chance experiment. An *event* is any collection of outcomes. The *sample space* is the collection of all the possible outcomes.

In a formal treatment of probability these concepts are expressed in set-theory terms. The sample space is a universal set, the outcomes are the elements of this set, and the events are the subsets of this set. In the following example events and outcomes are described using the language of sets.

EXAMPLE 1.1 ● A box contains three balls marked a, b, and c. You will select one ball "at random" from this box; that is, you will select in such a way that every ball has the same chance of being chosen (see Figure 1.1). List the events and sample space for this experiment.

SOLUTION The sample space or set of all outcomes can be denoted by $\Omega = \{a, b, c\}$. The events are the subsets of Ω and are denoted as follows:

$$\varnothing,\ \{a\},\ \{b\},\ \{c\},\ \{a, b\},\ \{a, c\},\ \{b, c\},\ \{a, b, c\}$$ ●

Figure 1.1 Selecting a ball at random.

We shall use set-theory notation whenever it seems advantageous, but in general, our descriptions of outcomes and events will be more informal.

Random sampling

Many applications of probability involve the selection of elements from a collection in such a way that every element of the collection has the same chance of being selected. This is known as **selecting at random.** A set of elements that have been selected at random is called a **random sample.** The collection from which a random sample is drawn is often referred to as the **population,** even when the objects being selected are not people.

Although taking a random sample is easy in principle, difficulties often arise in practice. For example, if you wanted to take a random sample of 100 students from a college population of 5,000 students, you could write the name of each student on a slip of paper, put the slips of paper in a large container, shake the container, and select 100 names without looking. This method would yield a random sample, but it is not very practical. An alternate method, such as standing in front of the cafeteria and selecting the first 100 students who came by, might not yield a random sample. (Do you see why?)

In order to take random samples more efficiently, statisticians have compiled lists of numbers selected at random. These numbers are called **random numbers** and the resulting list is called a **random number table.** If each element in the population is coded with a number (such as social security number or student identification number), numbers can be read off a random number table and the corresponding elements of the population can be selected. Most campus computer systems are equipped with **random number generators,** which enable one to obtain random numbers directly from the computer. These random number generators are actually programs that read numbers from a random number table that has been stored in the memory of the computer.

Probability

Probability measures the likelihood of the occurrence of an event. The probability of an event is the fraction of the time that the event can be expected to occur if the experiment is repeated over and over, a large number of times. Since probabilities are fractions, they are numbers between 0 and 1. If an event has probability 0, it has no chance of occurring, and if it has probability 1, it is certain to occur. Since the sample space is the event consisting of all the possible outcomes, it is certain to occur and thus has probability 1. For experiments such as tossing coins, rolling dice, and random sampling, in which every outcome has the same chance of occurring, there is an easy formula for obtaining probabilities. We shall call this the **basic probability formula.**

Basic probability formula
For an experiment in which every outcome has the same chance of occurring, the probability of an event is the fraction of the total number of outcomes that cause the event to occur. That is, for any event A, the probability of A is given by the formula

$$P(A) = \frac{\text{number of outcomes in } A}{\text{total number of outcomes}}$$

Here are six examples to illustrate this basic formula.

EXAMPLE 1.2 ● There are 5 tickets in a box, 3 marked "loser," and 2 marked "winner." The box is shaken, and without looking, you select a ticket. What is the probability that this ticket is a winner?

SOLUTION The event that a ticket marked winner is selected consists of 2 outcomes, and there are 5 possible outcomes in all. According to the basic probability formula it follows that the probabilty of selecting a ticket marked winner is 2/5. That is, $P(\text{winner}) = 2/5$. ●

The next example shows that if you know the percentage of the outcomes that cause a given event to occur, you can compute the probability of the event without knowing the population size.

EXAMPLE 1.3 ● In a large city, 70 percent of the registered voters favor passage of the equal rights amendment. If one person is selected at random from this population, find the probability that the selected person favors passage.

SOLUTION Although we are not given the total number of registered voters in the city or the total number who favor passage of the equal rights amendment, we are given the percentage who favor passage. To compute the probability that the selected person favors passage, we convert this percentage to a decimal and conclude that $P(\text{selected person favors passage}) = 0.7$. ●

EXAMPLE 1.4 ● A roulette wheel is divided into 38 sections, numbered 1 through 36, 0, and 00 (see Figure 1.2). Half the numbers from 1 through 36 are colored red, the other half black, while 0 and 00 are green. In the gambling game of roulette, players place bets on colors or numbers. Then the wheel is spun and a ball is tossed into it. When the wheel stops, the ball comes to rest in one of the sections, and the winning bets are paid off. Find the probablities of the following events.

(a) The ball lands on a red number.
(b) The ball lands on a black number.
(c) The ball lands on a green number.
(d) The ball lands on number 16.
(e) The ball lands on a number between 1 and 12, inclusive.

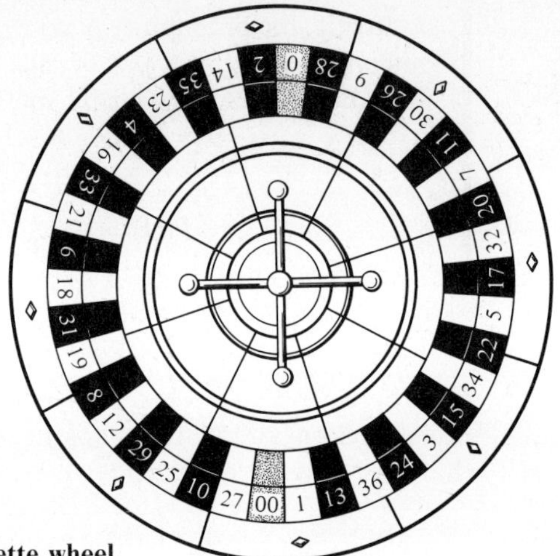

Figure 1.2 A roulette wheel.

SOLUTION (a) Since 18 of 38 places are red, the probability that the ball lands on a red number is 18/38. That is, $P(\text{red}) = 18/38$.
(b) Since 18 of the 38 places are black, $P(\text{black}) = 18/38$.
(c) There are only 2 green places, 0 and 00, and so $P(\text{green}) = 2/38$.
(d) There is only 1 number 16, so $P(16) = 1/38$.
(e) There are 12 numbers from 1 through 12, so the probability that the ball lands on a number between 1 and 12 is 12/38. ●

EXAMPLE 1.5 ● An evenly balanced coin is tossed 3 times. Find the probabilities of the following events.

(a) Exactly 2 heads come up.
(b) At least 1 head comes up.
(c) Three heads come up.

SOLUTION There are 8 equally likely outcomes:

HHH HHT HTH THH HTT THT TTH TTT

where HHH means heads comes up on all three tosses, HHT means heads comes up on the first two tosses and tails comes up on the third toss, and so on.

(a) By looking at the list of outcomes we see that there are 3 ways to get exactly 2 heads: HHT, HTH, THH. Thus, $P(\text{exactly 2 heads}) = 3/8$.

(b) There are 7 ways to get at least 1 head (all except TTT) so P(at least 1 head) $= 7/8$.
(c) There is only 1 way to get 3 heads, so P(3 heads) $= 1/8$. ●

EXAMPLE 1.6 ● A pair of dice is rolled (Figure 1.3) and the sum of the numbers on the upturned faces is recorded. Find the associated probabilities.

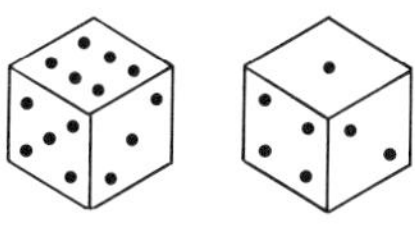

Figure 1.3 A pair of dice is rolled.

SOLUTION Assume for clarity of discussion that one die is red (R) and one die is green (G). Since there are 6 possible outcomes for each die, there are $6(6) = 36$ possible combinations, as indicated in the following table.

Outcomes when two dice are rolled

RG	RG	RG	RG	RG	RG
11	21	31	41	51	61
12	22	32	42	52	62
13	23	33	43	53	63
14	24	34	44	54	64
15	25	35	45	55	65
16	26	36	46	56	66

If the dice are evenly balanced, these outcomes are all equally likely. To find the probabilities of the various sums, we simply count the number of ways the sum can occur and divide this number by 36. The results are summarized in the following table:

Sum of upturned faces	Outcomes	Probability
2	11	1/36
3	12, 21	2/36
4	13, 22, 31	3/36
5	14, 23, 32, 41	4/36
6	15, 24, 33, 42, 51	5/36
7	16, 25, 34, 43, 52, 61	6/36
8	26, 35, 44, 53, 62	5/36
9	36, 45, 54, 63	4/36
10	46, 55, 64	3/36
11	56, 65	2/36
12	66	1/36

Notice that the most likely sum is 7, with $P(7) = 6/36$, while the least likely are 2 and 12, with $P(2) = P(12) = 1/36$. ●

EXAMPLE 1.7 ● Three digits are generated at random by a computer to form a number from 000 through 999, inclusive. (For example, 004 would denote the number 4, 037 would denote the number 37, and so on.)

(a) Find the probability that the generated number is odd.
(b) Find the probability that the generated number is a multiple of 9.
(c) Find the probability that the generated number is greater than 749.

SOLUTION (a) Half the 1,000 numbers between 000 and 999 are odd, so $P(\text{odd}) = 500/1{,}000 = 0.5$.
(b) The multiples of 9 are 9, 18, 27, 36, and so on. In fact, $1{,}000/9 = 111.11$, so there are 111 multiples of 9 (Do you see why?). Hence $P(\text{multiple of } 9) = 111/1{,}000 = 0.11$.
(c) There are 250 numbers greater than 749, so $P(\text{greater than } 749) = 250/1{,}000 = 0.25$. ●

Contingency tables

The next example shows how information from a population that is **cross-classified** according to two sets of characteristics can be efficiently displayed in a table called a **contingency table.**

EXAMPLE 1.8 ● In a group of 400 college seniors, there are 170 science majors and 230 nonscience majors. Fifty-five of the science majors are women, while 120 of the nonscience majors are women. One student is selected at random from this group.

(a) Find the probability that the selected student is a science major.
(b) Find the probability that the selected student is a woman.
(c) Find the probability that the selected student is neither a woman nor a science major.

SOLUTION Since each student is classified by two categories, major and sex, we call the given information cross-classified data. To construct a contingency table for this information, we first draw a 2×2 array labeled according to the various categories, and with marginal places reserved for totals.

	Science majors	Nonscience majors	Total
Women			
Men			
Total			

There are 170 science majors and 230 nonscience majors, and we fill in the column totals accordingly.

	Science majors	Nonscience majors	Total
Women			
Men			
Total	170	230	400

Since 55 of the science majors are women, the other 115 science majors must be men, and since 120 of the nonscience majors are women, the other 110 must be men. This information and corresponding row totals complete the table, and we can now easily obtain the desired probabilities.

	Science majors	Nonscience majors	Total
Women	55	120	175
Men	115	110	225
Total	170	230	400

(a) There are 170 science majors out of 400 students altogether, so $P(\text{science major}) = 170/400$.
(b) There are 175 women out of 400 students altogether, so $P(\text{woman}) = 175/400$.
(c) There are 110 students who are neither women nor science majors, so $P(\text{neither woman nor science major}) = 110/400$. ●

The law of large numbers

A famous mathematical theorem known as the **law of large numbers** asserts that the basic probability formula accurately describes what will happen in repeated trials of an experiment. In particular, the law states that if an experiment is performed a large number of times, the fraction of the trials that result in a particular event will be approximately the same as the theoretical probability of that event. Consequently, a gambling casino does not know whether a particular person will win or lose on a given play of a game, but it can calculate with accuracy how much money will be won (or lost) when the game is played over and over. A life insurance company does not know whether a claim will be filed on a particular policy, but if it knows the probability that a claim will be filed, it can predict with accuracy how many claims will be filed in the future. A business that does mail-order advertising does not know whether a particular person will respond favorably to an advertisement, but if it knows the overall response probabilities, it can calculate accurately how large a mailing must be in order to ensure a given volume of sales.

The next example is an experiment for you to perform to investigate the law of large numbers.

EXAMPLE 1.9 ● If you toss a coin twice, the possible outcomes are HH, HT, TH, TT. Thus, $P(\text{no heads}) = 0.25$, $P(1 \text{ head}) = 0.5$, and $P(2 \text{ heads}) = 0.25$. Perform this experiment 300 times and record the number of heads that occur each time.

When you are finished, calculate the fraction of the trials that resulted in each event and compare these fractions with the theoretical probabilities.

SOLUTION The law of large numbers asserts that the observed fractions will be close to the theoretical probabilities, but not necessarily exactly the same. In Figure 1.4 we have displayed what happened when we performed this experiment.

Event	Tallies	Total	Fraction	Probability
0 heads	𝍸 𝍸 𝍸 𝍸 𝍸 𝍸 𝍸 𝍸 𝍸 𝍸 𝍸 𝍸 𝍸 𝍸 𝍸 𝍸 IIII	84	0.28	0.25
1 head	𝍸 II	147	0.49	0.50
2 heads	𝍸 𝍸 𝍸 𝍸 𝍸 𝍸 𝍸 𝍸 𝍸 𝍸 𝍸 𝍸 𝍸 IIII	69	0.23	0.25

Figure 1.4 Results of a coin-tossing experiment.

Our results show close agreement between the observed fractions and theoretical probabilities. The event that was the farthest off was 0 heads, with an actual fraction of 0.28 and a theoretical probability of 0.25. ●

Probability models and simulation

Probability theory was invented by mathematicians who were studying gambling games. In this context, chance events are easily described, and probabilities, while sometimes difficult to compute, are usually well understood theoretically. Also, games of chance can be played over and over, and empirical probability estimates can be calculated. In real-life situations, however, this is not always the case.

For example, a weatherman estimating the probability of rain cannot repeat today's exact weather conditions over and over like a gambling game. An investment analyst trying to assess the probability that the price of a particular stock will go up has no clear-cut list of outcomes to look at, and no well-defined game that can be played over and over. A geologist trying to predict the likelihood of an earthquake or an economist trying to estimate the probability of a recession are also dealing with chance phenomena that are not as clearly formulated as most gambling games.

One way to deal with a situation like this is to build a **mathematical model** containing the essential features of the problem. Many of the probability experiments we shall discuss serve as models for more complicated situations. For example, the experiment of drawing tickets from a box can be used as a model for many random sampling problems, and tossing coins is sometimes used as a model in genetics.

Unfortunately, there are some situations that are too complicated for a simple model to describe adequately. In recent years, advances in computer technology have made possible a new type of modeling technique known as **simulation.** In simulation, the basic features of the situation under study are incorporated in a computer program. The computer then performs thousands of trials of the simulated situation and empirical probability estimates are obtained.

For example, it is not hard to simulate a gambling game on a computer. The computer can play thousands of games in a few seconds and can test many strategies, some of which may be too complicated to deal with theoretically. Using such an approach a mathematics professor named Edward Thorp simulated the card game of twenty-one (blackjack). The strategy he developed consistently won in the Nevada casinos. Professor Thorp later made his methods public in a fascinating book called *Beat the Dealer* (Vintage, 1964). In the book, Thorp describes how he used a computer to simulate the game of twenty-one and analyzes a variety of strategies. He also tells of his experiences in the casinos when the management discovered that someone actually had a system that worked.

Simulation models have proved to be important in areas like physics, genetics, economics, and the social sciences. In some cases, the study of simulation models has given the key to development of new theoretical results, while in other cases the simulation results have been ends in themselves. Simulation is a rapidly developing area, and as computer technology advances it will become a major tool in applied probability and statistics.

Problems

1 A box contains 4 tickets numbered 1 through 4. A ticket is selected at random from the box and a die is rolled.
   (a) List the elements of the sample space.
   (b) List the outcomes that make up the event that the sum of the numbers is even.
   (c) List the outcomes that make up the event that the number on the ticket is the same as the number on the die.

2 A number between 50 and 99, inclusive, is selected at random. Find the number of outcomes that make up each of the following events.
   (a) The selected number is less than 76.
   (b) The first digit of the selected number is the same as its second digit.
   (c) The sum of the digits of the selected number is 13.

3 How would you select 4 people at random from a class containing 35 people?

4 How would you select 100 students at random from a student body of 3,000?

5 How would you select 400 registered voters at random from a city containing 40,000 registered voters?

6 On a certain block there are 2 houses on the north side of the street and 3 houses on the south side. How could you select one of these houses at random, by tossing an evenly balanced coin?

7 A box contains 18 losing tickets and 2 winning tickets. A ticket is selected at random. Find the probability that a winning ticket is selected.

8 In a certain large band of pirates, 40 percent have patches over their left eyes. If a pirate from the band is selected at random, find the probability that the selected pirate has a patch over his left eye.

9 A coin is tossed 4 times.
(a) List all the possible outcomes.
(b) Find the probability that no heads turn up.
(c) Find the probability that exactly one head turns up.
(d) Find the probability that at least three heads turn up.

10 A large box contains 365 tickets, one for each day of the year. One ticket is selected at random.
(a) Find the probability that the selected day is in March.
(b) Find the probability that the selected day is in the winter.
(c) Find the probability that the selected day is your birthday.
(d) If the year starts on Sunday, find the probability that the selected day is on a weekend.

11 A pair of dice is rolled. If 2, 3, or 12 comes up, you win. Otherwise you lose. Find the probability that you win. (In the casino game of craps, this event is known as "any craps.")

12 A card is selected at random from an ordinary deck of playing cards.
(a) Find the probability that the card is a diamond.
(b) Find the probability that the card is a six.
(c) Find the probability that the card is the six of diamonds.

13 A number from 0 through 99, inclusive, is randomly generated by a computer.
(a) Find the probability that the number is odd.
(b) Find the probability that the number is greater than 74.
(c) Find the probability that the number is a perfect square.

14 A magician claiming to have ESP powers puts 3 cards numbered from 1 through 3 face down on a table. While he is looking the other way, you move the cards around so he no longer knows which is which. The magician then tries to guess the identity of each card. He cannot guess the same number more than once and he cannot look at the cards until he has made all 3 guesses. Assume that the magician is guessing at random.
(a) Find the probability that all 3 guesses are correct.
(b) Find the probability that exactly 2 guesses are correct.
(c) Find the probability that exactly 1 guess is correct.
(d) Find the probability that no guesses are correct.

15 A stream contains 40 percent trout, 25 percent perch, and 35 percent bass. A fish is selected at random from the stream.
(a) Find the probability that the fish is a trout.
(b) Find the probability that the fish is not a trout.
(c) Find the probability that the fish is either a trout or a perch.

16 In a certain population, 30 percent are allergic to poison oak, 25 percent have hay fever, and 10 percent are both allergic to poison oak and have hay fever. One person is selected at random from this population.

(a) Find the probability that the selected person is allergic to poison oak.
(b) Find the probability that the selected person is not allergic to poison oak.
(c) Find the probability that the selected person is allergic to poison oak or has hay fever or both.

17 Box I contains 5 tickets numbered from 1 through 5. Box II contains 3 tickets numbered from 4 through 6. One ticket is selected at random from each box.

(a) List the elements of the sample space.
(b) Find the probability that both tickets have the same number.
(c) Find the probability that the tickets do not have the same number.
(d) Find the probability that the sum of the numbers on the tickets is odd.
(e) Find the probability that the sum of the numbers on the tickets is less than 8.

18 In a certain aquarium there are 8 green turtles and 7 spotted turtles. Five of the green turtles are females and 2 of the spotted turtles are females. One turtle is selected at random.

(a) Display the given information in a contingency table.
(b) Find the probability that the selected turtle is a female.
(c) Find the probability that the selected turtle is a spotted male.
(d) Find the probability that the selected turtle is neither green nor a male.

19 In a population of middle-aged people, 40 percent are smokers and 60 percent are nonsmokers. Sixty percent of the smokers have high blood pressure, while 20 percent of the nonsmokers have high blood pressure. One of these people is selected at random.

(a) Construct a contingency table for the given information, assuming that there are 750 in the population.
(b) Find the probability that the selected person has high blood pressure.
(c) Find the probability that the selected person is neither a smoker nor has high blood pressure.
(d) Find the probability that the selected person is both a smoker and has high blood pressure.
(e) Suppose that 1,000 people are in the population instead of 750. Does this change the results of parts (b) through (d)? Explain your answer.

20 In a certain population 70 percent own two cars, 30 percent own a boat, and 20 percent own two cars and a boat. One person is selected at random from this population.

(a) Find the probability that the selected person owns two cars and a boat.
(b) Find the probability that the selected person owns either two cars or a boat or both.
(c) Find the probability that the selected person owns neither two cars nor a boat.

21 In a certain experiment the probability that event A occurs is 0.2. What is the probability that A doesn't occur?

22 If the probability that event A occurs is p, what is the probability that A doesn't occur?

23 Put 1 red marble and 2 green marbles in a bag. Select 300 marbles at random, replacing each selected marble in the bag before the next marble is drawn. Find the fraction of the selected marbles that are red and compare with the theoretical probability of selecting a red marble.

24 Take an ordinary deck of cards and select 200 cards at random, replacing each selected card before a new card is drawn. Find the fraction of the selected cards that are diamonds and compare with the theoretical probability of selecting a diamond.

25 Roll a pair of dice 300 times and compare the fraction of the rolls for which the dice add to 7 with the theoretical probability of rolling a 7.

26 If the computer on your campus has a program to generate random numbers, use it to obtain 500 random digits and compare the numbers of digits of each type.

27 If the computer on your campus has any game-playing programs, play a game 200 times to obtain an empirical estimate of the probability of winning. If you know the theoretical probability of winning, compare this with the actual fraction of wins.

2 • LAWS OF PROBABILITY

It will be helpful at this time to introduce convenient notation and state some basic laws of probability. Since events are subsets of the sample space, we may define the union, intersection, and complement of events as in set theory. In particular, the **union** of two or more events is the event consisting of all outcomes that are in at least one of the given events. The **intersection** of two or more events is the event consisting of all outcomes that are in all of the given events. As in set theory, the symbols $\cup$ and $\cap$ are used to denote union and intersection, respectively. The **complement** A' of an event A is the collection of all outcomes in the sample space that are not in A. Here is an example to illustrate these ideas.

EXAMPLE 2.1 ● A box contains tickets marked 1, 2, and 3, and a coin has one side marked 1 and the other side marked 2. A ticket is selected at random from the box and then the coin is tossed. The sample space is

$$\Omega = \{11, 12, 21, 22, 31, 32\}$$

where the second number in each pair is the number on the coin. Let A denote the event that an odd-numbered ticket is selected from the box, B the event that the sum of the numbers on the selected ticket and coin is 3, and C the event that the sum of the numbers on the selected ticket and coin is less than 4. Describe the following events by listing their outcomes.

(a) $A \cup B$ (b) $A \cap C$ (c) A'

SOLUTION (a) $A \cup B = \{11, 12, 21, 31, 32\}$
(b) $A \cap C = \{11, 12\}$
(c) $A' = \{21, 22\}$ ●

As in set theory, two events A and B are said to be **mutually exclusive** if they have no outcomes in common, that is, if $A \cap B$ cannot occur. For example, if A denotes the event that at least 2 heads come up in 3 tosses of a coin and B denotes the event that tails comes up on every toss, then $A \cap B$ cannot occur and so A and B are mutually exclusive. In general, we say that two or more events are mutually exclusive if no pair of them has any outcomes in common.

We are now ready to state some basic probability laws. The first law concerns the union of mutually exclusive events.

The probability of the union of mutually exclusive events
If A and B are mutually exclusive, then

$$P(A \cup B) = P(A) + P(B)$$

In general, the probability of the union of a collection of mutually exclusive events is the sum of their individual probabilities.

This law holds for any probability experiment. In the case of equally likely outcomes, this law has the following interpretation: If A and B have no outcomes in common, then the fraction of the outcomes that are either in A or B equals the fraction of the outcomes that are in A plus the fraction of the outcomes that are in B. The next example illustrates this law.

EXAMPLE 2.2 ● The probability that the value of a certain house will increase more than 20 percent over the next year is 0.30. The probability that the house will go down in value is 0.01. If A denotes the event that the value of the house increases more than 20 percent in the next year and B the event that the value of the house decreases, find $P(A \cup B)$.

SOLUTION Since the value cannot increase more than 20 percent and also decrease, the events A and B are mutually exclusive. Therefore,

$$P(A \cup B) = P(A) + P(B) = 0.31$$ ●

The next law concerns complementary events. It is a consequence of the law for mutually exclusive events and the fact that $P(\Omega) = 1$.

The probability of a complement
For any event A,

$$P(A) = 1 - P(A')$$

For experiments with equally likely outcomes the law for complementary events can be explained as follows: Since the fraction of the outcomes in A plus the fraction of the outcomes not in A equals 1, the fraction of the outcomes in A is 1 minus the fraction of the outcomes not in A. The next example illustrates this law.

EXAMPLE 2.3 ● If you have a head-on collision going more than 5 miles per hour, there is a 0.05 chance that your safety air bag will not inflate. What is the probability that it will inflate?

SOLUTION Letting A denote the event that the air bag will inflate, we have $P(A') = 0.05$, and so

$$P(A) = 1 - P(A') = 0.95$$ ●

The final probability law that we will discuss in this section enables us to find probabilities for the union of events that are not necessarily mutually exclusive.

The probability of a union
For any events A and B,

$$P(A \cup B) = P(A) + P(B) - P(A \cap B)$$

For experiments with equally likely outcomes, this law has the following interpretation: If you add the fraction of the outcomes in A and the fraction of the outcomes in B, you are adding the fraction of the outcomes in $A \cap B$ twice, so you must subtract this fraction when finding the fraction of outcomes in $A \cup B$. The Venn diagrams in Figure 2.1 illustrate this law, as does the next example.

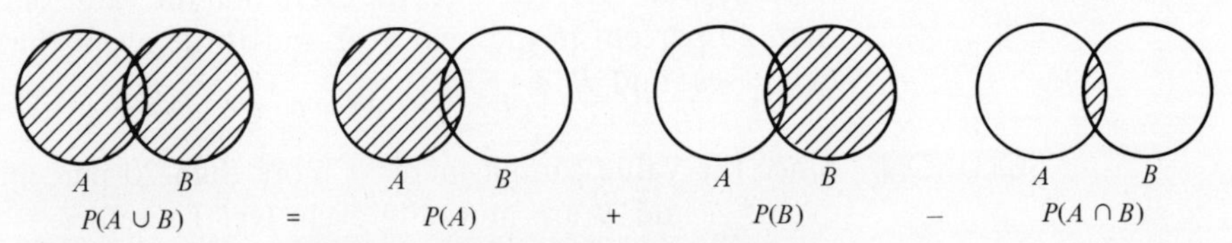

Figure 2.1 The probability of the union of two events.

EXAMPLE 2.4 ● It will rain in Fresno tomorrow with probability 0.3. It will rain in Merced tomorrow with probability 0.4. It will rain in both cities with probability 0.2. What is the probability that it will rain in at least one of the cities?

SOLUTION Let F denote the event that it will rain in Fresno tomorrow and M denote the event that it will rain in Merced. We wish to find $P(F \cup M)$. The general law for unions yields

$$\begin{aligned} P(F \cup M) &= P(F) + P(M) - P(F \cap M) \\ &= 0.3 + 0.4 - 0.2 = 0.5 \end{aligned}$$

●

Problems

1 A computer generates a random digit between 5 and 9, inclusive, and a researcher selects a ticket at random from a box containing 4 tickets marked 0, 1, 8, and 9. Let A denote the event that the sum of the two selected numbers equals 9, B denote the event that the sum of the selected numbers is greater than 8, and C denote the event that the number generated by the computer is odd. List the outcomes that make up the following events.
(a) $A \cup B$ (b) $A \cup C$ (c) $A \cap B$
(d) $A \cup B \cup C$ (e) B' (f) $B' \cap C'$

2 Five cards are selected at random from an ordinary deck. Let R denote the event that all the selected cards are red, B denote the event that 2 of the selected cards are black, and K denote the event that 3 of the selected cards are kings. Which of the following pairs of events are mutually exclusive?
(a) R and B (b) R and K (c) B and K

3 In a certain city, 45 percent of the families own only one car and 30 percent own exactly two cars. One family is selected at random. Let A denote the event that the selected family owns only one car, and B denote the event that the selected family owns exactly two cars.
(a) Find $P(A \cup B)$
(b) Find $P(A')$
(c) Find $P(A \cap B)$

4 In a large commune, 40 percent are vegetarians, 30 percent meditate, and 25 percent are both vegetarians and meditate. Suppose one person is selected at random from this commune. Let V denote the event that the selected person is a vegetarian and M denote the event that the selected person meditates.
(a) Find $P(V)$
(b) Find $P(V \cap M)$
(c) Find $P(V \cup M)$
(d) Find $P(V' \cap M')$ (*Hint:* Use De Morgan's laws.)

5 In a certain community, 40 percent of the houses need termite work, 55 percent need a new roof, and 35 percent need termite work and a new roof. One house is selected at random. Let T denote the event that the selected house needs termite work and R denote the event that the selected house needs a new roof.
(a) Find $P(T \cap R)$
(b) Find $P(T \cup R)$
(c) Find $P(T' \cup R')$ (*Hint:* Use De Morgan's laws.)

6 A box contains 10 red balls numbered from 1 through 10 and 15 green balls numbered from 1 through 15. A ball is selected at random. Let A denote the event that the selected ball is red and B denote the event that the number on the selected ball is odd.

(a) Find $P(A)$
(b) Find $P(B)$
(c) Find $P(A \cup B)$
(d) Find $P(A \cap B)$
(e) Find $P(A')$

7 A coin is tossed and a die is rolled.

(a) List the outcomes in the sample space.
(b) Find the probability that the coin comes up heads and the number on the die is odd.
(c) Find the probability that the coin comes up heads, or the number on the die is odd, or both.
(d) Find the probability that the number on the die is odd.

8 Suppose $P(A) = 0.2$, $P(B) = 0.4$, and $P(A \cap B) = 0.15$.

(a) Find $P(A \cup B)$ (b) Find $P(A')$
(c) Find $P(A' \cap B')$ (d) Find $P(A' \cup B')$

9 For any events A and B, show that $P(B) = P(B \cap A) + P(B \cap A')$. (*Hint:* Draw a Venn diagram to show that $B = (B \cap A) \cup (B \cap A')$ and then apply the law for mutually exclusive events.)

10 This problem generalizes the preceding problem. Suppose A, B, and C are mutually exclusive events, with $A \cup B \cup C = \Omega$. Show that for any event D, $P(D) = P(D \cap A) + P(D \cap B) + P(D \cap C)$.

11 For events A and B, the event $A - B$ is defined to be the collection of outcomes that are in A but not in B. (This event is called the **relative complement** of B in A.) Show that $P(A - B) = P(A) - P(A \cap B)$, and that the probability law for complements is a special case of this formula.

12 A certain restaurant offers 15 seafood dishes, 10 chicken dishes, and 20 beef dishes. A restaurant critic recommends 4 of the seafood dishes, 2 of the chicken dishes, and 5 of the beef dishes. Suppose you go to this restaurant and select a dish at random. Let S, C, B, and R denote the events that the selected dish is seafood, chicken, beef, and recommended, respectively.

(a) Find $P(R)$ (b) Find $P(S')$
(c) Find $P(S' \cap R')$
(d) Find $P(R \cap S)$, $P(R \cap C)$, and $P(R \cap B)$ and verify that $P(R) = P(R \cap S) + P(R \cap C) + P(R \cap B)$ as asserted in Problem 10.

3 • CONDITIONAL PROBABILITY

Sometimes the occurrence of one event affects the probability of another. For example, the probability that an individual who smokes cigarettes will have high blood pressure is higher than the probability in the population as a whole. A probability whose calculation is based on the assumption that a particular event occurs is called a **conditional probability.** We shall use the symbol $P(A \mid B)$

(read "the probability of A, given B") to denote the probability of A computed under the assumption that B occurs. For example, if A denotes the event that an individual has high blood pressure, and B the event that an individual smokes, then $P(A \mid B)$ denotes the probability that a smoker will have high blood pressure.

The next example illustrates how contingency tables can be used in the calculation of conditional probabilities.

EXAMPLE 3.1 ● The table below shows blood pressure levels and smoking habits for a group of 300 middle-aged men.

	Nonsmoker	Moderate smoker	Heavy smoker	Total
Normal blood pressure	81	84	27	192
High blood pressure	21	51	36	108
Total	102	135	63	300

Suppose someone is selected at random from this group.

(a) Find P(heavy smoker).
(b) Find P(high blood pressure).
(c) Find P(high blood pressure $\cap$ heavy smoker).
(d) Find P(high blood pressure | heavy smoker).

SOLUTION (a) There are 63 heavy smokers out of 300 so P(heavy smoker) $= 63/300$.
(b) There are 108 with high blood pressure out of 300 so P(high blood pressure) $= 108/300$.
(c) There are 36 who have high blood pressure and are heavy smokers, so P(high blood pressure $\cap$ heavy smoker) $= 36/300$.
(d) We can interpret P(high blood pressure | heavy smoker) as the fraction of heavy smokers who have high blood pressure. Thus, P(high blood pressure | heavy smoker) $= 36/63 = 0.57$, which is greater than the unconditional probability $108/300 = 0.36$ in part (b). ●

The conditional probability formula

For experiments with equally likely outcomes, the conditional probability of A given B is the fraction of outcomes in B that are also in A. To obtain this fraction we divide the number of outcomes in B that are also in A by the total number of outcomes in B. That is,

$$P(A \mid B) = \frac{\text{number of outcomes in } B \text{ that are also in } A}{\text{number of outcomes in } B}$$

Since the collection of outcomes in B that are also in A is simply the event $A \cap B$, we obtain the following formula, which is valid even for experiments in which the outcomes are not equally likely.

The conditional probability formula
For any events A and B,

$$P(A|B) = \frac{P(A \cap B)}{P(B)}$$

That is, the probability of A given B is the ratio of the probability of $A \cap B$ to the probability of B.

If we multiply both sides of the conditional probability formula by $P(B)$, we get the following formula for the probability of an intersection.

The probability of an intersection
For any events A and B,

$$P(A \cap B) = P(A|B)P(B)$$

In the following examples, we shall demonstrate how these formulas are used.

EXAMPLE 3.2 ● A survey of attitudes about one's job yielded the data below.

	Happy	Unhappy	Total
Bus drivers	50	75	125
Lawyers	40	35	75
Total	90	110	200

A person from this group is selected at random. Given that the selected person is a bus driver, find the probability that he or she is happy.

SOLUTION We let H denote the event that a happy person is selected and B denote the event that a bus driver is selected. We wish to find $P(H|B)$. The conditional probability formula tells us that $P(H|B) = P(H \cap B)/P(B)$. From the table we obtain $P(H \cap B) = 50/200 = 0.25$, $P(B) = 125/200 = 0.625$, and so $P(H|B) = 0.25/0.625 = 0.4$. Notice that we could have obtained this directly by computing the fraction of bus drivers who are happy. ●

Here is a more complicated example.

EXAMPLE 3.3 ● There are 6 boxes, 2 boxes of type A and 4 boxes of type B. Type A boxes each contain 2 green marbles and 3 red marbles. Type B boxes each contain 1 green marble and 3 red marbles. A box is selected at random and then a marble is drawn at random from the selected box. Letting R denote the event a red marble is selected, A denote the event that a type A box is selected, and B denote the event that a type B box is selected, find:

(a) $P(R)$ (b) $P(A \mid R)$

SOLUTION

(a) We note that R is the union of the mutually exclusive events $R \cap A$ and $R \cap B$ and use the formula

$$P(R) = P(R \cap A) + P(R \cap B)$$

We then use the intersection formula to obtain

$$P(R \cap A) = P(R \mid A)P(A) = \frac{3}{5}\left(\frac{1}{3}\right) = \frac{1}{5}$$

$$P(R \cap B) = P(R \mid B)P(B) = \frac{3}{4}\left(\frac{2}{3}\right) = \frac{1}{2}$$

Thus, we find

$$P(R) = \frac{1}{5} + \frac{1}{2} = \frac{7}{10}$$

(b) Using the result of part (a) we apply the conditional probability formula and obtain

$$P(A \mid R) = \frac{P(A \cap R)}{P(R)} = \frac{1/5}{7/10} = \frac{2}{7}$$ ●

Bayes' formula In the preceding example, we used the facts that for any events A and B,

$$B = (B \cap A) \cup (B \cap A')$$

and that $(B \cap A)$ and $(B \cap A')$ are mutually exclusive. We then applied the formula for the probability of the union of mutually exclusive events and the formula for the probability of an intersection to obtain

$$\begin{aligned} P(B) &= P(B \cap A) + P(B \cap A') \\ &= P(B \mid A)P(A) + P(B \mid A')P(A') \end{aligned}$$

Finally, we inserted this expression in the denominator of the conditional probability formula and obtained a useful formula for computing conditional probabilities. This formula is known as **Bayes' formula** (after the eighteenth century English clergyman and mathematician Thomas Bayes).

Bayes' formula
For any events A and B,

$$P(A|B) = \frac{P(A \cap B)}{P(B)} = \frac{P(B|A)P(A)}{P(B|A)P(A) + P(B|A')P(A')}$$

The next example illustrates the use of Bayes' formula.

EXAMPLE 3.4 ● In a certain population, 1 percent are heroin users. It is known that 95 percent of the heroin users also use marijuana, while 60 percent of those who don't use heroin use marijuana. A person is selected at random from this population. Find the probability that the selected person is a heroin user, given that he or she is a marijuana user.

SOLUTION Let H denote that the selected person is a heroin user and M denote that the selected person is a marijuana user. Using this notation in Bayes' formula we obtain

$$P(H|M) = \frac{P(M|H)P(H)}{P(M|H)P(H) + P(M|H')P(H')}$$

From the statement of the problem we know that $P(M|H) = 0.95$, $P(M|H') = 0.60$, $P(H) = 0.01$, and $P(H') = 0.99$. Putting these numbers in the formula we get

$$P(H|M) = \frac{0.95(0.01)}{0.95(0.01) + 0.60(0.99)} = 0.016$$ ●

Problems

1 The following contingency table gives the breakdown by race and position of employees in a certain company.

	Managerial	Nonmanagerial	Total
White	54	685	739
Nonwhite	12	244	256
Total	66	929	995

One employee is selected at random.

(a) Find the probability that the selected employee is in a managerial position.

(b) Find the probability that the selected employee is in a managerial position, given that he or she is nonwhite.

2 The following table gives the success record of a certain weather forecaster in predicting rain over a 700-day period.

	Predicted rain	Predicted no rain	Total
It rained	180	120	300
It didn't rain	190	210	400
Total	370	330	700

One day from this period is selected at random.

(a) Find the probability that it rained on the selected day.

(b) Find the probability that it rained on the selected day, given that the forecaster predicted rain.

(c) Do the probabilities in parts (a) and (b) indicate that the forecaster is doing a good job? Explain.

3 Following a major earthquake, a team of engineers visited a city to assess whether or not buildings constructed under the new earthquake safety guidelines had fared better than older buildings. Every building in a six-square-block downtown area was inspected, and the following results were obtained.

	New construction	Old construction	Total
No significant damage	18	75	93
Moderate damage	26	57	83
Severe damage	24	62	86
Total	68	194	262

One of these buildings is selected at random.

(a) Find the probability that the selected building has no significant damage.

(b) Find the probability that the selected building has no significant damage, given that it is of new construction.

(c) Find the probability that the selected building has severe damage.

(d) Find the probability that the selected building has severe damage, given that it is of new construction.

(e) Do these results indicate that the new construction is effective in preventing severe earthquake damage?

4 There are 2 large bags and 8 small bags. Each large bag contains 3 red marbles and 17 green marbles. Each small bag contains 2 red marbles and 18 green marbles. A bag is selected at random and then a marble is drawn from the selected bag. Find the probability that a small bag was selected, given that a red marble is drawn.

5 At a certain university, 60 percent of the marijuana smokers are cigarette smokers, while 40 percent of the students who don't smoke marijuana are cigarette smokers. Also, a total of 30 percent of the students are marijuana smokers. A student is selected at random.

(a) Find the probability that the selected student smokes marijuana, given that he or she smokes cigarettes.

(b) Find the probability that the selected student smokes marijuana, given that he or she doesn't smoke cigarettes.

(c) Suppose that there are 5,000 students at this university. Construct a contingency table consistent with the given information and recompute parts (a) and (b) using the table.

6 In pond A, 40 percent of the fish are trout and 60 percent are perch. In pond B, 70 percent are trout and 30 percent are perch. You are on your way to pond A but get lost. When you finally reach a pond you're not sure which pond it is, but you think there's a 75 percent chance that it's pond A. In any event, you've planned to go fishing, so that's what you do. Suppose you catch one fish and that each fish in the pond has the same chance of being caught.

(a) Find the probability that you catch a perch.

(b) Find the probability you're at pond A, given that you catch a perch.

(c) Find the probability you're at pond A, given that you catch a trout.

7 A medical diagnostic test for a certain disease (analogous to the skin test for tuberculosis) will yield either a positive or negative reaction. If you have the disease, there is a 0.99 chance that the test result will be positive, while if you don't have the disease, there is a 0.90 chance that the test will be negative. It is estimated that 0.003 of the population has the disease.

(a) Find the probability that you have the disease, given that you have a positive reaction.

(b) Find the probability that you don't have the disease, given that you have a negative reaction.

8 There are 5 coins in a box. One of the coins has 2 heads and the other 4 are ordinary coins. A coin is selected at random from the box and tossed, and you are told the outcome of the toss.

(a) Find the probability that an ordinary coin was selected, given that heads was the result of the toss.

(b) Find the probability that the 2-headed coin was selected, given that heads was the result of the toss.

(c) Find the probability that an ordinary coin was selected, given that tails was the result of the toss.

9 It is known that 1 percent of pet turtles smell bad. There are two types of pet turtles, green turtles and brown turtles. A pet-store owner discovers that 90 percent of his smelly turtles are green, whereas only 50 percent of the other turtles are green.

(a) Find the overall fraction of green turtles.

(b) If 1 turtle is selected at random, find the probability that it is smelly, given that it is green.

10 Box A has 4 green marbles and 5 red marbles. Box B has 3 green marbles and 2 red marbles. A marble is selected at random from box B and put in box A. Then a marble is randomly selected from box A.
(a) Find the probability that the marble selected from box A is green.
(b) Find the probability that the marble selected from box B was green, given that the marble selected from box A is green.

11 You mail a letter to a friend. There is a 10 percent chance that the letter will get lost on the way to the post office. Given that the letter gets to the post office, there is a 20 percent chance that it will get destroyed by the cancelling machine. Given that the letter makes it safely through the cancelling machine, there is a 10 percent chance that the mailman will deliver it to the wrong house. Given that your friend doesn't receive the letter, what is the probability that the cancelling machine destroyed it?

12 Each of 2 tennis players arrives on the court with a can of 3 balls. One has Wilsons and the other Penns. One player accidently hits a ball over the fence and it is lost. The other player wishes to determine (discretely) whose ball was lost. At the moment there are 3 balls on his side of the net, 2 of which are Penns and 1 of which is a Wilson. Given this information, what is the probability that the missing ball is a Penn?

13 Use the conditional probability formula to show that if A, B, and C are events and A and B are mutually exclusive, then $P(A \cup B \mid C) = P(A \mid C) + P(B \mid C)$. (*Hint:* $(A \cup B) \cap C = (A \cap C) \cup (B \cap C)$.)

14 Use the result of Problem 13 to show that if A and B are events, then $P(A \mid B) = 1 - P(A' \mid B)$. (*Hint:* $\Omega = A \cup A'$.)

4 • COUNTING AND PROBABILITY

In many experiments it is impossible or impractical to list all the outcomes of relevant events. In such situations, the selection formulas of Chapter 4 can often be used to calculate the desired probabilities. Here are some examples.

EXAMPLE 4.1 ● A box contains 8 green balls and 6 red balls. Five balls are selected at random.

(a) Find the probability that all are green.
(b) Find the probability that exactly 3 are red.
(c) Find the probability that at least 1 is red.
(d) Find the probability that 2 or 3 are red.

SOLUTION (a) There are 14 balls altogether and $\binom{14}{5} = 14(13)(12)(11)(10)/5! = 2{,}002$ ways to select 5 balls from 14. There are $\binom{8}{5} = 8(7)(6)(5)(4)/5! = 56$ ways to select 5 green balls from 8. Therefore, $P(\text{all green}) = 56/2{,}002$.

(b) If exactly 3 of the 5 chosen balls are red, the remaining 2 must be green. There are $\binom{6}{3} = 20$ ways to select 3 red balls from 6, and $\binom{8}{2} = 28$ ways to select 2 green balls from 8. By the joint selections formula, there are $\binom{6}{3}\binom{8}{2} = 20(28) = 560$ ways to select 5 balls, 3 of which are red and 2 of which are green. Therefore, $P(\text{exactly 3 red}) = 560/2{,}002$.

(c) The easiest way to find $P(\text{at least 1 red})$ is to use the formula for complementary events. The complement of the event that at least 1 of the selected balls is red is the event that all are green. In part (a) we found that $P(\text{all green}) = 56/2{,}002$. It follows that $P(\text{at least 1 red}) = 1 - P(\text{all green}) = 1 - 56/2{,}002 = 1{,}946/2{,}002$.

(d) In part (b) we found $P(\text{exactly 3 red}) = 560/2{,}002$. In a like manner, we can show that $P(\text{exactly 2 red}) = \binom{6}{2}\binom{8}{3}\Big/\binom{14}{5} = 840/2{,}002$. Since the events that exactly 2 reds are selected and exactly 3 reds are selected are mutually exclusive, it follows that $P(2 \text{ or } 3 \text{ red}) = P(2 \text{ red}) + P(3 \text{ red}) = 560/2{,}002 + 840/2{,}002 = 1{,}400/2{,}002$. ●

EXAMPLE 4.2 ● A jury of 12 people is selected, presumably at random, from a pool of 30 women and 20 men. After the selection is made, the resulting jury includes 2 women and 10 men. The defense attorney objects, claiming that the fraction, $2/12 = 0.167$, of women on the jury is not consistent with the fraction, $30/50 = 0.6$, of women in the pool. If the selection really was random, find the probability that at most 2 women are chosen.

SOLUTION Using the law for mutually exclusive events, we obtain

$$P(\text{at most 2 women}) = P(\text{exactly 2 women}) + P(\text{exactly 1 woman}) + P(\text{no women})$$

Using the selection formula, we obtain

$$P(\text{exactly 2 women}) = \frac{\binom{30}{2}\binom{20}{10}}{\binom{50}{12}} = 0.00066$$

$$P(\text{exactly 1 woman}) = \frac{\binom{30}{1}\binom{20}{11}}{\binom{50}{12}} = 0.00004$$

$$P(\text{no women}) = \frac{\binom{30}{0}\binom{20}{12}}{\binom{50}{12}} = 0.000001$$

Hence, $P(\text{at most 2 women}) = 0.00066 + 0.00004 + 0.000001 = 0.000701$, a figure that supports the defense attorney's contention. ●

Games of chance We are now able to compute probabilities for some popular games of chance. A discussion of odds and expected winnings for such games will be taken up in the next chapter. In the next example, we analyze a typical **lottery.**

EXAMPLE 4.3 ● Suppose you buy 10 tickets to a lottery in which there are to be 5 prizes awarded on the basis of a random drawing. Suppose that there are a total of 150 tickets sold. What are your chances of winning

(a) No prizes (b) At least 1 prize
(c) Exactly 3 prizes (d) All 5 prizes

SOLUTION (a) Since there are 150 tickets altogether and 5 prizes to be awarded, there are $\binom{150}{5} = 591{,}600{,}030$ possible sets of winning tickets. There are 140 tickets that are not yours, and so there are $\binom{140}{5} = 416{,}965{,}528$ ways to award 5 prizes so that none go to you. Therefore,

$$P(\text{no prizes}) = \frac{416{,}965{,}528}{591{,}600{,}030} = 0.705$$

(b) If you don't win at least 1 prize, you win no prizes, so

$$P(\text{at least 1 prize}) = 1 - P(\text{no prizes}) = 1 - 0.705 = 0.295$$

(c) There are $\binom{10}{3} = 120$ ways to select 3 of your 10 tickets for prizes and $\binom{140}{2} = 9{,}730$ ways to award 2 prizes to the other ticket holders. Thus,

$$P(\text{exactly 3 prizes}) = \frac{\binom{10}{3}\binom{140}{2}}{\binom{150}{5}} = \frac{120(9{,}730)}{591{,}600{,}030} = 0.002$$

(d) There are $\binom{10}{5} = 252$ ways to choose 5 of your tickets, so

$$P(\text{all 5 prizes}) = \frac{252}{591{,}600{,}030} = 0.0000004$$

which is not much of a chance. ●

In most lotteries many more than 150 tickets are sold, and your chances of winning prizes are even less than in this example, unless you buy an enormous number of tickets.

Keno is a popular casino game. One of the reasons it is so popular is that for a very small bet (say \$1) it is possible for you to win a very large amount of money (up to \$25,000). Needless to say, the chances of doing this are very low. Keno is played as follows. There are 80 balls numbered from 1 through 80. Each time the game is played, 20 of the balls are selected at random. You can bet on blocks of from 1 to 12 numbers, and you win or lose depending on how many of the numbers that you bet on are selected. In the next example we shall use the selection formulas to calculate probabilities for winning some typical keno bets.

EXAMPLE 4.4 ● Suppose you are playing keno.

(a) If you bet on 1 number, what is the probability that your number is selected?

(b) If you bet on 4 numbers, what is the probability that exactly 3 of your numbers are selected? That all 4 are selected?

(c) If you bet on 10 numbers, what is the probability that all 10 of your numbers are selected?

SOLUTION (a) There is $\binom{1}{1} = 1$ way to select your 1 number, $\binom{79}{19}$ ways to select the other 19 numbers from the 79 you didn't pick, and there are $\binom{80}{20}$ ways altogether to select 20 numbers from 80. It follows that

$$P(\text{your number selected}) = \frac{\binom{1}{1}\binom{79}{19}}{\binom{80}{20}} = 0.25$$

(If you make a \$1 bet on one number and that number is selected, you get \$3 back, for a net gain of \$2.)

(b) There are $\binom{4}{3}$ ways to select 3 numbers from 4 and $\binom{76}{17}$ ways to select the other 17 numbers from the 76 you didn't bet on, so

$$P(\text{3 of your 4 numbers selected}) = \frac{\binom{4}{3}\binom{76}{17}}{\binom{80}{20}} = 0.043$$

Similarly,

$$P(\text{all 4 of your numbers selected}) = \frac{\binom{4}{4}\binom{76}{16}}{\binom{80}{20}} = 0.003$$

(If you make a $1 bet on four numbers, you get $4 if three of your four numbers are selected and $180 if all four of your numbers are selected.)

(c) If you bet $1 on 10 numbers and all 10 are selected, you win $25,000 (or whatever is left in the betting pool for that game). Your chances of winning are $\binom{10}{10}\binom{70}{10}\Big/\binom{80}{20} = 0.0000001$, that is, 1 in 10 million! ●

Poker is a card game for two or more players in which the winner is the player with the best five-card hand. There is a hierarchy of types of hands, and in the next example we shall compute the probabilities of a few of these hands. We assume you are familiar with the basic rules of poker.

EXAMPLE 4.5 ● Suppose you are dealt 5 cards from a well-shuffled deck. Find the probability of getting each of the following poker hands:

(a) A **straight flush,** that is, 5 cards in sequence, all of the same suit (see Figure 4.1).
(b) A **full house,** that is, 3 of one denomination and 2 of another.
(c) A **flush,** that is, 5 cards of the same suit that are not all in sequence.
(d) A **pair,** that is, 2 cards of one denomination and 3 of three other denominations.

Figure 4.1 A straight flush.

SOLUTION (a) By applying the selection formula, we find that there are $\binom{52}{5} = 2{,}598{,}960$ ways to select 5 cards from a deck. If we count aces as either high or low, there are 10 types of "straight sequences": A2345, 23456, 34567, . . . , 10JQKA. Since there are 4 suits possible for each sequence, there are a total of $10(4) = 40$ possible straight flush hands, and so $P(\text{straight flush}) = 40/2{,}598{,}960 = 0.000015$ (that is, 15 chances in one million).

(b) There are 4 cards of each denomination, and thus $\binom{4}{3}$ ways to get 3 of a particular denomination. Since there are 13 denominations, there are $13\binom{4}{3}$ ways to get 3 cards of the same denomination. Once the 3-card denomination has been determined, there are 12 denominations left from which to get a pair and there are $\binom{4}{2}$ ways to select a pair of cards from 4 cards. Hence there are $13\binom{4}{3}(12)\binom{4}{2}$ possible ways to get a full house, and

$$P(\text{full house}) = \frac{13\binom{4}{3}(12)\binom{4}{2}}{\binom{52}{5}} = 0.00144$$

(c) There are 13 cards of each suit and 4 different suits, and so there are $4\binom{13}{5}$ ways to get 5 cards of the same suit. But this includes the 40 straight flushes, and so,

$$P(\text{flush}) = \frac{4\binom{13}{5} - 40}{\binom{52}{5}} = 0.00197$$

(d) There are 13 possible denominations with which to get a pair and $\binom{4}{2}$ ways to get a pair of a particular denomination. Once the denomination of the pair has been determined, there are $\binom{12}{3}$ ways to select 3 other denominations for the other 3 cards, since they must all be different. Once these 3 denominations have been determined, there are $\binom{4}{1}\binom{4}{1}\binom{4}{1}$ ways to select 1 card from each. Hence

$$P(\text{pair}) = \frac{13\binom{12}{3}\binom{4}{2}\binom{4}{1}\binom{4}{1}\binom{4}{1}}{\binom{52}{5}} = 0.42257$$

●

The probabilities we have obtained apply when five cards are selected at random. If there are more cards to choose from, as in draw poker or seven-card stud, the probabilities change, but the hierarchy of hands remains the same.

That is, the better the hand, the lower the probability of obtaining it. Here is a list of the hierarchy of poker hands from highest to lowest, along with the corresponding probabilities. The probabilities we have not derived will be assigned in the problems at the end of this section.

Poker hands

Hand	Probability
Straight flush	0.000015
Four of a kind	0.00024
Full house	0.00144
Flush	0.00197
Straight	0.00392
Three of a kind	0.02113
Two pair	0.04754
Pair	0.42257

Problems

1 From 20 light bulbs, 4 of which are burnt out, you select 3 bulbs at random.
  (a) Find the probability that all are burnt out.
  (b) Find the probability that none are burnt out.
  (c) Find the probability that at least 1 is burnt out.
  (d) Find the probability that exactly 1 is burnt out.

2 You and two of your friends are in a geology class with 30 people altogether. Five people are selected at random to go on a field trip to a nearby quarry.
  (a) Find the probability that you are selected.
  (b) Find the probability that you are selected but your friends aren't.
  (c) Find the probability that you and your friends are selected.
  (d) Find the probability that neither you nor your friends are selected.
  (e) Find the probability that your friends are selected but you aren't.
  (f) Find the probability that you and just one of your friends are selected.

3 There are 15 people in a police lineup, 3 of whom were involved in a robbery. A witness is trying to identify the criminals, but recognizes only one of them. Unwilling to admit his inability to identify the other two, he selects the one he recognizes along with the two standing on either side of him. (The one he recognizes is not on the end of the line.) Suppose the people in the lineup have been arranged in random order.
  (a) Find the probability that all three criminals are correctly identified.
  (b) Find the probability that the recognized person is the only one of the three criminals correctly identified.
  (c) Find the probability that at least two of the three criminals are correctly identified.

4 A box contains 15 shoes of the same size and style, including 10 right shoes and 5 left shoes. Eight shoes are selected at random from this box.
  (a) Find the probability that exactly 4 pairs of shoes are chosen.
  (b) Find the probability that at least 2 pairs are chosen.
  (c) Find the probability that no pairs are chosen.

5 Find the probability that all 13 cards selected at random from a deck of cards are red.

6 In a carnival game, 3 marbles are hidden under 8 cups so that no cup covers more than 1 marble. You can choose 3 cups and you win a prize if there are marbles under 2 or 3 of them. (You have to make all 3 choices before looking under any of the cups.) Suppose you choose 3 cups at random.
  (a) Find the probability that all 3 have marbles under them.
  (b) Find the probability that exactly 2 have marbles under them.
  (c) Find the probability that you win a prize.

7 On the TV game show "Frenzy," there are 10 doors. Behind 2 of the doors are valuable prizes. Behind the other 8 doors are a variety of worthless items. After the audience is worked into a frenzy by being shown pictures of the lavish prizes given away in recent weeks, the contestant is asked to select 3 doors at random from the 10 doors. The contestant then wins the item of highest value behind the 3 selected doors. Find the probability that the contestant wins a valuable prize.

8 Three of 25 newly built cars are selected at random to be checked for steering defects. Suppose 6 of the 25 cars have steering defects.
  (a) Find the probability that none of the selected cars is defective.
  (b) Find the probability that at least 1 of the selected cars is defective.
  (c) Find the probability that all 3 of the selected cars are defective.

9 Each week, a professor known for telling jokes writes 15 jokes on slips of paper and puts them in his pocket. Each class meeting he takes 3 pieces of paper from his pocket, tells the jokes written on them, and puts them back in his pocket. The class meets 3 times per week.
  (a) Find the probability that no joke is repeated in the first 2 classes.
  (b) Find the probability that no joke is repeated during the week.

10 Five cards are dealt from a well-shuffled deck. Find the probability that the following poker hands are dealt:
  (a) Four of a kind
  (b) Straight
  (c) Three of a kind
  (d) Two pair

11 You buy 4 tickets to a lottery in which 10 prizes are to be awarded on the basis of a random drawing. Suppose 200 other tickets are sold.
  (a) Find the probability that you win no prizes.
  (b) Find the probability that you win at least 1 prize.
  (c) Find the probability that you win 2 prizes.
  (d) Find the probability that you win 4 prizes.

12 Five cards are dealt from a well-shuffled deck.
  (a) Find the probability that all are face cards (J, Q, or K).
  (b) Find the probability that 3 are face cards and 2 are tens.
  (c) Find the probability that at least 3 are face cards.
  (d) Find the probability that none is a face card.

13 Suppose you bet on 5 numbers in keno.
  (a) Find the probability that exactly 4 of your numbers are selected.
  (b) Find the probability that all 5 of your numbers are selected.

14 In horse racing, if you bet on a horse to "show," you win if the horse comes in first, second, or third. If you bet on a horse to "place," you win if the horse comes in either first or second, and if you bet on a horse to "win," you win only if the horse comes in first. In a certain race you bet on 2 horses to show, you bet on 3 other horses to place, and you bet on still another horse to win. Assume that there are 10 evenly matched horses in the race.
(a) Find the probability that both of your bets to show pay off.
(b) Find the probability that exactly 1 of your bets to place pays off.
(c) Find the probability that your bet to win pays off.

5 • INDEPENDENCE

In Section 3, we saw that if there is a relationship between events A and B (as there is between high blood pressure and smoking), then the unconditional probability $P(A)$ will be different from the conditional probability $P(A|B)$. In this section we shall study unrelated or **independent** events. We say that two events A and B are independent if $P(A|B) = P(A)$. For example, if you toss a coin repeatedly, the result of one toss does not affect the result of another toss and so the individual tosses are independent.

The multiplication rule

When A and B are independent events, the formula $P(A \cap B) = P(A|B)P(B)$ for the probability of an intersection can be rewritten more simply as $P(A \cap B) = P(A)P(B)$, since $P(A|B) = P(A)$. This simplified formula is sometimes called the multiplication rule.

The multiplication rule
If A and B are independent events, then

$$P(A \cap B) = P(A)P(B)$$

That is, the probability of the intersection of two (or more) independent events is the product of their individual probabilities.

If probabilities are viewed as fractions, the multiplication rule says that if A and B are independent, the fraction of the time that A and B both occur is the product of the individual fractions. For example, if A occurs $\frac{1}{3}$ of the time and B occurs $\frac{1}{2}$ of the time, and if A and B are independent, then the fraction of the time that both A and B occur will be $\frac{1}{3}$ of $\frac{1}{2}$, or $\frac{1}{6}$. The next three examples illustrate the use of the multiplication rule.

EXAMPLE 5.1 ● A deep-sea diver has two independent oxygen supply systems, so that even if one breaks down, the diver still gets oxygen. Suppose the probability that system 1 works is 0.9, while the probability that system 2 works is 0.8.

(a) Find the probability that neither system breaks down.

(b) Find the probability that both systems break down.
(c) Find the probability that at least one system works.

SOLUTION (a) We let S_1 denote the event that system 1 works and S_2 denote the event that system 2 works. Then $P(S_1) = 0.9$ and $P(S_2) = 0.8$, and we wish to find $P(S_1 \cap S_2)$. Since S_1 and S_2 are independent events, we apply the multiplication rule and obtain

$$P(S_1 \cap S_2) = P(S_1)P(S_2) = 0.9(0.8) = 0.72$$

(b) In this case we wish to find $P(S_1' \cap S_2')$. Noting that $P(S_1') = 0.1$ and $P(S_2') = 0.2$, we apply the multiplication rule and obtain

$$P(S_1' \cap S_2') = P(S_1')P(S_2') = 0.1(0.2) = 0.02$$

(c) The event that at least one system works is $S_1 \cup S_2$, which is the complement of the event that neither works, Therefore,

$$P(S_1 \cup S_2) = 1 - P(S_1' \cap S_2') = 1 - 0.02 = 0.98$$

An alternative approach is to use the general rule for unions to obtain

$$\begin{aligned} P(S_1 \cup S_2) &= P(S_1) + P(S_2) - P(S_1 \cap S_2) \\ &= 0.9 + 0.8 - 0.72 = 0.98 \end{aligned}$$ ●

EXAMPLE 5.2 ● A pair of dice is rolled three times.
(a) Find the probability that the dice add up to 7 each time.
(b) Find the probability that the dice add up to 7 at least once.

SOLUTION (a) Let S_1, S_2, and S_3 denote the events that the dice add up to 7 on the first, second, and third rolls, respectively. We wish to find $P(S_1 \cap S_2 \cap S_3)$. We have seen earlier (in Example 1.6) that on any roll, $P(7) = \frac{1}{6}$. Since successive rolls of the dice are independent, we use the multiplication rule to obtain $P(S_1 \cap S_2 \cap S_3) = \frac{1}{6}(\frac{1}{6})(\frac{1}{6}) = \frac{1}{216}$.
(b) The complement of the event that the dice add up to 7 at least once is the event that the dice do not add up to 7 on any of the rolls, that is, $S_1' \cap S_2' \cap S_3'$. Since $P(S_1) = P(S_2) = P(S_3) = \frac{1}{6}$, it follows that $P(S_1') = P(S_2') = P(S_3') = \frac{5}{6}$, and using the rule for complementary events and the multiplication rule, we obtain

$$\begin{aligned} P(\text{dice add up to 7 at least once}) &= 1 - P(S_1' \cap S_2' \cap S_3') \\ &= 1 - \frac{5}{6}\left(\frac{5}{6}\right)\left(\frac{5}{6}\right) = \frac{91}{216} \end{aligned}$$ ●

Sampling with replacement

When cards are dealt at random from a deck, the denomination of a particular card is dependent upon the cards that have already been dealt. In fact, strategies for card games usually involve remembering cards as they appear and choosing one's subsequent play accordingly. There is another type of random sampling in which the results of the selections are independent. This occurs when the sampling is done **with replacement,** that is, when each selected item is returned to the population before the next selection is made. The next example illustrates this concept.

EXAMPLE 5.3 ● A box contains 3 green marbles and 2 yellow ones. Two marbles are selected at random with replacement. In other words, the first marble is replaced in the box before the second marble is drawn.

(a) Find the probability that both selected marbles are green.
(b) Find the probability that the first marble is green and the second marble is yellow.
(c) Find the probability that at least 1 marble is yellow.

SOLUTION Since the marbles are drawn with replacement, the outcome of one selection does not affect the outcome of the other. That is, the selections are independent. It is convenient to introduce the following notation. We let G denote the selection of a green marble and Y the selection of a yellow marble. On each selection, $P(G) = \frac{3}{5}$ and $P(Y) = \frac{2}{5}$. The sample space can then be written as $\Omega = \{GG, GY, YG, YY\}$ where, for example, GY means that the first marble is green and the second is yellow. (More formally, the event GY is the *intersection* of the event that the first marble is green with the event that the second marble is yellow.)

(a) We wish to find $P(GG)$. Applying the multplication rule we obtain

$$P(GG) = \frac{3}{5}\left(\frac{3}{5}\right) = \frac{9}{25}$$

(b) We wish to find $P(GY)$. Applying the multiplication rule we obtain

$$P(GY) = \frac{3}{5}\left(\frac{2}{5}\right) = \frac{6}{25}$$

(c) The event that at least one yellow marble is selected is the complement of the event GG. Using the result of part (a) we obtain

$$P(\text{at least one yellow}) = 1 - P(GG) = \frac{16}{25}$$

●

Probabilities computed when sampling is done with replacement differ, in general, from the corresponding probabilities computed when sampling is done without replacement. The reason for this difference is clear. When sampling is done without replacement, each selection alters the makeup of the population from which subsequent selections are to be made. When sampling is done with replacement, the population is the same for each selection.

Sampling from large populations

In one important situation, there is almost no difference in the probabilities computed when sampling is done with replacement and without replacement. This is when the population size is large relative to the sample size. The reason is that when you select from a large population, the removal of a few members of the population has almost no effect on the population as a whole. The importance of this is that when dealing with samples from large populations, you may assume that the sampling is done with replacement, whether or not this is actually the case. There are several advantages to this approach. For example, certain events that would not be independent if sampling were done without replacement become independent if replacement is assumed before each selection. Moreover, when sampling is done without replacement, the calculations of many probabilities are based on the selection formulas. These formulas become unmanageable when the population is large, and cannot be used at all if percentages rather than absolute numbers are used to describe the frequency of events.

These ideas are illustrated in the next example.

EXAMPLE 5.4 ● In a certain state 55 percent will vote for Spiff for governor. Three people are selected at random.

(a) Find the probability that all 3 will vote for Spiff.
(b) Find the probability that exactly 2 will vote for Spiff.

SOLUTION Since the population is large we shall calculate probabilities assuming that sampling is done with replacement. We let S denote that a person will vote for Spiff and N that the person will not. On any selection, $P(S) = 0.55$ and $P(N) = 0.45$.

(a) We wish to find $P(SSS)$. Applying the multiplication rule we get

$$P(SSS) = (0.55)^3 = 0.166$$

(b) We break the event that exactly 2 will vote for Spiff into the 3 mutually exclusive events SSN, SNS, and NSS, and use the multiplication rule to obtain the probability of each. We get

$$\begin{aligned} P(\text{exactly 2 will vote for Spiff}) &= P(SSN) + P(SNS) + P(NSS) \\ &= 3(0.55)^2(0.45) \\ &= 0.408 \end{aligned}$$

●

Binomial probabilities

The preceding three examples were alike in the following ways. In each, a sequence of independent selections, or trials, was performed, and on each trial a particular event either occurred or did not occur. In Example 5.2 the event of interest was rolling 7 with a pair of dice; in Example 5.3 it was selecting a green marble; in Example 5.4 it was selecting someone who would vote for Spiff. Traditionally the event of interest is known as "success" and its complement is known as "failure." This is an arbitrary label and is used only to standardize the terminology.

Probabilities for obtaining a particular number of successes in n independent trials of an experiment are known as **binomial probabilities.** We shall state a general formula for these probabilities, but first we shall give an example to illustrate the reasoning that leads to the formula.

EXAMPLE 5.5 ● A certain operation is successful 80 percent of the time. If 4 people undergo this operation, find the probability of exactly 3 successes.

SOLUTION We let S denote success and F failure. On each trial (i.e., operation), $P(S) = 0.8$ and $P(F) = 0.2$. The sample space can be represented as the collection of all sequences of length 4 of S's and F's, where, for example, $SSSF$ denotes the outcome that the first 3 operations are successes and the 4th is a failure. The event that there are exactly 3 successes is the collection of all ordered sequences with 3 S's and 1 F. To compute the probability of this event, we calculate the probability of obtaining a sequence of 3 S's and 1 F *in a particular order* (each ordering has the same probability) and then multiply this probability by the number of possible rearrangements of the sequence. By the multiplication rule, each ordering has probability $(0.8)^3(0.2) = 0.1024$. By the selection formula, the number of sequences with 3 S's and 1 F is $\binom{4}{3} = 4$; namely $SSSF$, $SSFS$, $SFSS$, and $FSSS$. Therefore,

$$P(\text{exactly 3 successes}) = \binom{4}{3}(0.8)^3(0.2) = 0.4096$$ ●

The reasoning in Example 5.5 can be used to derive the general binomial probability formula.

Binomial probabilities
Suppose an experiment with two possible outcomes (traditionally called success and failure) is performed n times, and that each time the probability of success is p. Let X denote the total number of successes that occur. Then, for $0 \leq k \leq n$,

$$P(X = k) = \binom{n}{k} p^k (1 - p)^{n-k}$$

The next two examples will illustrate the use of the binomial probability formula.

EXAMPLE 5.6 ● In a large suburban area, 10 percent of adults have high blood pressure. Suppose you select 15 adults at random. Find the probability that 2 have high blood pressure.

SOLUTION Let X be the number in the sample who have high blood pressure. Since 15 are selected, $n = 15$. Since 10 percent have high blood pressure, $p = 0.1$ and $1 - p = 0.9$. Hence, $P(X = 2) = \binom{15}{2}(0.1)^2(0.9)^{13} = 0.2669$. ●

EXAMPLE 5.7 ● An evenly balanced coin is tossed 10 times.

(a) Find the probability that heads comes up 5 times.
(b) Find the probability that heads comes up less than 3 times.

SOLUTION Let X be the number of times heads comes up. Since the coin is tossed 10 times, we have $n = 10$, and since the coin is evenly balanced, we have $p = 0.5$.

(a) $P(X = 5) = \binom{10}{5}(0.5)^5(0.5)^5 = 0.2461$.

(b) $P(X < 3) = P(X = 0) + P(X = 1) + P(X = 2)$

$$= \binom{10}{0}(0.5)^0(0.5)^{10} + \binom{10}{1}(0.5)^1(0.5)^9 + \binom{10}{2}(0.5)^2(0.5)^8 = 0.0547.$$ ●

Since calculations like the one in part (b) of Example 5.7 can be rather tedious, a table of binomial probabilities has been included in the Appendix.

Problems

1 An electric component on a satellite has 3 independent circuits. The component will work if 1 or more of the circuits is working. The probability that each circuit is working is 0.9. Find the probability that the component is working.

2 There are 3 stocks, A, B, and C. On a given day there is a 0.3 chance that stock A will go up in price, a 0.5 chance that stock B will go up in price, and a 0.2 chance that stock C will go up in price. Assume that the stocks behave independently of one another.
(a) Find the probability that all 3 stocks will go up in price.
(b) Find the probability that only stock A will go up in price.
(c) Find the probability that exactly 1 of the stocks will go up in price.
(d) Find the probability that at least 1 of the stocks will go up in price.

3 A box contains 2 red balls and 6 blue balls. Two balls are selected at random, with replacement.
(a) Find the probability that both balls are red.
(b) Find the probability that at least 1 is red.
(c) Find the probability that exactly 1 is red.

4 A box contains 5 yellow marbles, 6 blue marbles and 4 clear marbles. Three marbles are selected at random, with replacement.
 (a) Find the probability that all are clear.
 (b) Find the probability that exactly 1 is clear.
 (c) Find the probability that 1 is clear and 2 are yellow.
 (d) Find the probability that 1 is clear, 1 is yellow, and 1 is blue.

5 A pair of dice is rolled 3 times. Find the probability that the dice add up to 9 at least twice.

6 A box contains 2 tickets marked "success" and 3 tickets marked "failure." Four tickets are selected at random, with replacement.
 (a) Find the probability that all 4 tickets are marked success.
 (b) Find the probability that at least 3 tickets are marked success.
 (c) Find the probability that at least 1 ticket is marked success.

7 A large population of crickets contains 30 percent who can't chirp. Suppose 4 crickets are selected at random from this population.
 (a) Find the probability that all can chirp.
 (b) Find the probability that at least 1 can chirp.
 (c) Find the probability that exactly 2 can chirp.

8 In a large herd of cows, 10 percent have hoof and mouth disease. Three cows are selected at random.
 (a) Find the probability that all 3 have the disease.
 (b) Find the probability that none have the disease.
 (c) Find the probability that at least 2 have the disease.

9 In a certain community, 40 percent are Republicans and 60 percent are Democrats. Ten voters are selected at random.
 (a) Find the probability that exactly 4 Republicans are selected.
 (b) Find the probability that at most 4 Republicans are selected.
 (c) Find the probability that at least 4 Republicans are selected.

10 On the average, 2 people out of 10 will respond favorably to a certain telephone sales pitch. Suppose 15 people are called at random.
 (a) Find the probability that exactly 3 respond favorably.
 (b) Find the probability that at most 1 responds favorably.
 (c) Find the probability that at least 3 respond favorably.

11 The probability that a certain vaccine is effective is 0.8. The vaccine is given to 8 people selected at random.
 (a) Find the probability that the vaccine is effective for at least 6 of these people.
 (b) Find the probability that the vaccine is effective for fewer than 3 of these people.
 (c) Find the probability that the vaccine is effective for all 8 of these people.

12 A packaging machine underfills 15 percent of the pretzel packages it turns out. Suppose you select 10 packages at random.
 (a) Find the probability that at least 1 is underfilled.
 (b) Find the probability that exactly 3 are underfilled.

(c) Find the probability that at most 2 are underfilled.

13 The average weight of newborn babies is 3 kilograms, with 20 percent weighing more than 3.4 kilograms and 20 percent weighing less than 2.6 kilograms. Suppose 12 babies are selected at random.
  (a) Find the probability that exactly 4 weigh more than 3.4 kilograms.
  (b) Find the probability that exactly 3 weigh less than 2.6 kilograms.
  (c) Find the probability that exactly 6 weigh between 2.6 and 3.4 kilograms.

14 There is a 50 percent chance that a certain type of lightbulb will burn for more than 400 hours. Suppose 8 lightbulbs are selected at random and tested.
  (a) Find the probability that all burn more than 400 hours.
  (b) Find the probability that at least 1 burns more than 400 hours.
  (c) Find the probability that exactly 4 burn more than 400 hours.

15 Each morning for 10 days, the weather forecaster rolls a die. If it comes up 1, 2, or 3, the forecaster predicts rain and if it comes up 4, 5, or 6 the forecaster predicts no rain.
  (a) Find the probability that the weather forecaster predicts rain twice.
  (b) Find the probability that the weather forecaster predicts rain at least once.

6 • MARKOV CHAINS

In the preceding section, we discussed situations involving sequences of independent events. Many practical problems involve sequences in which the events are *not* independent. If the probability of each event in a sequence is determined by the event immediately preceding it, such a sequence of dependent events is called a **Markov chain** (named after the nineteenth century Russian mathematician Andrei Markov).

Markov chains are used to model situations, processes, or systems that are observed at regular intervals. At the time of each observation, the **state** of the system is noted. The sequence of these states is the Markov chain. The goal in a typical problem is to compute the probability that the system will be in a particular state at a specified time.

The probabilities associated with a Markov chain can be summarized conveniently in a matrix. The construction of such a matrix is illustrated in the next example.

EXAMPLE 6.1 ● A sociologist has developed a model of class mobility that describes the socioeconomic class movement of a family from generation to generation as a Markov chain. In the model, a family is classified as being in one of three states: upper class, middle class, or lower class. If the parents are upper class, the probability that an offspring will be upper class is 0.7, the probability that an offspring will be middle class is 0.2, and the probability that an offspring will be lower class is 0.1. If the parents are middle class, the probability that an offspring will be upper class is 0.4, the probability that an offspring will be middle class is 0.5, and the probability that an offspring will be lower class is 0.1. If the parents are lower class, the probability that an offspring will be upper class is 0.1, the probability that an offspring will be middle class is 0.6, and the probability that an offspring will be lower class is 0.3. Represent these probabilities in a matrix.

SOLUTION Letting state 1 be upper class, state 2 be middle class, and state 3 be lower class, we construct the matrix P, where the entry p_{ij} is the probability of going from state i to state j.

$$
\begin{array}{c} \\ P = \begin{array}{r} \text{Upper} \\ \text{Middle} \\ \text{Lower} \end{array} \end{array}
\begin{array}{c} \begin{array}{ccc} \text{Upper} & \text{Middle} & \text{Lower} \end{array} \\ \begin{bmatrix} 0.7 & 0.2 & 0.1 \\ 0.4 & 0.5 & 0.1 \\ 0.1 & 0.6 & 0.3 \end{bmatrix} \end{array}
$$

●

Transition probabilities In the preceding example, we used a matrix to display the probabilities of changing from one state to another. Since these probabilities refer to the transition of states, they are called **transition probabilities** and the matrix P is called the **transition matrix.**

Transition matrix
For a Markov chain with m states, the *transition matrix* P is the $m \times m$ matrix in which the entry p_{ij} is the probability of going from state i to state j.

Using the transition probabilities, it is possible to obtain probabilities for the state of a system more than one stage ahead. The next example introduces this technique.

EXAMPLE 6.2 ● A meteorologist has developed the following model for predicting the weather. If it is raining today, the probability is 0.6 that it will rain again tomorrow and the probability is 0.4 that it will not. If it is not raining today, the probability is 0.2 that it will rain tomorrow and the probability is 0.8 that it will not. (This is a simplified version of an actual rainfall model.)

(a) Find the transition matrix for this Markov chain.
(b) If it is raining today, what is the probability that it will be raining the day after tomorrow?
(c) If it is not raining today, what is the probability that it will be raining the day after tomorrow?

SOLUTION (a) Using the meteorologist's transition probabilities and letting "rain" be state 1 and "no rain" be state 2, we get the transition matrix

$$P = \begin{array}{r} \\ \text{Rain} \\ \text{No rain} \end{array} \begin{array}{c} \begin{array}{cc} \text{Rain} & \text{No rain} \end{array} \\ \begin{bmatrix} 0.6 & 0.4 \\ 0.2 & 0.8 \end{bmatrix} \end{array}$$

(b) There are two possible transition sequences in which it is raining today and raining the day after tomorrow. They are shown, along with the associated transition probabilities, in the **tree diagram** in Figure 6.1.

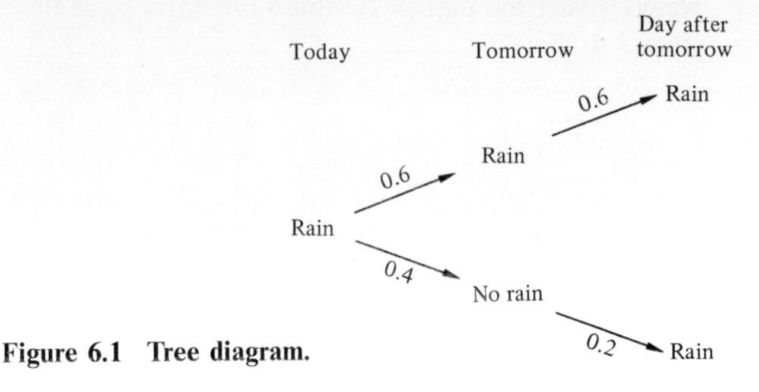

Figure 6.1 Tree diagram.

Because the probability of each state depends only on the preceding state, we can find the probability of each sequence by multiplying the individual transition probabilities. (This can be justified formally by means of the conditional probability formula.) We get

$$P(\text{rain} \rightarrow \text{rain} \rightarrow \text{rain}) = 0.6(0.6) = 0.36$$

$$P(\text{rain} \rightarrow \text{no rain} \rightarrow \text{rain}) = 0.4(0.2) = 0.08$$

The probability that it will be raining the day after tomorrow, given that it is raining today, is the sum

$$\begin{aligned} P(\text{rain} \rightarrow \text{rain} \rightarrow \text{rain}) + P(\text{rain} \rightarrow \text{no rain} \rightarrow \text{rain}) &= 0.36 + 0.08 \\ &= 0.44 \end{aligned}$$

(c) From the tree diagram in Figure 6.2 we see that

$$P(\text{no rain} \rightarrow \text{rain} \rightarrow \text{rain}) = 0.2(0.6) = 0.12$$

$$P(\text{no rain} \rightarrow \text{no rain} \rightarrow \text{rain}) = 0.8(0.2) = 0.16$$

Figure 6.2 Tree diagram.

and we conclude that if it is not raining today, the probability that it will be raining the day after tomorrow is $0.12 + 0.16 = 0.28$. ●

The probability that a system will be in a particular state after n transitions is called an ***n*-step transition probability.** The probabilities we computed in Example 6.2 with the aid of tree diagrams were 2-step transition probabilities. You can also use matrix multiplication to find these probabilities. Here's how.

Suppose that

$$P = \begin{array}{c} \text{State 1} \\ \text{State 2} \end{array} \overset{\begin{array}{cc} \text{State 1} & \text{State 2} \end{array}}{\begin{bmatrix} p_{11} & p_{12} \\ p_{21} & p_{22} \end{bmatrix}}$$

is the transition matrix for a two-state Markov chain. The corresponding tree diagrams for the 2-step transitions are shown in Figure 6.3.

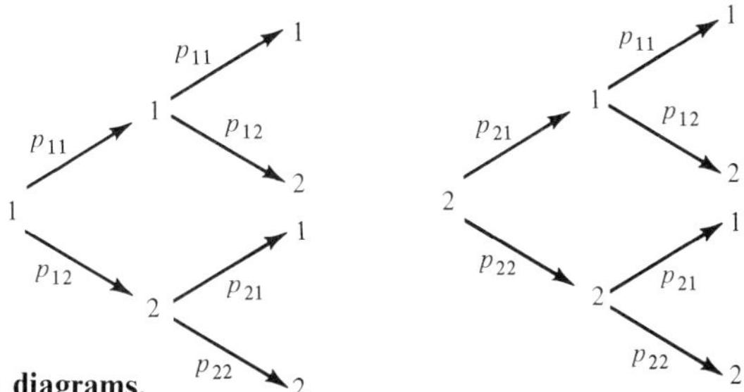

Figure 6.3 Tree diagrams.

The 2-step probabilities can be summarized in the matrix

$$\begin{array}{c} \text{State 1} \\ \text{State 2} \end{array} \overset{\begin{array}{cc} \text{State 1} & \text{State 2} \end{array}}{\begin{bmatrix} p_{11}p_{11} + p_{12}p_{21} & p_{11}p_{12} + p_{12}p_{22} \\ p_{21}p_{11} + p_{22}p_{21} & p_{21}p_{12} + p_{22}p_{22} \end{bmatrix}}$$

Notice that this matrix is precisely

$$P^2 = \begin{bmatrix} p_{11} & p_{12} \\ p_{21} & p_{22} \end{bmatrix} \begin{bmatrix} p_{11} & p_{12} \\ p_{21} & p_{22} \end{bmatrix}$$

which is the product of the transition matrix with itself. This result generalizes to any number of steps.

The *n*-step transition matrix

For any integer n, the matrix

$$P^n = P \times P \times \cdots \times P \qquad (n \text{ factors})$$

gives the n-step transition probabilities for the system. In particular, p_{ij}^n is the probability that the system will be in state j after n transitions if it is currently in state i.

Here is an example.

EXAMPLE 6.3 ● In a certain model of child behavior, a child's behavior at school on a given day is classified as either "good" or "bad." If a certain child is good today, there is a 0.9 chance that he will be good tomorrow, while if the child is bad today, there is a 0.3 chance that he will be bad tomorrow. Given that the child is good today, find the probability that he will be good 4 days from now.

SOLUTION The child's behavior can be described by a Markov chain with two states, "good" (state 1) and "bad" (state 2). The transition matrix for this chain is

$$P = \begin{array}{c} \\ \text{Good} \\ \text{Bad} \end{array} \begin{array}{c} \begin{array}{cc} \text{Good} & \text{Bad} \end{array} \\ \begin{bmatrix} 0.9 & 0.1 \\ 0.7 & 0.3 \end{bmatrix} \end{array}$$

The probability that the child will be good 4 days from now is the probability that the child will be in state 1 after 4 transitions. Since the child started in state 1 this probability is the entry p_{11}^4 in the product matrix P^4. Using matrix multiplication we get

$$P^2 = \begin{bmatrix} 0.9 & 0.1 \\ 0.7 & 0.3 \end{bmatrix} \begin{bmatrix} 0.9 & 0.1 \\ 0.7 & 0.3 \end{bmatrix} = \begin{bmatrix} 0.88 & 0.12 \\ 0.84 & 0.16 \end{bmatrix}$$

and

$$P^4 = P^2 \times P^2 = \begin{bmatrix} 0.88 & 0.12 \\ 0.84 & 0.16 \end{bmatrix} \begin{bmatrix} 0.88 & 0.12 \\ 0.84 & 0.16 \end{bmatrix} = \begin{bmatrix} 0.875 & 0.125 \\ 0.874 & 0.126 \end{bmatrix}$$

Thus, the probability that the child will be good 4 days from now is $p_{11}^4 = 0.875$.

Notice that the probability that the child will be bad 4 days from now is $p_{12}^4 = 0.125$ and that the two probabilities 0.875 and 0.125 add up to 1 as expected. ●

Distribution of states

The probability that a system will be in a particular state at a specified time is a number between 0 and 1. If there are n possible states, the sum of the corresponding n probabilities must be 1 since one of these states has to occur. These probabilities can be summarized in a $1 \times n$ matrix

$$[q_1 \quad q_2 \quad \cdots \quad q_n]$$

where q_i is the probability that the system is in state i at the time in question. This matrix, said to describe the **distribution of states** at the time in question, is an example of a **probability vector.** In general, a probability vector is an ordered n-tuple whose entries are nonnegative numbers that add up to 1.

The calculation of probability vectors describing distributions of states is illustrated in the next example.

EXAMPLE 6.4 ● Suppose that before evidence linking smoking to respiratory disease was discovered, 40 percent of adult males were smokers and 60 percent were nonsmokers. One year after the evidence was made public, 30 percent of the smokers had stopped smoking, while 10 percent of the nonsmokers had started smoking.

(a) Represent the shift in smoking patterns as a transition matrix for a Markov chain with two states.
(b) Represent the original distribution of smokers and nonsmokers as a probability vector.
(c) Find the probability vector describing the distribution of smokers and nonsmokers after one year.
(d) Assuming that the trend continues, find the probability vector describing the distribution of smokers and nonsmokers after two years.

SOLUTION (a) The states are "smoker" (state 1) and "nonsmoker" (state 2) and the transition matrix is

$$P = \begin{array}{l} \text{Smoker} \\ \text{Nonsmoker} \end{array} \overset{\begin{array}{cc} \text{Smoker} & \text{Nonsmoker} \end{array}}{\begin{bmatrix} 0.7 & 0.3 \\ 0.1 & 0.9 \end{bmatrix}}$$

(b) The initial distribution of 40 percent smokers and 60 percent nonsmokers can be represented by the probability vector

$$[0.4 \quad 0.6]$$

(c) A fraction of 0.4 of the original population were smokers and 0.7 of these were still smoking one year later. Hence the fraction of the population who were smoking originally and were still smoking at the end of one year is $0.4(0.7) = 0.28$. Similarly, 0.6 of the original population were nonsmokers and 0.1 of these started smoking during the year. Hence the fraction of the population who were originally nonsmokers and had started smoking by the end of one year is $0.6(0.1) = 0.06$. It follows that the total fraction of the population smoking at the end of one year is the sum $0.28 + 0.06 = 0.34$. The fraction not smoking at the end of one year is $1 - 0.34 = 0.66$. The distribution of states at the end of one year can be represented by the probability vector

$$[0.34 \quad 0.66]$$

(d) The 2-step transition matrix is

$$P^2 = \begin{bmatrix} 0.52 & 0.48 \\ 0.16 & 0.84 \end{bmatrix}$$

The same reasoning used in part (c) shows that if the trend continues, the distribution of states after two years is given by the probability vector

$$[0.4(0.52) + 0.6(0.16) \quad 0.4(0.48) + 0.6(0.84)] = [0.304 \quad 0.696] \qquad \bullet$$

Notice that in the preceding example we obtained the probability vector representing the distribution of smokers and nonsmokers after one year by multiplying the initial probability vector by the transition matrix to get

$$[0.4 \quad 0.6] \begin{bmatrix} 0.7 & 0.3 \\ 0.1 & 0.9 \end{bmatrix} = [0.34 \quad 0.66]$$

and we obtained the distribution after two years by multiplying the initial probability vector by the 2-step transition matrix to get

$$[0.4 \quad 0.6] \begin{bmatrix} 0.52 & 0.48 \\ 0.16 & 0.84 \end{bmatrix} = [0.304 \quad 0.696]$$

This method works in general.

> ***n*-Step distribution of states**
> If the initial distribution of states of a Markov chain with transition matrix P is represented by the probability vector X, then the distribution of states after n transitions is represented by the probability vector XP^n.

Regular Markov chains

The ith row of the n-step transition matrix P^n is a probability vector giving the distribution of states after n transitions for a system that was originally in state i. If all the rows of P^n happen to be equal, it follows that the distribution of states after n transitions is independent of the initial state. There are very few Markov chains for which the rows of P^n are equal. However, there are many Markov chains that arise in practice for which the rows of P^n are *approximately* equal for large n, and for which the differences between rows approach zero as n increases. Markov chains with this property are said to be **regular.** The Markov chain in Example 6.4 is regular. To illustrate the behavior of P^n for regular Markov chains, here are some powers of the transition matrix from Example 6.4. (To make the calculations simpler, entries are rounded off to the nearest thousandth.)

$$P = \begin{bmatrix} 0.7 & 0.3 \\ 0.1 & 0.9 \end{bmatrix} \qquad P^2 = \begin{bmatrix} 0.52 & 0.48 \\ 0.16 & 0.84 \end{bmatrix} \qquad P^4 = \begin{bmatrix} 0.347 & 0.653 \\ 0.218 & 0.782 \end{bmatrix}$$

$$P^8 = \begin{bmatrix} 0.263 & 0.737 \\ 0.246 & 0.754 \end{bmatrix} \qquad P^{12} = \begin{bmatrix} 0.252 & 0.748 \\ 0.250 & 0.750 \end{bmatrix}$$

In fact, as n increases, the matrix P^n gets closer and closer to the matrix

$$\begin{bmatrix} 0.25 & 0.75 \\ 0.25 & 0.75 \end{bmatrix}$$

in which each row is the probability vector

$$Q = [0.25 \quad 0.75]$$

This vector Q is known as the **limiting distribution** of the states of the Markov chain.

The limiting distribution of states
A Markov chain is said to be *regular* if as n grows large, the rows of the n-step transition matrix P^n get closer and closer to a fixed probability vector Q. This vector Q is known as the *limiting distribution* of the states of the Markov chain.

Fortunately, we do not have to compute powers of the transition matrix to discover the limiting distribution of a regular Markov chain. Here is a more efficient procedure.

Computation of limiting distributions
If P is the transition matrix for a regular Markov chain, then the limiting distribution Q can be obtained by solving the vector equation

$$QP = Q$$

The vector QP represents the distribution of states after one transition for a system whose initial distribution is Q. The equation $QP = Q$ says that if a system's distribution of states is its limiting distribution, then the system will have the same distribution of states one transition later and consequently, after all future transitions. This stability property characterizes limiting distributions of states.

The use of the equation $QP = Q$ in the calculation of limiting distributions is illustrated in the next example.

EXAMPLE 6.5 ● Find the limiting distribution of states for the Markov chain in Example 6.4 and interpret the result.

SOLUTION The transition matrix is

$$P = \begin{bmatrix} 0.7 & 0.3 \\ 0.1 & 0.9 \end{bmatrix}$$

To find the limiting distribution $Q = [q_1 \quad q_2]$, we solve the vector equation $QP = Q$ or

$$[q_1 \quad q_2]\begin{bmatrix} 0.7 & 0.3 \\ 0.1 & 0.9 \end{bmatrix} = [q_1 \quad q_2]$$

Carrying out the required multiplication, we obtain

$$0.7q_1 + 0.1q_2 = q_1$$
$$0.3q_1 + 0.9q_2 = q_2$$

which we rewrite as

$$-0.3q_1 + 0.1q_2 = 0$$
$$0.3q_1 - 0.1q_2 = 0$$

Since the second of these two linear equations can be obtained by multiplying both sides of the first by -1, this linear system reduces to the single equation

$$-0.3q_1 + 0.1q_2 = 0$$

or, equivalently,

$$q_2 = 3q_1$$

We now use the fact that $[q_1 \quad q_2]$ is a probability vector and must satisfy the condition $q_1 + q_2 = 1$.

Letting $q_2 = 1 - q_1$, we substitute into the equation $q_2 = 3q_1$ to obtain

$$1 - q_1 = 3q_1 \qquad \text{or} \qquad q_1 = 0.25$$

We get $q_2 = 0.75$ by substituting $q_1 = 0.25$ into the equation $q_2 = 3q_1$, and conclude, as expected, that the limiting distribution is

$$Q = [0.25 \quad 0.75]$$

In the context of Example 6.4, this says that if the trend observed in the first year continues, there will eventually be about 25 percent smokers and 75 percent nonsmokers. Moreover, this will happen regardless of the initial distribution of smokers and nonsmokers. ●

In Example 6.5, we found Q by solving the system of linear equations

$$-0.3q_1 + 0.1q_2 = 0$$
$$0.3q_1 - 0.1q_2 = 0$$
$$q_1 + q_2 = 1$$

using elementary algebraic operations. If you prefer, you could solve this linear system using the matrix-reduction technique introduced in Chapter 2, Section 3.

If you try to find a probability vector Q that satisfies $QP = Q$ when the corresponding Markov chain is not regular, you will get a system of linear equations that has no solution. There is also a simple test that you can apply to determine in advance whether a given Markov chain is regular.

A test for regularity

A Markov chain with N states is regular if and only if there is some integer n between 1 and $(N - 1)^2 + 1$, inclusive, for which none of the entries in the matrix P^n is zero.

For most regular Markov chains, you will only have to compute a few powers of P before getting a matrix containing only nonzero entries. In many cases, the original transition matrix P itself will have this property.

Here is a final example.

EXAMPLE 6.6 ● The company that manufactures Novabrite toothpaste mounts a large advertising campaign. A survey taken 6 months later reveals that 5 percent of Novabrite users have switched to a rival brand, while 20 percent of those who used rival brands have switched to Novabrite. If this trend continues, what percentage will eventually be using Novabrite?

SOLUTION These percentages can be represented by a transition matrix for a two-state Markov chain whose states are "Novabrite user" (state 1) and "rival brand user" (state 2).

$$\begin{array}{r} \\ \text{Novabrite} \\ \text{Rival brand} \end{array} \begin{array}{c} \begin{array}{cc} \text{Novabrite} & \text{Rival brand} \end{array} \\ \begin{bmatrix} 0.95 & \qquad 0.05 \\ 0.20 & \qquad 0.80 \end{bmatrix} \end{array}$$

Since all its entries are nonzero, the transition matrix is regular and we can solve the appropriate matrix equation to get the limiting distribution. We have

$$[q_1 \quad q_2] \begin{bmatrix} 0.95 & 0.05 \\ 0.20 & 0.80 \end{bmatrix} = [q_1 \quad q_2]$$

from which we get

$$0.95q_1 + 0.20q_2 = q_1$$
$$0.05q_1 + 0.80q_2 = q_2$$

or, equivalently,

$$q_2 = 0.25q_1$$

Combining this with the fact that $q_1 + q_2 = 1$ we get

$$q_1 = 0.80 \quad \text{and} \quad q_2 = 0.20$$

We conclude that if the present trend continues, about 80 percent of the population will eventually be using Novabrite and 20 percent will be using a rival brand. ●

Problems

1 Suppose that each year 5 percent of registered Democrats change their party affiliation to Republican while 10 percent of Republicans change their party affiliation to Democrat. Represent this process as a two-state Markov chain and give the transition matrix.

2 An employer has found that if a worker arrives on time today, the probability is 0.8 that he will arrive on time tomorrow, while if he is late today, the probability is 0.5 that he will be late tomorrow.

(a) Represent this as a two-state Markov chain and give the transition matrix.

(b) If the worker arrives on time today, find the probability that he will be late the day after tomorrow.

3 A large city undertakes a vigorous inner-city renovation program. One year after the program is completed, a survey indicates that 5 percent of those living in the suburbs have moved to the city, while 3 percent of those living in the city have moved to the suburbs.

(a) Represent this as a two-state Markov chain and give the transition matrix.

(b) If the transition probabilities remain unchanged, find the probability that someone living in the city now will be living in the suburbs in two years.

(c) Find the two-step transition matrix P^2.

4 A gambler who has \$5 plays a game in which he wins \$1 with probability 0.4 and loses \$1 with probability 0.6. The gambler will keep playing this game until he has \$10 or goes broke, whichever happens first.

(a) Represent the gambler's successive fortunes as a Markov chain and give the transition matrix. (*Hint:* There are 11 possible states corresponding to 0, 1, 2, . . . , and 10 dollars.)

(b) Find the probability that the gambler will have \$6 after three plays of the game.

5 After each song, a folk singer checks to see whether or not her guitar is in tune. If it is in tune, the probability is 0.9 that it will be in tune after the next song. If it is not in tune, the probability is 0.8 that she will notice and be able to adjust it so that it will be in tune after the next song.

(a) Represent this as a two-state Markov chain and give the transition matrix.

(b) If the guitar is currently in tune, find the probability that it will be in tune after three more songs.
(c) Find the three-step transition matrix P^3.

6 Decide which of the following matrices could be transition matrices for a Markov chain.

(a) $\begin{bmatrix} 0.2 & 0.8 \\ 0.5 & 0.5 \end{bmatrix}$ (b) $\begin{bmatrix} 0.1 & 0.6 \\ 0.9 & 0.4 \end{bmatrix}$

(c) $\begin{bmatrix} 0.2 & 0.5 & 0.3 \\ 0.4 & 0.4 & 0.2 \end{bmatrix}$ (d) $\begin{bmatrix} 0.6 & 0.3 & 0.1 \\ 0.2 & 0 & 0.8 \\ 0.1 & 0.2 & 0.7 \end{bmatrix}$

7 A trailer rental agency has outlets in three cities. If a trailer is rented in City 1, there is a 70 percent chance that it will be returned to City 1, a 20 percent chance that it will be returned to City 2, and a 10 percent chance that it will be returned to City 3. If a trailer is rented in City 2, there is a 50 percent chance that it will be returned to City 2, a 30 percent chance that it will be returned to City 1, and a 20 percent chance that it will be returned to City 3. If a trailer is rented in City 3, there is a 40 percent chance that it will be returned to City 3, a 40 percent chance that it will be returned to City 2, and a 20 percent chance that it will be returned to City 1.
(a) Give the transition matrix for the Markov chain that represents the successive locations of a trailer as it moves from city to city.
(b) If a trailer is currently in City 1, find the probability that it will be in City 3 after it has been rented two times.
(c) If a trailer is currently in City 3, find the probability that it will be back in City 3 after it has been rented four times.

8 Suppose that 60 percent of voters are Democrats and 40 percent are Republicans and that party changeovers take place according to the transition matrix in Problem 1.
(a) Find the percentage of Democrats after 1 year.
(b) Find the percentage of Democrats after 2 years.
(c) Find the percentage of Democrats after 5 years.

In Problems 9 through 12, determine whether the given transition matrix is regular.

9 $P = \begin{bmatrix} 0.2 & 0.3 & 0.5 \\ 0.1 & 0.6 & 0.3 \\ 0.4 & 0.4 & 0.2 \end{bmatrix}$

10 $P = \begin{bmatrix} 0.3 & 0.7 & 0 \\ 0.2 & 0 & 0.8 \\ 0.2 & 0.2 & 0.6 \end{bmatrix}$

11 $P = \begin{bmatrix} 0.5 & 0.5 \\ 1 & 0 \end{bmatrix}$

12 $P = \begin{bmatrix} 0.1 & 0.9 & 0 \\ 0.3 & 0.7 & 0 \\ 0.3 & 0.4 & 0.3 \end{bmatrix}$

13 Determine which of the transition matrices in Problems 1 through 5 are regular.

14 Assume that the transition probabilities in Problem 3 remain unchanged. What fraction of the population will eventually be living in the city?

15 Suppose that the folk-singer in Problem 5 performs for several hours and that the transition probabilities remain unchanged. Approximately what percentage of the time will her guitar be in tune?

16 Assume that the transition probabilities in Problem 7 remain unchanged. What fraction of the trailers will eventually be at each outlet?

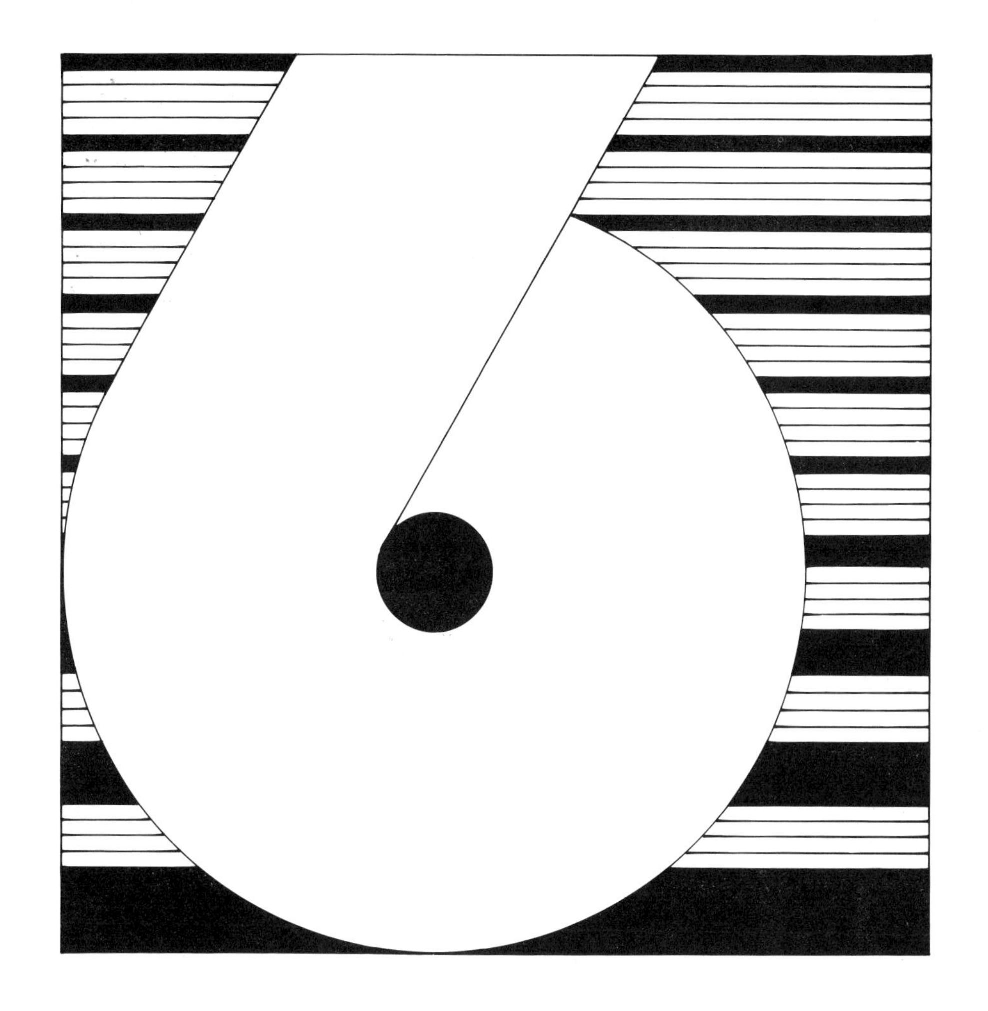

DISTRIBUTIONS

1 • RANDOM VARIABLES

In many situations, we may be interested in numerical values that are associated with the outcomes of a random experiment. For example, if a coin is tossed twice, we may be interested in the number of heads that come up. If two dice are rolled, we may be interested in the sum of the numbers shown on their faces. When we assign numbers to the outcomes of an experiment, we are actually defining a function that associates a real number to each outcome. Such a function is called a **random variable.** Here is the formal definition.

A random variable
A *random variable* is a function that assigns a number to each outcome of a random experiment.

It will be convenient to express events in terms of the values assigned by a random variable. For example, we shall write $(X = 2)$ to denote the set of all outcomes to which the random variable X assigns the value 2. The next two examples illustrate these ideas.

EXAMPLE 1.1 ● A coin is tossed twice. Let X denote the number of times heads comes up.

(a) List all the possible outcomes of this experiment and the value that the random variable X assigns to each.
(b) List the outcomes that make up the event $(X = 1)$.

SOLUTION (a) The possible outcomes can be represented as HH, HT, TH, and TT. The value of X for each of these is shown in the following table.

Outcome	HH	HT	TH	TT
X	2	1	1	0

(b) The event $(X = 1)$ is the set of outcomes for which $X = 1$, that is, for which exactly 1 head comes up. Thus, $(X = 1)$ denotes the event {HT, TH}. ●

EXAMPLE 1.2 ● In a certain experiment, one mouse is selected at random from a collection of five mice whose weights (in grams) are summarized in the following table.

Mouse	A	B	C	D	E
Weight	65	47	59	31	47

Let X denote the weight of the selected mouse. List the outcomes that make up the following events:

(a) $(X = 65)$ (b) $(X = 47)$ (c) $(X \geq 30)$

SOLUTION (a) Mouse A is the only mouse that weighs 65 grams, so $(X = 65) = \{A\}$.
(b) Only mice B and E weigh 47 grams, so $(X = 47) = \{B, E\}$.

(c) All the mice weigh 30 grams or more, so $(X \geq 30) = \{A, B, C, D, E\}$. ●

Since they are real-valued functions, random variables can be combined using the algebraic operations of addition, subtraction, and multiplication. For example, if X_1 and X_2 are random variables, the number associated with an outcome by the random variable $X_1 + X_2$ is the sum of the numbers associated with that outcome by X_1 and X_2 individually. Subtraction and multiplication of random variables are defined similarly.

Distributions

Each value assumed by a random variable corresponds to a particular event in the sample space, namely the collection of all outcomes that are assigned that value. Hence each value has a probability associated with it, namely, the probability of the corresponding event. The list or formula giving these values and their probabilities is called the **distribution** of the random variable. Here is an example.

EXAMPLE 1.3 ● A box contains 3 red balls and 2 green balls. Two balls are selected at random, with replacement. Find the distribution of the random variable X, where X is the number of red balls selected.

SOLUTION Letting R denote that a red ball is selected and G that a green ball is selected, we write the sample space as $\Omega = \{RR, RG, GR, GG\}$. Since the selections are made with replacement, we can calculate probabilities using the multiplication rule:

$$P(RR) = \tfrac{3}{5}(\tfrac{3}{5}) = \tfrac{9}{25} \qquad P(RG) = \tfrac{3}{5}(\tfrac{2}{5}) = \tfrac{6}{25}$$

$$P(GR) = \tfrac{2}{5}(\tfrac{3}{5}) = \tfrac{6}{25} \qquad P(GG) = \tfrac{2}{5}(\tfrac{2}{5}) = \tfrac{4}{25}$$

Since X is the number of red balls selected, the possible values of X are 0, 1, and 2. To find the distribution of the random variable X we must compute the probabilities $P(X = 0)$, $P(X = 1)$, and $P(X = 2)$. There is only one outcome for which $X = 0$, namely GG. Therefore, $P(X = 0) = \frac{4}{25}$. There are two outcomes for which $X = 1$, RG and GR. Therefore, using the probability law for mutually exclusive events, $P(X = 1) = P(RG) + P(GR) = \frac{6}{25} + \frac{6}{25} = \frac{12}{25}$. Finally, $X = 2$ only if both selected balls are red, so $P(X = 2) = P(RR) = \frac{9}{25}$. Here is the resulting distribution for X.

x	0	1	2
$P(X = x)$	4/25	12/25	9/25

It follows from the laws of probability that the probabilities of the values of a random variable must add up to 1. This provides a quick way to check for errors in the distribution table. ●

Distribution of a random variable
The *distribution of a random variable* is a list or formula giving the values of the variable along with their probabilities.

The binomial distribution

The next example illustrates an important distribution that involves binomial probabilities.

EXAMPLE 1.4 ● Ten percent of the items produced on a certain assembly line are defective. Twenty-five items are selected at random. Find the distribution of X, where X denotes the number of defective items in the sample.

SOLUTION Probabilities for the number of defective items in the sample are binomial probabilities, which we studied in Chapter 5. Applying the binomial probability formula with $n = 25$ and $p = 0.1$, we see that $P(X = k) = \binom{25}{k}(0.1)^k(0.9)^{25-k}$ for $k = 0, 1, \ldots, 25$. Since this formula can be used to obtain the probability of every value assumed by X, there is no need to write a distribution table. ●

The distribution of the random variable in the preceding example was obtained from the binomial probability formula and is called a **binomial distribution.**

The binomial distribution
Let X denote the number of successes in n independent trials of an experiment for which the probability of success is p on each trial. Then X is said to have a *binomial distribution* and, for $k = 0, 1, \ldots, n$,

$$P(X = k) = \binom{n}{k}p^k(1 - p)^{n-k}$$

As you may have noticed, binomial probabilities are difficult to compute, even with a calculator. A table of binomial probabilities for some small values of n is included in the Appendix. For large values of n, there is an extremely accurate approximation to the binomial distribution that we shall study later in this chapter.

Problems

1 A coin is tossed three times. Find the distribution of the random variable X, where X denotes the number of times heads comes up.

2 A ticket is selected at random from a box containing two tickets marked 1, three tickets marked 2, and five tickets marked 3. Let X denote the number on the selected ticket.
(a) Find the distribution of X.
(b) Find $P(X \leq 2)$.

3 Two dice are rolled. Let X_1 denote the number showing on the first die and X_2 the number showing on the second.
(a) Find the distributions of X_1 and X_2.
(b) Find the distribution of $X_1 + X_2$.

4 In a certain card game, you are dealt two cards from a standard deck. Suppose that the denominations ace through five are worth 1 point, denominations six through nine are worth 2 points, and denominations ten through king are worth 3 points. Find the distribution of X, where X denotes your point total in this game. (*Hint:* Use the selection formula.)

5 Two tickets are selected at random, with replacement, from the box in Problem 2. Let X_1 denote the number on the first ticket selected and X_2 the number on the second ticket selected.
(a) Find the distributions of X_1 and X_2.
(b) Find the distribution of $X_1 + X_2$.

6 Decide whether the following statements are true or false and justify your answers.
(a) It is possible to have a random variable X with $P(X = 1) = 0.3$, $P(X = 4) = 0.3$, and $P(X = 7) = 0.5$.
(b) It is possible to have a random variable X with $P(X \geq 1{,}000) = 0.8$ and $P(X = 0) = 0.2$.
(c) It is possible to have a random variable X with $P(X \leq -17) = 1$.
(d) It is possible to have a random variable X with $P(X = 2) = 0.6$ and $P(X \leq 2) = 0.4$.

7 A cereal-packaging machine underfills boxes about 10 percent of the time. Suppose 20 boxes of cereal packaged by this machine are selected at random and let X denote the number of these that are underfilled.
(a) Find $P(X = 2)$.
(b) Find $P(X \leq 2)$.

8 A baseball player has a batting average of .300. If X denotes the number of hits in 10 times at bat, find $P(X \geq 4)$. (Assume that X has a binomial distribution.)

9 A newspaper delivery boy throws his papers toward customers' porches while riding his bicycle down the street. Suppose that on any throw the probability that the paper will land on the porch is 0.6. Each week, each customer on this route gives the boy a tip if the paper lands on his porch at least 5 out of the 7 days.
(a) Let X denote the number of days in a particular week that the paper lands on a certain customer's porch and find $P(X \geq 5)$.
(b) Suppose there are 6 customers on the route. Let Y denote the number of tips the boy gets in a particular week and find $P(Y = 4)$.

10 In a certain large school district, 40 percent of the children have colds. If X denotes the number of children with colds in a random sample of 8 students, find the distribution of X. Express your answer as a formula for $P(X = k)$ for $k = 1, 2, \ldots, 8$.

11 Fifteen marbles are placed at random in three boxes marked A, B, and C.

Find the distribution of X, where X is the number of balls put in box B. Express your answer as a formula.

12 The probability is 0.4 that you will win a certain game. Suppose you play this game repeatedly until you win. Find the distribution of X, where X is the total number of games you play. Express your answer as a formula. (*Hint:* Find the probability of a sequence of $k - 1$ losses followed by a win.)

13 On each drilling, the probability that an oil prospector will strike oil is 0.1. Find the distribution of X, where X is the number of the attempt on which the prospector first strikes oil. Express your answer as a formula.

14 A personnel officer knows that about 20 percent of the applicants for a certain position are suitable for the job. A group of applicants is interviewed in sequence. Let X denote the number of the interview on which the first suitable applicant is found.

(a) Find $P(X = 6)$.
(b) Find $P(X \geq 5)$.

15 Suppose that 15 percent of the items produced at a certain plant are defective. Items are tested at random as they come off the assembly line. Let X denote the number of the test on which the fourth defective item is found.

(a) Find $P(X = 18)$.
(b) Find a formula for $P(X = k)$ for $k = 4, 5, \ldots$.

16 Suppose the result of an experiment can be classified as either success or failure, with success probability p, and suppose the experiment is repeated over and over until the rth success occurs. Let X denote the number of the trial on which the rth success occurs, and find the distribution of X. (This is known as the **negative binomial distribution.**)

17 If you know a programming language, write a program to calculate binomial probabilities for selected values of n and p, so that if you supply n and p, the computer will print out the probabilities $P(X = k)$, for $k = 0, 1, 2, \ldots, n$, where X is a random variable with the appropriate binomial distribution.

2 • EXPECTED VALUE

In many practical situations we may wish to know the "average" value of a random variable. For example, an investor may be interested in the average yearly earnings of a particular stock over a given period of time. A sociologist may want to know the average time a person convicted of a felony spends in jail. A consumer may want to know the average gas mileage obtained by a particular type of car.

One way to obtain the average of a random variable whose possible values are $x_1, x_2, \ldots, x_n$ would be to compute the numerical average

$$\frac{x_1 + x_2 + \cdots + x_n}{n} = \frac{1}{n}x_1 + \frac{1}{n}x_2 + \cdots + \frac{1}{n}x_n$$

This average would be meaningful if each value occurred $1/n$th of the time. In general, however, some of the values may be more likely to occur than others.

To take this into account, multiply each possible value of the random variable by its probability and then add the results. This *weighted average* is called the **expected value,** or **mean.** Defined in this way, expected value is a "long-run average" in the sense that if repeated trials are made of an experiment with a random variable X, the numerical average of the observed values of X will get closer and closer to the expected value of X as the number of repetitions gets large.

For random variables that assume only a finite number of values, the following formula can be used to compute the expected value.

Expected value of a random variable
If X is a random variable that assumes values $x_1, x_2, \ldots, x_n$ with probabilities $p_1, p_2, \ldots, p_n$, respectively, the *expected value* of X, denoted by $E(X)$, is the sum

$$E(X) = x_1p_1 + x_2p_2 + \cdots + x_np_n$$

In other words, to compute $E(X)$, multiply each possible value of X by its probability and then add the results.

The next two examples illustrate the use of the expected-value formula.

EXAMPLE 2.1 ● Find the expected value of a random variable X that has the following distribution.

x	0	4	2	-3
$P(X = x)$	0.2	0.2	0.5	0.1

SOLUTION Applying the formula for expected value, we obtain

$$E(X) = 0(0.2) + 4(0.2) + 2(0.5) + (-3)(0.1) = 1.5$$

(Notice that the numerical average of the possible values of X is $(0 + 4 + 2 - 3)/4 = 0.75$.) ●

EXAMPLE 2.2 ● A bag contains three \$1 bills, two \$5 bills, and one \$10 bill. One bill is selected at random. If X denotes the denomination of the selected bill, find $E(X)$.

SOLUTION First we display the distribution of X.

x	1	5	10
$P(X = x)$	1/2	1/3	1/6

Applying the formula for expected value, we obtain

$$E(X) = 1(\tfrac{1}{2}) + 5(\tfrac{1}{3}) + 10(\tfrac{1}{6}) = \tfrac{23}{6}$$ ●

Notice that if we list the denominations of the bills in the bag in Example 2.2, listing each denomination as many times as there are bills of that denomination, we get the numbers 1, 1, 1, 5, 5, 10 whose numerical average

$$\frac{1 + 1 + 1 + 5 + 5 + 10}{6} = \frac{23}{6}$$

is precisely the expected value of X.

Listing the possible values of X in this way is the same as weighting each value by its probability and shows the connection between a weighted average and a numerical average.

The next example illustrates the notion of long-run average.

EXAMPLE 2.3 ● You buy one ticket in a lottery that has a first prize of \$1,000, three second prizes of \$500 each, and five third prizes of \$100 each. In all, 10,000 tickets are sold. Let X denote the value of your ticket. Find $E(X)$ and interpret your result.

SOLUTION You should be able to show that the distribution of X is as follows.

x	0	100	500	1,000
$P(X = x)$	0.9991	0.0005	0.0003	0.0001

From the distribution table we see that

$$E(X) = 0(0.9991) + 100(0.0005) + 500(0.0003) + 1{,}000(0.0001) = 0.3$$

This means that if the lottery were repeated many times you would average about 30 cents per ticket in prize money. For this reason, 30 cents is said to be the "fair" ticket price. ●

Properties of expected value

Here are two useful algebraic properties of expected value.

The expected value of a sum
For any random variables $X_1, X_2, \ldots, X_n$,

$$E(X_1 + X_2 + \cdots + X_n) = E(X_1) + E(X_2) + \cdots + E(X_n)$$

That is, the expected value of a sum of random variables is the sum of the individual expected values.

The expected value of a constant multiple
For any number c,

$$E(cX) = cE(X)$$

The next two examples illustrate these properties.

EXAMPLE 2.4 ● A die is rolled twice. Find the expected value of the sum of the numbers that come up.

SOLUTION Let X denote the sum of the numbers that come up. We wish to find $E(X)$. If we let X_1 denote the upturned number on the first roll and X_2 denote the upturned number on the second roll, then $X = X_1 + X_2$. The first property states that to find $E(X)$ we need only find $E(X_1)$ and $E(X_2)$ and add. Since a die has faces numbered from 1 through 6 with probabilities 1/6 for each, we obtain the distribution and expected value of each X_i as follows.

x	1	2	3	4	5	6
$P(X_i = x)$	1/6	1/6	1/6	1/6	1/6	1/6

$$E(X_i) = 1(\tfrac{1}{6}) + 2(\tfrac{1}{6}) + 3(\tfrac{1}{6}) + 4(\tfrac{1}{6}) + 5(\tfrac{1}{6}) + 6(\tfrac{1}{6}) = 3.5$$

Thus,

$$E(X) = E(X_1) + E(X_2) = 7$$

Check this answer by computing $E(X)$ directly from the distribution of X. ●

In light of the previous example, it is clear that each additional time you roll the die you add another 3.5 to the expected sum of the rolls. This illustrates the basic idea behind the formula for the expected value of a sum.

The next example demonstrates that multiplication by a constant changes the expected value by the same factor.

EXAMPLE 2.5 ● Suppose that X is the weight in pounds of a randomly selected chicken. If $E(X) = 3.5$ pounds, find the expected weight in kilograms.

SOLUTION Let $Y =$ denote the weight in kilograms. Since 1 kilogram is 2.2 pounds, $Y = (1/2.2)X$ and

$$E(Y) = E\left(\frac{1}{2.2}X\right) = \frac{1}{2.2}E(X) = \frac{3.5}{2.2} = 1.59 \text{ kilograms}$$ ●

The expected value of a binomial random variable

Here is a convenient formula you can use to compute the expected value of a binomial random variable.

The expected value of a binomial random variable
Let X be the binomial random variable denoting the number of successes in n trials of an experiment for which the success probability on each trial is p. Then

$$E(X) = np$$

To understand the formula for the expected value of a binomial variable X, think of X as the sum $X_1 + X_2 + \cdots + X_n$, where each X_i is the number of successes (0 or 1) on the ith trial. For each i, $E(X_i) = p$ and so, by the formula for the expected value of a sum

$$E(X) = E(X_1) + E(X_2) + \cdots + E(X_n) = np$$

The next two examples further illustrate this formula.

EXAMPLE 2.6 ● A coin is tossed three times. Find the expected number of heads directly and then find it using the formula for the expected value of a binomial random variable.

SOLUTION Let X denote the number of heads that come up. Then X is binomial with $n = 3$ and $p = 0.5$. Using the formula

$$P(X = k) = \binom{n}{k} p^k (1 - p)^{n-k}$$

with $n = 3$ and $p = 0.5$ we get

k	0	1	2	3
$P(X = k)$	1/8	3/8	3/8	1/8

from which we see that

$$E(X) = 0(\tfrac{1}{8}) + 1(\tfrac{3}{8}) + 2(\tfrac{3}{8}) + 3(\tfrac{1}{8}) = 1.5$$

Using the formula $E(X) = np$ with $n = 3$ and $p = 0.5$, we get

$$E(X) = 3(0.5) = 1.5$$ ●

EXAMPLE 2.7 ● Ten percent of the items coming off a certain assembly line are defective. If 25 items are selected at random, find the expected number of defective items.

SOLUTION Let X denote the number of defective items selected. Then X has a binomial distribution with $n = 25$ and $p = 0.1$. Using the formula $E(X) = np$ we get $E(X) = 25(0.1) = 2.5$. ●

Fair games and fair odds An interesting application of expected value involves the notion of a **fair game** in gambling. A game is said to be fair if, on the average, gains equal losses. That is, if X denotes a player's winnings, a fair game is one in which $E(X) = 0$.

In casinos, payoffs for winning bets of various types are expressed as **odds.** For example, odds of 1 to 1 indicate that a winning player will gain \$1 for every dollar bet, and odds of 3 to 2 or, equivalently, $\frac{3}{2}$ to 1, indicate that a winning player will gain \$1.50 for every dollar bet. In general, odds of n to 1 indicate that a winning player will gain n dollars for every dollar bet. That is, if the player bets \$1 and wins, he or she will receive n dollars in addition to the dollar that was bet. **Fair odds** on a given bet are odds that make the bet a fair game and can be computed using the following formula.

Fair odds
If the probability of winning a particular bet is p, the *fair odds* are $(1 - p)/p$ to 1.

To derive this formula, suppose the probability of winning the bet is p and the odds are n to 1. Let X denote the winnings of a player who bets \$1. The random variable X can take on one of two values: n if the player wins (the probability of which is p) and -1 if the player loses (the probability of which is $1 - p$). Hence, the distribution table for X is

x	n	-1
$P(X = x)$	p	$1 - p$

and the corresponding expected value is

$$E(X) = np + (-1)(1 - p) = np - 1 + p$$

To determine the fair odds, we must find the value of n that makes $E(X)$ equal to zero. Setting the expression for $E(X)$ equal to zero and solving for n we get

$$np - 1 + p = 0$$

$$n = \frac{1 - p}{p}$$

We conclude that the fair odds are $(1 - p)/p$ to 1.

The following examples illustrate the use of this formula.

EXAMPLE 2.8 • The probability that you win a certain gambling game is $\frac{1}{4}$. What are the fair odds?

SOLUTION The probability that you win is $p = \frac{1}{4}$. Hence, $(1 - p)/p = \frac{3}{4}/\frac{1}{4} = 3$, and the fair odds are 3 to 1. This means that if the game is fair you should win \$3 for each successful \$1 bet. ●

EXAMPLE 2.9 ● What are the fair odds for a bet on red in roulette?

SOLUTION The probability that red comes up is $p = \frac{18}{38}$. Hence, $(1 - p)/p = \frac{20}{38}/\frac{18}{38}$ and the fair odds are $\frac{20}{18}$ to 1. This means that for the bet to be fair, you should win $\frac{20}{18}$ dollars or \$1.11 for each successful \$1 bet on red. ●

If the odds pay less than fair odds, they favor the casino. If the odds pay more than fair odds, they favor the player. If the odds favor one side even slightly, then in repeated play that side will eventually accumulate arbitrarily large winnings. For example, the sub-fair odds of 1 to 1 (instead of $\frac{20}{18}$ to 1) offered by casinos for bets on red in roulette result in an expected gain of 5.3 cents for the casinos for every dollar bet. (Do you see why?)

Problems

In Problems 1 through 4, find the expected value of the random variable with the given distribution.

1

x	3	5	7
$P(X = x)$	1/3	1/3	1/3

2

x	3	5	7
$P(X = x)$	1/2	3/8	1/8

3

x	−4	0	1	40
$P(X = x)$	1/8	1/8	1/8	5/8

4

x	0	1
$P(X = x)$	$1 - p$	p

5 A machine that gives change for a \$1 bill will give \$1 in change with probability 0.9, \$1.50 in change with probability 0.01, and 75 cents in change with probability 0.09. If you put a dollar bill into this machine, how much change can you expect to get back on the average?

6 In the coming year the probability is 0.3 that the price of a certain house will go up 20 percent, the probability is 0.5 that the price will go up 10 percent, the probability is 0.1 that the price will remain the same, and the probability is 0.1 that the price will go down 10 percent. Find the expected percentage increase.

7 If a first offender is convicted of a certain crime, the probability is 0.1 that the judge will sentence him to 3 years in jail, the probability is 0.2 that the judge will sentence him to 1 year in jail, and the probability is 0.7 that the judge will place him on probation. What is the expected length of time a first offender will be sentenced to jail?

8 In the casino game of craps, the probability that the shooter wins is 0.493. The casino pays 1 to 1 odds for bets on the shooter, so that if you bet \$1, you win \$1 if the shooter wins and lose your dollar if the shooter loses. Find your expected winnings for a \$1 bet on the shooter.

9 In the casino game of roulette, there are 18 red numbers, 18 black numbers, and 2 green numbers, so that the probability of red is $\frac{18}{38}$. The casino pays 1 to 1 odds for bets on red. Find your expected winnings for a \$1 bet on red.

10 In the casino game of craps, two dice are rolled. One version of the "field bet" is as follows. If you bet \$1, you win \$1 if 3, 4, 9, 10, or 11 comes up on the next roll of the dice. You win \$2 if 2 comes up and \$3 if 12 comes up. You lose your dollar if anything else comes up. Find your expected winnings if you bet \$1 in this way.

11 In the game of "Chuck-A-Luck" three dice are rolled. If you bet \$1 you win \$1 for every 6 that comes up. (For example, if each die shows a 6, you win \$3 and if no 6's come up you lose your dollar.) Find your expected winnings for a \$1 bet in this game.

12 You have 9 keys and exactly 1 of them fits a certain lock. You try the keys at random, setting aside keys that don't work. How many keys can you expect to try until you find the right one?

13 You toss a coin repeatedly until heads occurs or until four tosses have been made, whichever comes first. Let X denote the total number of tosses you make and find $E(X)$.

14 You have \$7 dollars to gamble in a game in which the probability of winning is 0.5, and which pays \$1 for each dollar bet. You decide to use the strategy known as double or nothing. That is, you begin by betting \$1. If you win, you quit; if you lose, you bet \$2 and play again. If you win the second game, you quit; if you lose, you bet \$4 and play again. If you win, you quit; and if you lose, you must also quit since you will have lost all your money. Find your expected winnings if you use this strategy.

15 Suppose $E(X) = 7$, $E(Y) = 3$, and $E(Z) = -2$.

(a) Find $E(X + Y)$.
(b) Find $E(X + Y + Z)$.
(c) Find $E(4X + 9Y - 3Z)$.

16 In a certain lottery there are 5 first prizes worth \$100 each, 10 second prizes worth \$50 each, and 20 third prizes worth \$25 each. Suppose 10,000 tickets are sold.

(a) Find the expected value of 1 ticket.
(b) Find the expected value of 10 tickets.

17 A box contains five tickets marked 1, 2, 4, 4, 4. Two tickets are selected at random, with replacement. Let X_1 denote the number on the first ticket selected and X_2 denote the number on the second ticket.

(a) Find $E(X_1)$ and $E(X_2)$.
(b) Find $E(X_1 + X_2)$.
(c) Find $E(X_1X_2)$.

18 Do Problem 17 again, this time assuming that the tickets are selected without replacement.

19 Five percent of the packages of potato chips produced at a certain factory are underfilled. How many packages do you expect to find underfilled in a random sample of 200 packages?

20 A coin is tossed 100 times. Find the expected number of times heads will come up.

21 A certain type of chemotherapy is supposed to be 40 percent effective in causing tumor remission. If is it used on 200 patients, how many remissions can be expected?

22 In a certain population 40 percent will vote for Spiff for mayor. If 100 people are selected at random, find the expected number in the sample who will vote for Spiff.

In Problems 23 and 24, find fair odds for a game in which the probability of winning is p.

23 $p = \frac{2}{5}$. **24** $p = 0.261$.

In Problems 25 through 27, assume that the given odds are fair and find the corresponding probability of winning.

25 Odds are 10 to 1.

26 Odds are $\frac{6}{5}$ to 1.

27 Odds are 1.674 to 1.

28 In the game of roulette, if you bet on an individual number, the casino odds are 35 to 1. The probability that you win such a bet is $\frac{1}{38}$. Find the fair odds and decide whether the casino odds are favorable to the casino, favorable to the bettor, or fair.

29 In the game of craps, the probability that the shooter wins is 0.493. Casino odds for bets on the shooter are 1 to 1. Find the fair odds and decide whether the casino odds are favorable to the casino, favorable to the bettor, or fair.

3 • THE NORMAL DISTRIBUTION

The most important distribution in statistics is the **normal distribution.** It arises frequently in connection with natural phenomena and scientific experiments, and also serves as an accurate approximation to the distributions of many random variables that arise in random-sampling problems. For example, the distribution of the heights of the people in a given population is usually normal. So is the distribution of the recorded weights obtained when an object is weighed over and over on the same scale. And, in many cases, the *averages* of data obtained from random samples are normally distributed, even if the data themselves are not.

Continuous random variables and density functions

Random variables representing such quantities as the height of an individual or the weight of an object can take on any value in a continuous range. Random variables with this property are said to be **continuous,** and differ from the random variables we have encountered previously that could take on only a finite number of possible values.

The values of a continuous random variable cannot be put in a list. Hence the definition of the distribution of a random variable given in Section 1 of this chapter cannot be used for continuous random variables. To define the distribution of a continuous random variable, we use the concept of a **density function.** A density function is a nonnegative function whose graph is a curve with the property that the total area between the curve and the x axis is 1. When the distribution of a random variable is defined by such a function, the probability that the variable assumes a value in a given interval $a \leq x \leq b$ is the area between the curve and the x axis from $x = a$ to $x = b$. The situation is illustrated in Figure 3.1.

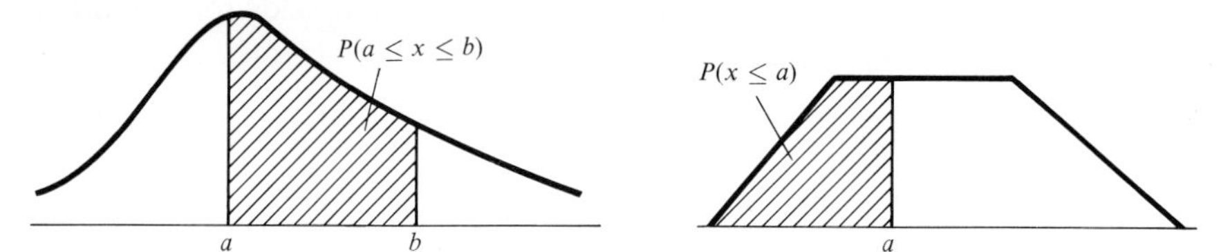

Figure 3.1 Probabilities represented as areas under density functions.

The expected value of a continuous random variable

The expected value of a random variable is its average value. A rigorous definition of this for continuous random variables requires the use of integral calculus. To get an intuitive understanding of expected value in this context, think of the number line as a beam and of the density function as assigning "weights" to the points on this beam. The expected value of the corresponding random variable is then the balancing point of the beam. If the graph of the density function is symmetric, the expected value is the point of symmetry, as illustrated in Figure 3.2.

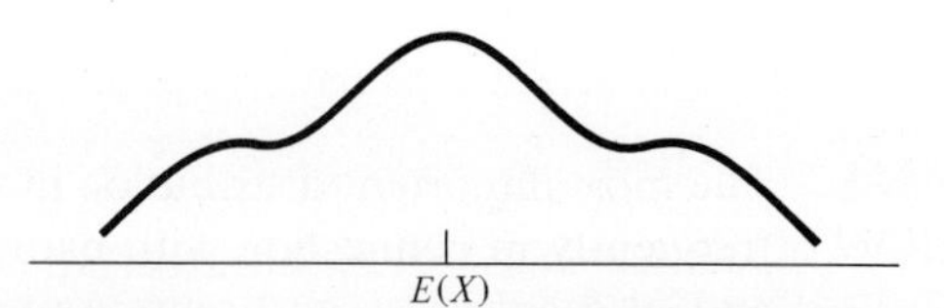

Figure 3.2 The expected value of a symmetric random variable.

Standard deviation

In many situations it is useful to know the variation or spread of a random variable: that is, the tendency of the values of the variable to cluster about the expected value. The **standard deviation** of a random variable, which we shall denote by the Greek letter σ (sigma), is the most commonly used measure of the variation of a random variable. It is approximately the average distance between the values of the variable and its expected value. The standard deviation of a random variable is always nonnegative, large values of σ indicating more variation than small values. If $\sigma = 0$, then there is *no* variation and the random variable is a constant. Except in a few special cases, we shall not give the formula for calculating standard deviation, but shall instead supply the appropriate values where needed.

The standard normal distribution

For the rest of this chapter, we shall study random variables that have normal distributions. There are many normal distributions. Their density functions are all bell-shaped curves whose formulas are determined by their expected values and standard deviations. Fortunately, it is not necessary to study all these distributions individually, because by a simple transformation, any normal random variable can be "standardized" so that its distribution is the **standard normal distribution** determined by the **standard normal density** whose graph is shown in Figure 3.3.

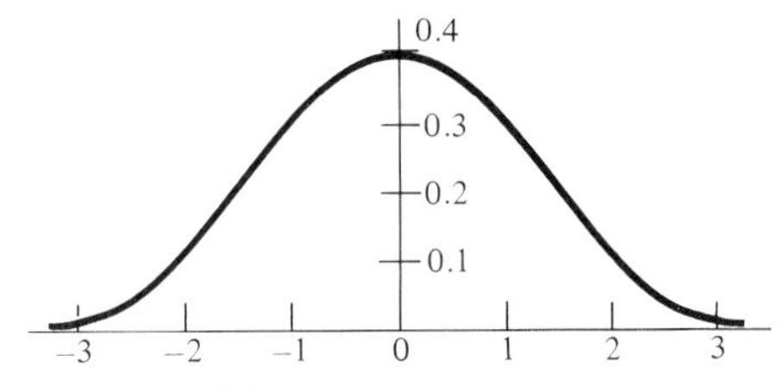

Figure 3.3 The standard normal density.

A random variable with standard normal distribution is usually denoted by the letter Z. The graph of the standard normal density is symmetric about 0, so $E(Z) = 0$. The standard deviation of Z is 1. It is customary to refer to the expected value of a normal random variable as the **mean** and to denote it by the Greek letter μ (mu). Thus, for a standard normal random variable Z, $\mu = 0$ and $\sigma = 1$.

How to use the normal table

Areas under the standard normal curve have been computed and tabulated. A table of such areas appears in the Appendix of this book. It gives areas under the standard normal curve to the left of specified positive numbers z, or, equivalently, probabilities of the form $P(Z \leq z)$. The use of this table will be illustrated in the next two examples. In these examples, we shall use the fact that $P(Z \leq z) = P(Z < z)$. This is true because there is no area under one point on the curve and so, for any z, $P(Z = z) = 0$.

EXAMPLE 3.1 ● Suppose that the random variable Z has the standard normal distribution. Find the following probabilities.

(a) $P(Z \leq 1)$ (b) $P(Z \geq 0.03)$
(c) $P(Z \leq -1.87)$ (d) $P(-1.87 \leq Z \leq 1)$
(e) $P(Z \leq 4.7)$

SOLUTION (a) $P(Z \leq 1)$ is the area under the normal curve to the left of $Z = 1$ (Figure 3.4). Looking up $Z = 1$ in the table we find that $P(Z \leq 1) = 0.8413$.

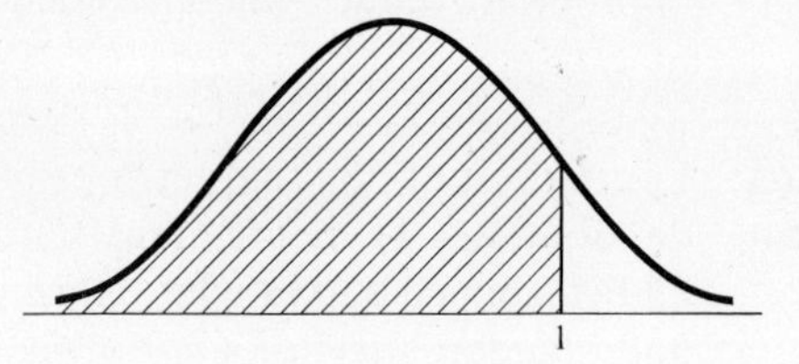

Figure 3.4 $P(Z \leq 1)$.

(b) $P(Z \geq 0.03)$ is the area under the curve to the right of 0.03 (Figure 3.5).

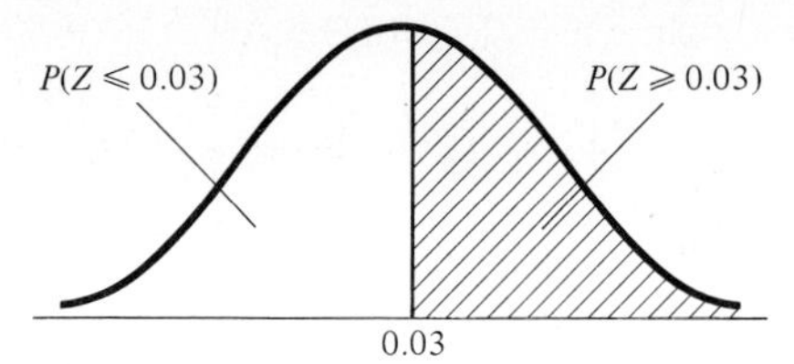

Figure 3.5 $P(Z \geq 0.03) = 1 - P(Z \leq 0.03)$.

Since the table only gives areas to the *left* of values of Z, we cannot get this area directly from the table. Instead, we use the law for complementary events to get

$$P(Z \geq 0.03) = 1 - P(Z \leq 0.03)$$

That is, the area to the right of 0.03 is 1 minus the area to the left of 0.03. From the table we get

$$P(Z \leq 0.03) = 0.5120$$

and so

$$P(Z \geq 0.03) = 1 - 0.5120 = 0.4880$$

(c) $P(Z \leq -1.87)$ is the area under the curve to the left of -1.87 (Figure 3.6*a*). Since the curve is symmetric, this is the same as the area to the right of 1.87 (Figure 3.6*b*). That is,

$$P(Z \leq -1.87) = P(Z \geq 1.87)$$

Proceeding as in part (b), we use the law for complementary events and the table to get

$$P(Z \geq 1.87) = 1 - P(Z \leq 1.87) = 1 - 0.9693 = 0.0307$$

Hence, $P(Z \leq -1.87) = 0.0307$.

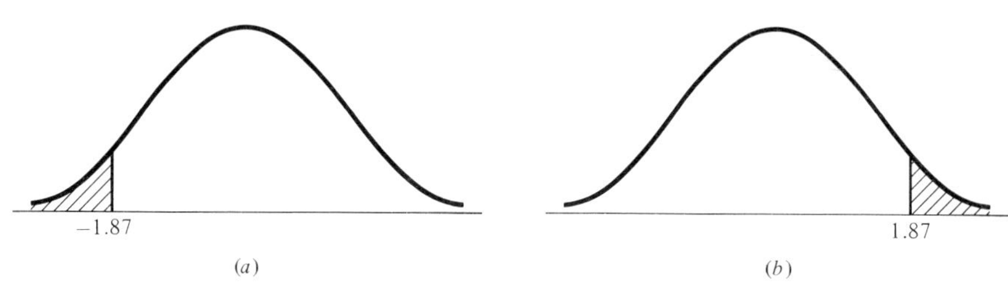

Figure 3.6 $P(Z \leq -1.87) = P(Z \geq 1.87) = 1 - P(Z \leq 1.87)$.

(d) $P(-1.87 \leq Z \leq 1)$ is the area under the curve between -1.87 and 1 (Figure 3.7*a*). This can be computed by subtracting the area to the left of -1.87 (Figure 3.7*c*) from the area to the left of 1 (Figure 3.7*b*). That is, we get

$$P(-1.87 \leq Z \leq 1) = P(Z \leq 1) - P(Z \leq -1.87)$$

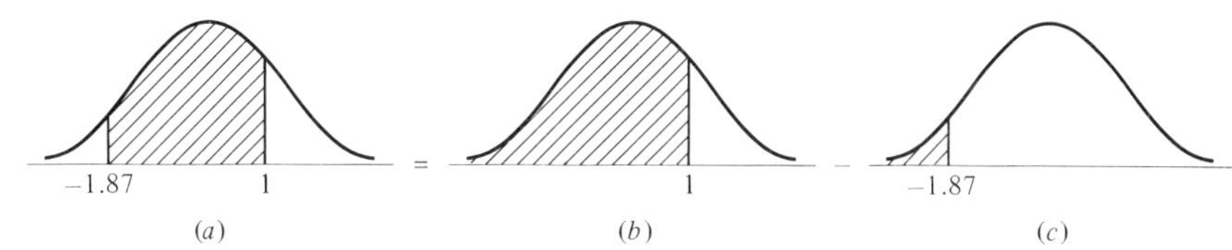

Figure 3.7 $P(-1.87 \leq Z \leq 1) = P(Z \leq 1) - P(Z \leq -1.87)$.

Using the results of parts (a) and (c) we get

$$P(-1.87 \leq Z \leq 1) = 0.8413 - 0.0307 = 0.8106$$

(e) $P(Z \leq 4.7)$ is the area under the curve to the left of 4.7. However, the table doesn't go this far. This is because most of the area under the curve is between -3 and 3 (Figure 3.8). Outside this interval, the curve is so close to the x axis that there is virtually no area under it. Since almost all the area under the curve is to the left of 4.7 we have, approximately, $P(Z \leq 4.7) = 1$.

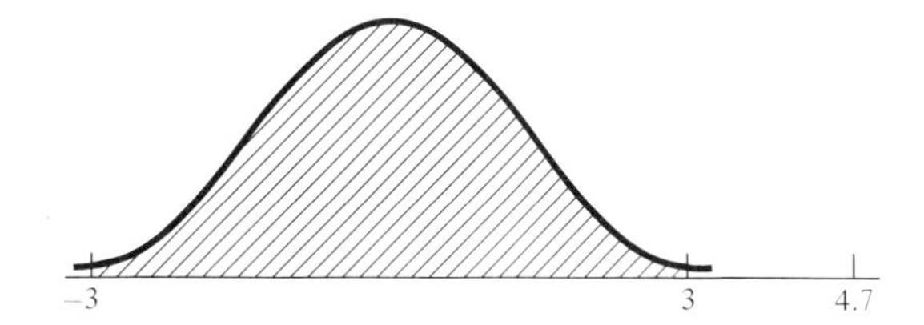

Figure 3.8 $P(Z \leq 4.7) \approx 1$. ●

The next example illustrates how to use the table to find Z values corresponding to given probabilities.

EXAMPLE 3.2 ●

(a) Find the value z such that $P(Z \leq z) = 0.9$.
(b) Find the value z such that $P(Z \geq z) = 0.01$.
(c) Find the value z such that $P(Z \leq z) = 0.1$.
(d) Find the positive number z such that $P(-z \leq Z \leq z) = 0.95$.

SOLUTION (a) We want the value of Z with the property that the area to the left of it is 0.9 (Figure 3.9*a*). Looking in the area columns of the normal table, we find no entry equal to 0.9, so we choose the entry closest to it, 0.8997. This corresponds to a Z value of 1.28. That is, rounded off to two decimal places, $P(Z \leq 1.28) = 0.9$, and so the desired value is (approximately) $z = 1.28$.

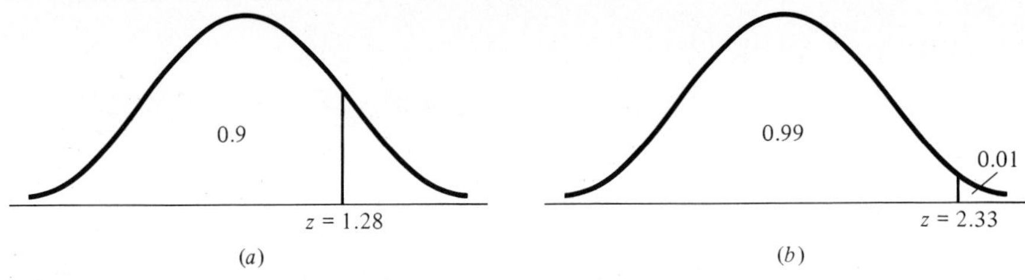

Figure 3.9 (*a*) $P(Z \leq z) = 0.9$; (*b*) $P(Z \geq z) = 0.01$.

(b) We want the value of Z with the property that the area to the right of it is 0.01, or equivalently, the area to the left of it is 0.99 (Figure 3.9*b*). Looking in the table, we find that 0.9901 is the area closest to 0.99 and so we take the corresponding value $z = 2.33$.

(c) We want the value of Z with the property that the area to the left of it is 0.1 (Figure 3.10*a*). Since the area under the curve to the left of 0 is 0.5, it follows that the desired value of Z is less than 0 and does not appear in the table. Proceeding indirectly, we first find the value of Z with the property that the area to the right of it is 0.1. This is also the value of Z for which the area to the left of it is 0.9. From part (a) we know that this value is 1.28. By symmetry, the value of Z we are looking for is the negative of this and so we conclude that $z = -1.28$.

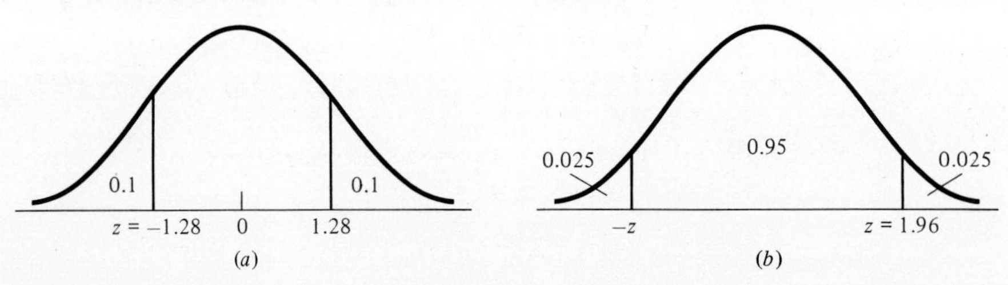

Figure 3.10 (*a*) $P(Z \leq z) = 0.1$; (*b*) $P(-z \leq Z \leq z) = 0.95$.

(d) From Figure 3.10*b* we see that the desired value of Z has the property that the area to the left of it is $0.025 + 0.95 = 0.975$. From the table we find that this value is $z = 1.96$. ●

Other normal distributions

If you know the mean and standard deviation of any normal random variable X, there is a simple formula you can use to transform X into the standard normal random variable Z.

Transforming to standardized values
If X has a normal distribution with mean μ and standard deviation σ, then the transformed random variable

$$Z = \frac{X - \mu}{\sigma}$$

has a standard normal distribution. It follows that for any number x,

$$P(X \leq x) = P\left(Z \leq \frac{x - \mu}{\sigma}\right)$$

The situation is illustrated in Figure 3.11. The transformation $Z = (X - \mu)/\sigma$ has the property that the area under the X curve between x_1 and x_2 is equal to the area under the Z curve between z_1 and z_2, where z_1 and z_2 are the Z values obtained from x_1 and x_2, respectively, using the formula $Z = (X - \mu)/\sigma$.

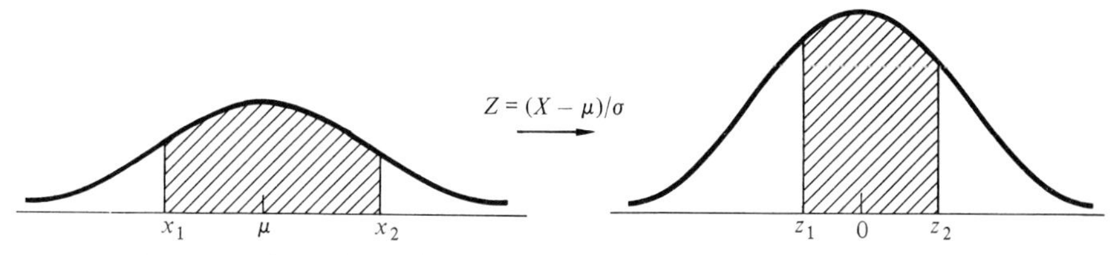

Figure 3.11 Transformation of X to Z.

The following example illustrates the use of this transformation.

EXAMPLE 3.3 ● Suppose X has a normal distribution with $\mu = 20$ and $\sigma = 4$. Find the following probabilities.

(a) $P(X \leq 26)$ (b) $P(X \geq 18)$ (c) $P(15 \leq X \leq 21)$

SOLUTION In each case, we apply the formula $Z = (X - \mu)/\sigma$ and then use the table of Z values in the Appendix.

(a) $P(X \leq 26) = P[Z \leq (26 - 20)/4] = P(Z \leq 1.5) = 0.9332$

(b) $P(X \geq 18) = P[Z \geq (18 - 20)/4] = P(Z \geq -0.5) = 0.6915$

(c) $P(15 \leq X \leq 21) = P[(15 - 20)/4 \leq Z \leq (21 - 20)/4]$
$= P(-1.25 \leq Z \leq 0.25)$
$= 0.5987 - 0.1056 = 0.4931$ ●

Problems

1 Suppose Z has the standard normal distribution. Find the following probabilities.
(a) $P(Z \leq 1.24)$ (b) $P(Z \geq 0.19)$
(c) $P(Z \leq -1.20)$ (d) $P(-1 \leq Z \leq 1)$
(e) $P(Z \leq 4.26)$

2 Suppose Z has the standard normal distribution. For each of the following probabilities, find the appropriate value z.
(a) $P(Z \leq z) = 0.8413$ (b) $P(Z \geq z) = 0.2266$
(c) $P(Z \leq z) = 0.8643$ (d) $P(-z \leq Z \leq z) = 0.9544$

3 Suppose X has a normal distribution with $\mu = 75$ and $\sigma = 5$. Transform the following values on the X scale to the corresponding values on the Z scale.
(a) $X = 78$ (b) $X = 75$ (c) $X = 98$
(d) $X = 70$ (e) $X = 50$

4 Suppose X has a normal distribution with $\mu = 60$ and $\sigma = 4$. Find the following probabilities.
(a) $P(X \geq 68)$ (b) $P(X \geq 60)$
(c) $P(56 \leq X \leq 64)$ (d) $P(X \leq 40)$

5 Suppose X has a normal distribution with $\mu = 4$ and $\sigma = 0.1$. Transform the following Z values to the corresponding values on the X scale.
(a) $Z = 4.2$ (b) $Z = 3.7$
(c) $Z = -1$ (d) $Z = 0$

6 Suppose X has a normal distribution with $\mu = 5$ and $\sigma = 0.6$. For each of the following probabilities find the appropriate value x.
(a) $P(X \geq x) = 0.0228$ (b) $P(X \leq x) = 0.1587$
(c) $P(5 - x \leq X \leq 5 + x) = 0.6826$ (d) $P(X \leq x) = 0.6$

4 • APPLICATIONS OF THE NORMAL DISTRIBUTION

In this section, you will see a variety of practical situations to which you can apply the normal distribution. The first example deals with random errors made by measuring devices.

EXAMPLE 4.1 ● On a certain scale, measurement errors are made that are normally distributed with $\mu = 0$ and $\sigma = 0.1$ ounces. If you weigh an object on this scale, what is the probability that the measurement error is no more than 0.15 ounces?

SOLUTION We let X denote the measurement error made when an object is weighed on this scale. Our goal is to find $P(-0.15 \leq X \leq 0.15)$. Since $\mu = 0$ and $\sigma = 0.1$, we transform to Z using the formula

$$Z = \frac{X - 0}{0.1}$$

and find that

$$P(-0.15 \leq X \leq 0.15) = P\left(\frac{-0.15 - 0}{0.1} \leq Z \leq \frac{0.15 - 0}{0.1}\right)$$
$$= P(-1.5 \leq Z \leq 1.5) = 0.8664$$
●

EXAMPLE 4.2 ● The number of loaves of bread that can be sold during a day by a certain supermarket is normally distributed with $\mu = 1{,}000$ loaves and $\sigma = 100$ loaves. If the market stocks 1,200 loaves on a given day, what is the probability that the loaves will be sold out before the day is over?

SOLUTION If X denotes the number of loaves that can be sold during a day, X is normally distributed with $\mu = 1{,}000$ and $\sigma = 100$. We wish to find $P(X \leq 1{,}200)$. Transforming to Z and using the table we obtain

$$P(X \geq 1{,}200) = P\left(Z \geq \frac{1{,}200 - 1{,}000}{100}\right) = P(Z \geq 2) = 0.0228$$

Thus, there is about a 2 percent chance that the market will run out of loaves before the day is over. ●

In the next example, the probability is known and the goal is to find the corresponding value of the random variable.

EXAMPLE 4.3 ● The lengths of trout in a certain lake are normally distributed with a mean of 7 inches and a standard deviation of 2 inches. If the fish and game department would like fishermen to keep only the largest 20 percent of the trout, what should the minimum size for "keepers" be?

SOLUTION Let X denote the length of a randomly selected trout and c the minimum size for a keeper. Since only the largest 20 percent are to be keepers, c must satisfy the equation

$$P(X \geq c) = 0.2$$

Using $\mu = 7$ and $\sigma = 2$, we transform X to Z and obtain

$$P\left(Z \geq \frac{c - 7}{2}\right) = 0.2 \qquad \text{or} \qquad P\left(Z \leq \frac{c - 7}{2}\right) = 0.8$$

From the table of Z values we see that

$$P(Z \leq 0.84) = 0.8$$

Hence,

$$\frac{c - 7}{2} = 0.84 \qquad \text{or} \qquad c = 8.68$$

We conclude that the fish and game department should set 8.68 inches as the minimum size for keepers. ●

The empirical rule

A convenient way to characterize the normal distribution is with the **empirical rule.** The empirical rule is often used by statisticians in describing normally distributed data.

> **The empirical rule**
> If a random variable X has a normal distribution, then
>
> **1** the probability is (approximately) 0.68 that X is within 1 standard deviation of the mean.
> **2** the probability is (approximately) 0.95 that X is within 2 standard deviations of the mean.
> **3** the probability is (approximately) 0.997 that X is within 3 standard deviations of the mean.

That is, approximately 68 percent of all values lie within 1 standard deviation of the mean, approximately 95 percent of all values lie within 2 standard deviations of the mean, and almost all values lie within 3 standard deviations of the mean.

To derive the empirical rule, we use the transformation $Z = (X - \mu)/\sigma$ and find that

$$P(\mu - \sigma \leq X \leq \mu + \sigma) = P(-1 \leq Z \leq 1) = 0.6826 \approx 0.68$$

$$P(\mu - 2\sigma \leq X \leq \mu + 2\sigma) = P(-2 \leq Z \leq 2) = 0.9544 \approx 0.95$$

$$P(\mu - 3\sigma \leq X \leq \mu + 3\sigma) = P(-3 \leq Z \leq 3) = 0.9974 \approx 0.997$$

The following example illustrates the use of the empirical rule.

EXAMPLE 4.4 ● The weights of cans of tomato sauce filled by a certain machine are normally distributed with a mean of $\mu = 6$ ounces and standard deviation of $\sigma = 0.05$ ounces. Using the empirical rule, describe this distribution of weights.

SOLUTION According to the empirical rule, the probability is 0.68 that a randomly selected can weighs between $6 - 0.05 = 5.95$ ounces and $6 + 0.05 = 6.05$ ounces. Equivalently, 68 percent of the cans weigh between 5.95 ounces and 6.05 ounces.

Similarly, about 95 percent of the cans weigh between $6 - 2(0.05) = 5.90$ and $6 + 2(0.05) = 6.10$ ounces and almost all (99.7 percent) of the cans weigh between $6 - 3(0.05) = 5.85$ and $6 + 3(0.05) = 6.15$ ounces. ●

The normal approximation to the binomial distribution

One of the most important results in probability theory is known as the **central limit theorem,** the first versions of which were proved over 200 years ago by the mathematicians De Moivre and Laplace. The following special case of this theorem states that under certain circumstances, the normal distribution can be used to approximate the binomial distribution.

Normal approximation to the binomial distribution
If X is a binomial random variable arising from an experiment with n trials and success probability p, and if n is large, then the distribution of X is approximately normal with

$$\mu = np \qquad \text{and} \qquad \sigma = \sqrt{np(1-p)}$$

It can be shown that if $np(1-p) > 5$, the error resulting from this approximation is insignificant.

The use of the normal distribution to approximate binomial probabilities is illustrated in the next example.

EXAMPLE 4.5 ● The probability that a certain surgical operation is successful is 0.8. If the operation is performed on 120 people, find the probability that 90 or more operations are successful.

SOLUTION We wish to find $P(X \geq 90)$, where X denotes the number of successful operations. Since n is large, we use the normal distribution with $\mu = 120(0.8) = 96$ and $\sigma = \sqrt{120(0.8)(0.2)} = 4.38$ to approximate the binomial distribution of X. We obtain

$$P(X \geq 90) = P\left(Z \geq \frac{90 - 96}{4.38}\right) = P(Z \geq -1.37) = 0.9147$$

(To see what a tedious job it would be to calculate this probability without the use of the normal curve, try to write down the *formula* for the required binomial probability.) ●

Problems

1 The incomes of industrial workers in a certain region are normally distributed with a mean of \$12,500 and a standard deviation of \$1,000. Find the probability that a randomly selected worker has an income between \$11,000 and \$14,000.

2 Suppose all fourth graders in a certain school are taught to read by the same teaching method and that at the end of the year they are tested for reading speed. If the average reading speed is 150 words per minute with standard deviation of 25 words per minute, find the percentage of students who read more than 180 words per minute. (Assume that reading speeds are normally distributed.)

3 A scale makes measurement errors that are normally distributed with $\mu = 0$ grams and $\sigma = 0.4$ grams.
  (a) If you weigh an object on this scale, what is the probability that the measurement error is more than 0.5 grams?
  (b) If you weigh an object on this scale, what is the probability that the recorded weight will be correct to within 1 gram?

4 A certain company produces glass jars. On the average the jars hold 1 quart but there is some variation among them. In fact, the volumes are normally distributed with standard deviation 0.01 quarts.

(a) Find the probability that a randomly chosen jar will hold between 0.98 and 1.02 quarts.

(b) Find the fraction of jars that hold less than 0.97 quarts.

5 A meat company produces hot dogs having lengths that are normally distributed with a mean length of 4 inches and standard deviation of 0.2 inches. Find the fraction of hot dogs that are less than 3.5 inches long.

6 An automobile manufacturer claims that its new cars get an average of 30 miles per gallon in city driving. Assume that the manufacturer's claim is correct and that the standard deviation is 2 miles per gallon. Assume also that gas mileages are normally distributed.

(a) Find the probability that a randomly selected car will get less than 25 miles per gallon.

(b) If you test two cars, what is the probability that both get less than 25 miles per gallon?

7 A certain type of rope has an average breaking strength of 1,200 pounds. If the standard deviation is 100 pounds, find the probability that a randomly selected piece of rope will break under a strain of less than 1,000 pounds.

8 A carrot farm is famous for growing carrots of almost identical length. If the mean length is 5 inches, what must the standard deviation be if 99 percent of the carrots are between 4.9 and 5.1 inches long?

9 The scores on a psychology exam are normally distributed with a mean of 75 and a standard deviation of 10.

(a) If a score of at least 60 is needed to pass, find the fraction of students who pass the exam.

(b) If the instructor wishes to pass 70 percent of those taking the test, what should the lowest passing score be?

(c) How high must someone's score be to be in the top 10 percent?

10 A certain club is having a meeting. There are 1,000 members who will attend, but only 200 seats. It is decided that the oldest members will get seats. If the average age of members is 40 with standard deviation of 5, find the age of the youngest person who will get a seat. (Assume the ages are normally distributed.)

11 A random variable X has a normal distribution with $\mu = 15$ and $\sigma = 4$. Using the empirical rule find:

(a) $P(11 \leq X \leq 19)$

(b) $P(7 \leq X \leq 23)$

(c) $P(3 \leq X \leq 27)$

12 The weights of adult basset hounds are normally distributed with $\mu = 50$ pounds and $\sigma = 5$ pounds. Use the empirical rule to describe this distribution of weights.

13 The scores on a certain test are normally distributed with $\mu = 78$ and $\sigma = 7$. Use the empirical rule to describe the distribution of these scores.

In Problems 14 through 19, use a normal distribution to approximate the appropriate binomial distribution.

14 Each question on a multiple-choice test has 5 choices, only 1 of which is correct. There are 100 questions altogether. The passing grade is 60. Find the probability that someone who guesses at random on each question will pass the test.

15 A coin is tossed 400 times. Let X denote the number of times that heads comes up and find the following probabilities.

(a) $P(X \geq 210)$ (b) $P(190 \leq X \leq 210)$ (c) $P(X \leq 280)$

16 The probability is 0.4 that you will win a certain gambling game. You play the game 75 times.

(a) What is the probability that you win at least 18 times?

(b) Use the empirical rule to find a range in which the number of games you win will lie with probability 0.95.

17 From a box containing 3 green balls and 1 red ball, you select 200 times, with replacement. Find the probability that a green ball is selected at least 160 times.

18 A self-proclaimed mind reader is given the following test. From a deck containing an equal number of red and green cards, the experimenter chooses a card at random and without showing it to the subject, concentrates on its color. The subject then tries to determine the color of the card by "reading the mind" of the experimenter. This process is repeated 100 times and the number of correct responses is recorded. Suppose the subject is really only guessing at random and has no psychic powers.

(a) Find the probability that at least 40 responses will be correct.

(b) Find the probability that at least 70 responses will be correct.

(c) Use the empirical rule to find a range in which the number of correct responses will lie with probability 0.95.

19 Because of a predictable number of "no-shows," most airlines have a policy of overbooking. That is, they make more reservations than there are seats available. Suppose that the no-show rate is about 10 percent and that there are 300 seats available on a particular flight.

(a) If an airline makes 325 reservations, find the probability that everyone will get a seat.

(b) If an airline wants the probability to be 0.99 that everyone will get a seat, what is the largest number of reservations it can make for this flight?

20 Decide whether the following statements are true or false. Explain your answer.

(a) If X has a normal distribution, then $P(X \geq 2.1) = P(X \leq -2.1)$.

(b) If Z has the standard normal distribution, then $P(Z \geq 2.1) = P(Z \leq -2.1)$.

(c) If X has a normal distribution with $\mu = 8$, then $P(X \geq 8.21) = P(X \leq 7.79)$.

(d) If X has a normal distribution, then the probability is about 0.87 that its value will be within 1.5 standard deviations of the mean.

21 A manufacturer needs washers whose thickness is between 0.18 and 0.22 inches; any other thickness is unusable. Machine shop A sells washers for \$1.00 per thousand. The thickness of its washers is normally distributed with $\mu = 0.2$ inches and $\sigma = 0.010$ inches. Machine shop B sells washers for \$0.90 per thousand. The thickness of its washers is normally distributed with $\mu = 0.2$ inches and $\sigma = 0.011$ inches. Using the price per usable washer as the criterion, which shop offers the manufacturer a better deal?

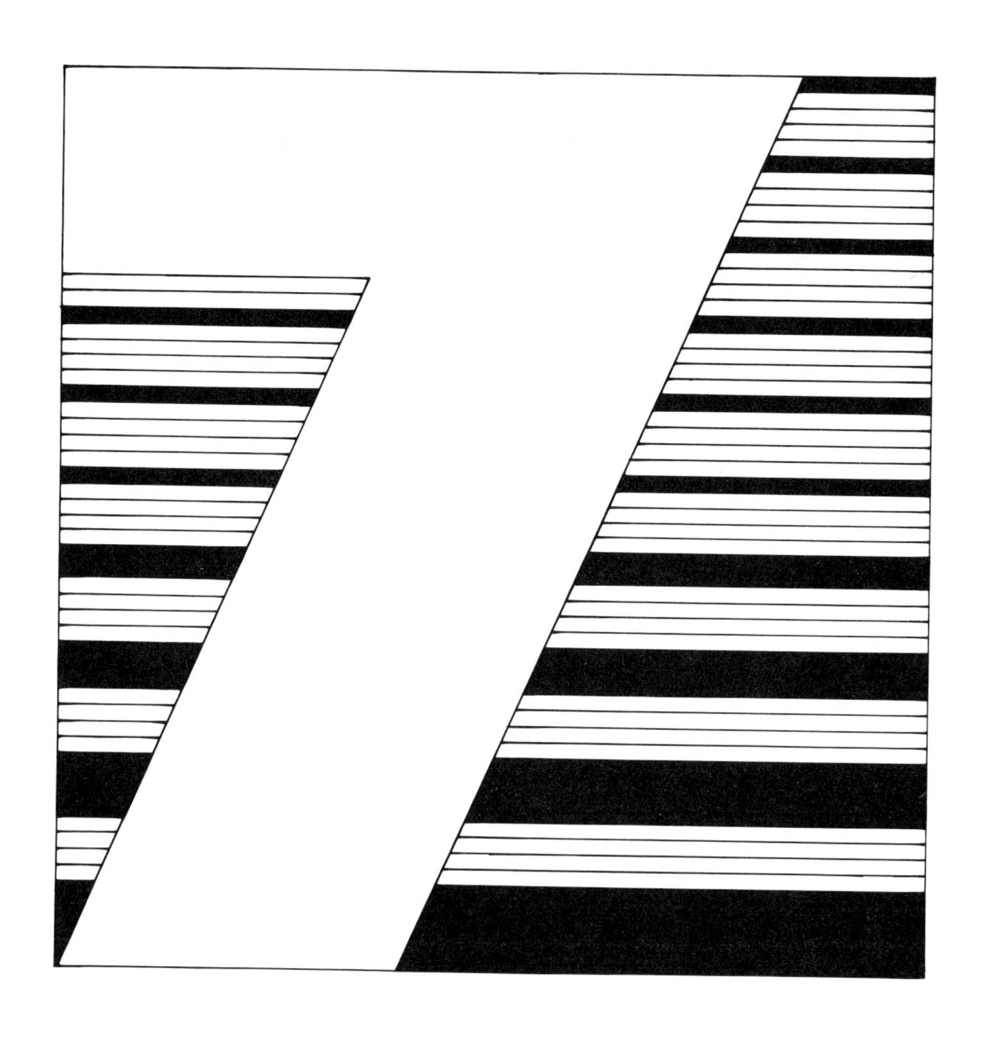

STATISTICAL INFERENCE

1 • RANDOM SAMPLING

There are many situations in which information about a population under study must be estimated from information obtained from a random sample. For example, a consumers' group may want to determine whether a certain car really averages 30 miles per gallon as the manufacturer claims. It would be impractical to test all the cars produced by the manufacturer, and so the group selects a number of the cars at random and tests them, drawing conclusions about the entire population of the cars from the information obtained from the sample.

In this chapter, you will learn how to use information from a random sample to estimate unknown parameters of a population such as its mean and standard deviation. You will also learn how to measure the reliability of your estimates. The process of drawing conclusions about the population from information obtained from a random sample is called **statistical inference.**

The sample mean and standard deviation

We have seen that the expected value and standard deviation are useful in describing the behavior of a random variable. Similarly, the **sample mean,** $\bar{X}$, and the **sample standard deviation,** s, are useful in describing the results of a random sample and can serve as estimates of the corresponding population parameters. The sample mean, $\bar{X}$, is simply the numerical average of the values obtained in a random sample.

Sample mean
If $X_1, X_2, \ldots, X_n$ are the values associated with the n members of a random sample, the *sample mean* $\bar{X}$ is the numerical average

$$\bar{X} = \frac{X_1 + X_2 + \cdots + X_n}{n} = \frac{\sum X_i}{n}$$

The sample standard deviation, s, is a measure of variation similar to the standard deviation of a random variable. If $X_1, X_2, \ldots, X_n$ denote the sample values, and $\bar{X}$ denotes the sample mean, the sample standard deviation is given by the formula

$$s = \sqrt{\frac{\sum (X_i - \bar{X})^2}{n - 1}}$$

Except for the fact that $n - 1$ rather than n appears in the denominator, the sample standard deviation is the square root of the average squared deviation of the sample values from the mean. The more spread out the sample values are, the greater the sample standard deviation will be.

An alternative way of writing the formula for s, which is easier to use for calculating purposes, is

$$s = \sqrt{\frac{\sum X_i^2 - \left(\sum X_i\right)^2/n}{n-1}}$$

where ΣX_i^2 tells you to add up the squares of each of the sample values, and $(\Sigma X_i)^2$ tells you to add up the sample values and then square the sum. We call this the **calculator formula** for s and shall use it in all our computations.

Sample standard deviation
The *sample standard deviation, s*, is a measure of the amount of variation in the data obtained in a random sample. A formula for s called the *calculator formula* is

$$s = \sqrt{\frac{\sum X_i^2 - \left(\sum X_i\right)^2/n}{n-1}}$$

The next two examples illustrate the calculation of $\bar{X}$ and s.

EXAMPLE 1.1 ● Find $\bar{X}$ and s for a random sample yielding the values -1, 1, 0, 2, and 3.

SOLUTION First we find $\bar{X}$. Since the sample size is $n = 5$, we obtain

$$\bar{X} = \frac{-1+1+0+2+3}{5} = 1$$

Next, we calculate s. To find s using the calculator formula, we must calculate the sum of the squared data values as well as the square of the sum of the values. It is convenient to do these calculations from a table, as follows.

X	-1	1	0	2	3	$\Sigma X_i = 5$
X^2	1	1	0	4	9	$\Sigma X_i^2 = 15$

From this we obtain $\Sigma X_i^2 = 15$ and $(\Sigma X_i)^2 = 5^2 = 25$. Putting these results in the formula, we get

$$s = \sqrt{\frac{\sum X_i^2 - \left(\sum X_i\right)^2/n}{n-1}} = \sqrt{\frac{15 - 25/5}{4}} = \sqrt{2.5} = 1.58$$ ●

EXAMPLE 1.2 ● In a survey of dental costs in a certain city, 6 dentists were selected at random and the amount each charged for a routine office visit was recorded. Find $\bar{X}$ and s for the resulting data: \$18, \$17, \$25, \$15, \$20, \$15.

SOLUTION First we fill in the table.

X	18	17	25	15	20	15	$\Sigma X_i = 110$
X^2	324	289	625	225	400	225	$\Sigma X_i^2 = 2{,}088$

Then we calculate $\bar{X}$:

$$\bar{X} = \frac{\sum X_i}{n} = \frac{110}{6} = \$18.33$$

Next we calculate s. From the table we obtain $\Sigma X_i^2 = 2{,}088$ and $(\Sigma X_i)^2 = 110^2 = 12{,}100$. Using the formula for s, we get

$$s = \sqrt{\frac{\sum X_i^2 - \left(\sum X_i\right)^2/n}{n-1}} = \sqrt{\frac{2{,}088 - 12{,}100/6}{5}} = \$3.78$$ ●

The distribution of $\bar{X}$

In many statistical studies in the social and managerial sciences, the goal is to find the mean μ of a particular population. Often, the population will be too large to permit direct calculation of its mean, and you will have to estimate μ using the information obtained from a random sample. The sample mean $\bar{X}$ is the natural estimate to use, but since different samples may have different means, the variability of $\bar{X}$ has to be taken into account.

To do this, think of $\bar{X}$ as a random variable that associates a number to each random sample of some particular size n. It can be shown that when n is large, the distribution of $\bar{X}$ is approximately normal, no matter what the distribution of the population itself is. This remarkable fact is a generalization of the central limit theorem for binomial random variables, which you saw in the preceding chapter. Here is a more precise statement of the situation.

The distribution of $\bar{X}$

Suppose samples of fixed size n are drawn from a population with mean μ and standard deviation σ. If n is large, the distribution of the sample mean, $\bar{X}$, is approximately normal with mean μ and standard deviation $\sigma/\sqrt{n}$. Hence, for any number c,

$$P(\bar{X} \leq c) = P\left(Z \leq \frac{c-\mu}{\sigma/\sqrt{n}}\right)$$

This approximation is good whenever $n \geq 30$, regardless of the distribution of the population from which the sample is taken.

As n increases, the amount of possible variation in $\overline{X}$ decreases. That is, the larger the sample size, the more likely it is that $\overline{X}$ will be close to μ. This is reflected in the formula $\sigma/\sqrt{n}$ for the standard deviation of $\overline{X}$, which gets smaller as n increases.

In the next section, you will learn how to use the normal approximation to the distribution of $\overline{X}$ to estimate an unknown population mean μ. Here is an example to illustrate how to use this approximation to find probabilities for $\overline{X}$ when the population mean μ is known.

EXAMPLE 1.3 ● The average weight of newborn babies in a certain city is $\mu = 6.8$ pounds. Suppose 50 newborn babies are to be selected at random, and the average, $\overline{X}$, of their weights is to be obtained.

(a) If it is known that the standard deviation of the weights of all the newborn babies in the city is $\sigma = 0.9$ pounds, find the probability that $\overline{X}$ will be less than or equal to 6.6 pounds.
(b) Use the empirical rule for normally distributed random variables to describe an interval in which $\overline{X}$ will lie with probability 0.95.

SOLUTION (a) We wish to find $P(\overline{X} \leq 6.6)$. We are given $\mu = 6.8$, $\sigma = 0.9$ and $n = 50$. Using the formula for the normal approximation to the distribution of $\overline{X}$ we obtain

$$P(\overline{X} \leq 6.6) = P\left(Z \leq \frac{6.6 - 6.8}{0.9/\sqrt{50}}\right) = P(Z \leq -1.57) = 0.0582$$

(b) The empirical rule states that the probability is 0.95 that $\overline{X}$ will be within two standard deviations of its mean. Since the standard deviation of $\overline{X}$ is $\sigma/\sqrt{n} = 0.13$ and the mean of $\overline{X}$ is $\mu = 6.8$, it follows that the probability is 0.95 that $\overline{X}$ will be between $6.8 - 2(0.13) = 6.54$ pounds and $6.8 + 2(0.13) = 7.06$ pounds. ●

Problems

1 Calculate $\overline{X}$ and s for the following sets of sample data.
(a) −1, 1, −2, 2, −3, 3 (b) 1, 2, 2, 3, 3, 6
(c) 12.0, 15.0, 17.1, 16.3, 11.4 (d) 8, 8, 8, 8, 8, 8, 8, 8

2 Seven randomly selected bags of flour labeled "5 pounds" were weighed and the following results obtained: 5.0, 5.1, 4.9, 4.9, 5.3, 5.0, 4.5. Find $\overline{X}$ and s.

3 Suppose you take a random sample of size $n = 100$ from a population with $\mu = 8$. Find the following probabilities for the sample mean $\overline{X}$ if $\sigma = 2$.
(a) $P(\overline{X} \geq 8.1)$ (b) $P(\overline{X} \leq 8)$ (c) $P(7.8 \leq \overline{X} \leq 8.2)$

4 Do the same calculations as in Problem 3, but this time assume that the sample size is 30 instead of 100.

5 Do the same calculations as in Problem 3, but this time assume that $\sigma = 0.2$ instead of 2.

6 Use the normal table to find intervals in which $\overline{X}$ will lie with probabilities 0.95 and 0.99 for Problems 3 and 4.

2 • ESTIMATION

Suppose we want to estimate the average weight μ of all the crates of oranges packed by a certain grower. We can't weigh all the crates, so we take a random sample, compute the sample mean $\overline{X}$, and use this as an estimate of the population mean μ. The problem with this approach is that we have no idea how close our estimate really is to the true value of μ. If, for example, our random sample happens to consist of unusually heavy crates, our estimate of μ could be much too high.

A more useful method of estimation is to use the sample mean $\overline{X}$ to construct an interval of values in which the true value of μ lies with high probability. In particular, if a probability p is given, we can calculate a number d such that the probability that μ lies within d units of $\overline{X}$ is equal to p. The interval from $\overline{X} - d$ to $\overline{X} + d$ is called a **confidence interval.** A 95 percent confidence interval is illustrated in Figure 2.1. For about 95 percent of all samples of size n drawn from the population the 95 percent confidence interval will contain the true value of μ.

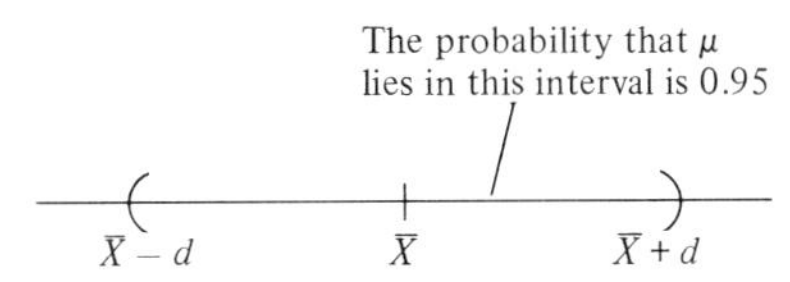

Figure 2.1 A 95 percent confidence interval for μ.

Here are the formulas for obtaining the 95 percent and 99 percent confidence intervals for μ. They are valid if $n \geq 30$ and (as you will see) are based on the fact that the distribution of $\overline{X}$ is approximately normal for large n. Similar formulas can be obtained for any level of confidence; however, 95 percent and 99 percent are the levels used most frequently.

Confidence intervals for μ

Suppose $\overline{X}$ is the mean and s the standard deviation of a random sample of size n drawn from a population with unknown mean μ. A *confidence interval* for μ is an interval of the form

$$\overline{X} \pm d$$

where

$$d = \frac{1.96s}{\sqrt{n}} \qquad \text{for 95 percent confidence}$$

Since the mean of $\overline{X}$ is μ and the standard deviation of $\overline{X}$ is $\sigma/\sqrt{n}$, it follows that

and

$$d = \frac{2.58s}{\sqrt{n}} \qquad \text{for 99 percent confidence}$$

The next two examples illustrate the use of these formulas.

EXAMPLE 2.1 ● Suppose that you want to estimate the average weight, μ, of the crates of oranges packed by a certain grower. Suppose also that a random sample of 75 crates yields a sample mean of $\bar{X} = 45$ pounds, and sample standard deviation of $s = 5.3$ pounds. Find the 95 percent confidence interval for μ.

SOLUTION Using $d = 1.96s/\sqrt{n}$ we find that the 95 percent confidence interval is

$$\bar{X} \pm \frac{1.96s}{\sqrt{n}} = 45 \pm \frac{1.96(5.3)}{\sqrt{75}} = 45 \pm 1.20$$

That is, we are 95 percent confident that the true average weight of the crates is between $45 - 1.20 = 43.80$ and $45 + 1.20 = 46.20$ pounds. ●

EXAMPLE 2.2 ● In a random sample of 250 automobile insurance claims paid out during a 2-year period by a large insurance company, the average payment was $300 and the sample standard deviation was $175. Find the 99 percent confidence interval for the mean claim payment μ.

SOLUTION Using $d = 2.58s/\sqrt{n}$, we find that the 99 percent confidence interval is

$$\bar{X} \pm \frac{2.58s}{\sqrt{n}} = 300 \pm \frac{2.58(175)}{\sqrt{250}} = 300 \pm 28.56$$

That is, we are 99 percent confident that the average payment is between $300 - 28.56 = \$271.44$ and $300 + 28.56 = \$328.56$. ●

Derivation of the formula

We shall now verify the formula for the 95 percent confidence interval. Intervals for other levels of confidence are obtained similarly. Our goal is to find d such that

$$P(\bar{X} - d \le \mu \le \bar{X} + d) = 0.95$$

From the standard normal table we find that

$$P(-1.96 \le Z \le 1.96) = 0.95$$

Since the mean of $\bar{X}$ is μ and the standard deviation of $\bar{X}$ is $\sigma/\sqrt{n}$, it follows that

that

$$Z = \frac{\overline{X} - \mu}{\sigma/\sqrt{n}}$$

and so

$$P\left(-1.96 \le \frac{\overline{X} - \mu}{\sigma/\sqrt{n}} \le 1.96\right) = 0.95$$

which can be rewritten as

$$P\left(\overline{X} - \frac{1.96\sigma}{\sqrt{n}} \le \mu \le \overline{X} + \frac{1.96\sigma}{\sqrt{n}}\right) = 0.95$$

This tells us that $d = 1.96\sigma/\sqrt{n}$. Since we usually do not know the value of σ, we use the sample standard deviation s in its place. (The effect of this substitution turns out to be negligible). We conclude that the 95 percent confidence interval for μ is $\overline{X} \pm d$, where $d = 1.96s/\sqrt{n}$.

Notice that since $s/\sqrt{n}$ is approximately the standard deviation of $\overline{X}$, the 95 percent confidence interval for μ is obtained by allowing 1.96 units of standard deviation on either side of $\overline{X}$. Similarly, the 99 percent confidence interval for μ is obtained by allowing 2.58 units of standard deviation on either side of $\overline{X}$.

Controlling the precision of an estimate

The more confident we wish to be that an interval contains the true value of μ, the more room we must allow for random variation. In other words, as we increase the reliability of our estimate, we decrease its precision. One way to increase both reliability and precision is to increase the sample size. In applied situations, it is usually too costly, too time consuming, or physically impossible to increase the sample size at will, so we would like to know the minimum sample size necessary to obtain a desired level of confidence and a desired interval size. Here is a useful formula.

How to control the precision of an estimate of μ
To be 95 percent confident that the error committed when using $\overline{X}$ to estimate μ is no more than d units, take a sample of size $n \ge (1.96s/d)^2$. To be 99 percent confident, take a sample of size $n \ge (2.58s/d)^2$.

To see why this works, look at the picture in Figure 2.2 showing a 95 percent confidence interval for μ.

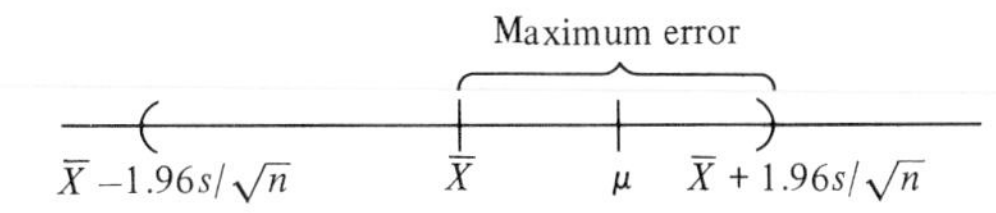

Figure 2.2 Possible error in a 95 percent confidence interval for μ.

Since we are 95 percent confident that μ lies in the interval, it follows that we are 95 percent confident that the error is no greater than $1.96s/\sqrt{n}$. Thus, we can be at least 95 percent confident that the error is no greater than d if

$$\frac{1.96s}{\sqrt{n}} \leq d$$

Solving this inequality for n we get

$$\sqrt{n} \geq \frac{1.96s}{d} \quad \text{or} \quad n \geq \left(\frac{1.96s}{d}\right)^2$$

Note that this formula requires us to have information about s, the sample standard deviation, that we won't know until after we've taken the sample. There are a variety of ways to get around this difficulty, one being to use the standard deviation from a preliminary sample, and another being to obtain information about s from other sources. For simplicity, we shall assume that an estimate of the standard deviation is known beforehand.

The next example illustrates the use of this formula.

EXAMPLE 2.3 ● Suppose that the standard deviation of the lengths of hot dogs produced by a certain machine is estimated to be 0.08 inches. How large a sample should we take to be 95 percent confident that the average length of all hot dogs produced by the machine will be within 0.01 inches of the sample mean?

SOLUTION Using $d = 0.01$ and $s = 0.08$ we find that

$$n \geq \left(\frac{1.96s}{d}\right)^2 = \left[\frac{1.96(0.08)}{0.01}\right]^2 = 245.86$$

and we conclude that the sample size should be at least 246. ●

Estimating proportions

Suppose you are conducting a voter preference poll (like a Gallup poll) to estimate the proportion of voters in the population who will vote for Smith for mayor. Suppose that out of a random sample of 100 voters, 55 say they will vote for Smith. Is it safe to conclude that a majority of voters would vote for Smith if the election were held today?

It is true that a majority of those in the sample (namely a proportion of $55/100 = 0.55$) will vote for Smith. However, this sample proportion, which we shall denote by $\hat{p}$, will vary from sample to sample and is at best only an estimate of the true population proportion, p, of people who will vote for Smith. The relationship between the sample proportion $\hat{p}$ and the population proportion p is similar to the relationship between the sample mean $\bar{X}$ and the population mean μ.

We shall now state a formula you can use to find confidence intervals for population proportions p.

Confidence intervals for a proportion p
Suppose $\hat{p}$ is the proportion of individuals with a certain characteristic in a random sample of size n drawn from a population in which the proportion p of individuals with this characteristic is unknown. A 95 percent confidence interval for the population proportion p is an interval of the form

$$\hat{p} \pm d$$

where

$$d = 1.96\sqrt{\frac{\hat{p}(1-\hat{p})}{n}} \quad \text{for 95 percent confidence}$$

and

$$d = 2.58\sqrt{\frac{\hat{p}(1-\hat{p})}{n}} \quad \text{for 99 percent confidence}$$

It can be shown that the term $\sqrt{\hat{p}(1-\hat{p})/n}$ is an estimate of the standard deviation of $\hat{p}$. The 95 and 99 percent confidence intervals thus allow 1.96 and 2.58 units of standard deviation respectively on either side of $\hat{p}$. A 95 percent confidence interval for p is illustrated in Figure 2.3.

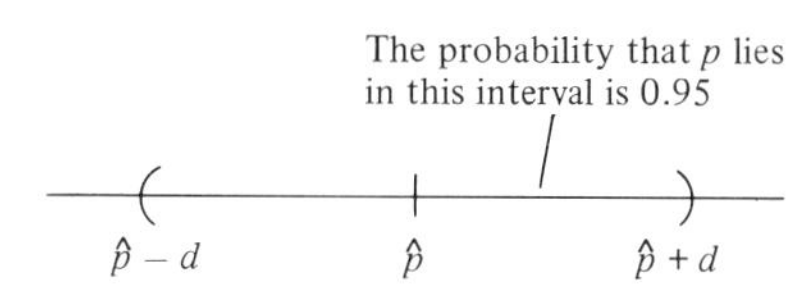

Figure 2.3 A 95 percent confidence interval for p.

The next two examples show how to apply these formulas.

EXAMPLE 2.4

● In a random sample of 100 voters it is found that 55 would vote for Smith for mayor if the election were held today. Find the 95 percent confidence interval for p, the proportion of voters in the population who would vote for Smith. Based on this, can you conclude with 95 percent confidence that a majority of the voters would vote for Smith if the election were held today?

SOLUTION Using the formula $d = 1.96\sqrt{\hat{p}(1-\hat{p})/n}$ with $n = 100$ and $\hat{p} = 0.55$, you find that the 95 percent confidence interval is

$$\begin{aligned}\hat{p} \pm 1.96\sqrt{\hat{p}(1-\hat{p})/n} &= 0.55 \pm 1.96\sqrt{0.55(0.45)/100} \\ &= 0.55 \pm 0.10\end{aligned}$$

That is, you can be 95 percent confident that the proportion who would vote for Smith is between $0.55 - 0.10 = 0.45$ and $0.55 + 0.10 = 0.65$. Since a portion of this confidence interval lies below 0.5, you cannot conclude that a majority would vote for Smith. ●

EXAMPLE 2.5 ● In a certain city, a random sample of 300 houses is taken to estimate the proportion p of houses with faulty wiring. Thirty-six of the houses in the sample are found to have faulty wiring. Find the 99 percent confidence interval for p.

SOLUTION Since 36 of 300 houses have faulty wiring, the sample proportion is $\hat{p} = 36/300 = 0.12$. Using $d = 2.58\sqrt{\hat{p}(1-\hat{p})/n}$ we find that the 99 percent confidence interval is

$$\hat{p} \pm 2.58\sqrt{\frac{\hat{p}(1-\hat{p})}{n}} = 0.12 \pm 2.58\sqrt{\frac{0.12(0.88)}{300}}$$
$$= 0.12 \pm 0.05$$

That is, we are 99 percent confident that the proportion of houses with faulty wiring is between $0.12 - 0.05 = 0.07$ and $0.12 + 0.05 = 0.17$. ●

Controlling the precision of an estimate

In estimating proportions, we again encounter the problem of trying to control both the reliability and precision of an estimate. For example, in the voter poll of Example 2.4, the 95 percent confidence interval was not narrow enough to indicate whether Smith would get a majority of the votes. This suggests that a larger sample should have been taken. Here is a rule you can use to determine the appropriate sample size.

> **How to control the precision of an estimate of p**
> To be 95 percent confident that the error committed when using $\hat{p}$ to estimate p is no more than d units, take a sample of size $n \geq (0.98/d)^2$. To be 99 percent confident, take a sample of size $n \geq (1.29/d)^2$.

The next example illustrates this technique.

EXAMPLE 2.6

● Suppose that you want to be 95 percent confident that your estimate of the proportion of voters who favor Smith will be "correct to within 3 percentage points." How large a sample should you take?

SOLUTION

For the estimate to be correct to within 3 percentage points, the maximum error must be no more than $d = 0.03$. Since 95 percent confidence is desired, you use the inequality $n \geq (0.98/d)^2$ to get

$$n \geq \left(\frac{0.98}{d}\right)^2 = \left(\frac{0.98}{0.03}\right)^2 = 1{,}067.11$$

and conclude that a sample of at least 1,068 people should be taken. ●

Problems

1 In the firing of ceramics, pyrometric cones are used to determine when the temperature in the kiln reaches the desired level. At this temperature the cones will melt. A random sample of 100 cones is chosen from a large supply. The average melting temperature of the sample is found to be $\overline{X} =$ 2,290°F with a standard deviation of $s = 8$°F. Find the 95 and 99 percent confidence intervals for the average melting point of the cones in the supply.

2 A random sample of 64 lightbulbs manufactured by a certain company is tested to estimate the average lifetime. For the sample bulbs it is found that the average lifetime is 440 hours and the standard deviation is 75 hours. Find the 95 and 99 percent confidence intervals for the average lifetime of the company's bulbs.

3 A fire department wants to know the average amount of time needed to extinguish fires. Over a 1-year period, 40 fires are selected at random and from the fire department's records it is found that for these fires the average time was 1.1 hours, with a standard deviation of 0.6 hours. Find the 95 and 99 percent confidence intervals for the average time needed to put out a fire.

4 A machine that fills 10-ounce cans with tomato sauce is checked by weighing a random sample of 100 cans out of every 2,000. The lot of 2,000 cans is rejected if the entire 95 percent confidence interval for the sample mean is less than 10 ounces. Would the lot be rejected if a sample has an average weight of 9.9 ounces and a standard deviation of 0.4 ounces?

5 Suppose that for a radar system, the measurement error in pinpointing the longitude of the position of an airplane several miles away has mean 0 and standard deviation 0.5 miles. How many observations are needed to ensure with probability 0.95 that the average of the observed positions is within 0.1 miles of the true position?

6 The standard deviation of weights of cereal boxes filled by a certain machine is known to be 0.4 ounces. How large a sample should you take to be 99 percent confident that the sample mean is within 0.05 ounces of the true average weight of all the boxes filled by this machine?

7 Show that the 90 percent confidence interval for μ is $\overline{X} \pm 1.65(s/\sqrt{n})$.

8 A random sample of 1,600 registered voters was taken and 825 said they would vote for Smiley. Basing your decision on a 95 percent confidence interval, can you say that Smiley would win if the election were held today?

9 The department of transportation wishes to estimate the proportion of commuters in a certain city who drive to work with at least one other person. A sample of 500 commuters yielded 210 who commute with at least one other person. Find the 95 and 99 percent confidence intervals for the proportion who drive to work with at least one other person.

10 A biologist found that in a sample of 450 insects selected at random from a certain insect colony, 298 were males. Find the 95 and 99 percent confidence intervals for the proportion of males in this insect colony.

11 A person received a ticket 15 out of 120 times he double parked.

(a) Find the 95 percent confidence interval for the proportion of the time one could expect to get a double-parking ticket.

(b) If double-parking tickets cost \$5 and it costs \$1 to park in a garage, which method of parking is cheaper in the long run?

12 A survey is to be taken to estimate the proportion of people who use a certain brand of aspirin. How large a sample should you take to be 95 percent confident that the proportion of those in the sample who use the aspirin is within 2 percentage points of the population proportion?

13 (a) In estimating the proportion of defective items in a batch of transistors, how many transistors should you sample to be 99 percent confident that $\hat{p}$ is within 0.02 of the true proportion?

(b) Suppose you take a sample of the size obtained in part (a) and you find that $\hat{p} = 0.13$. Find the 99 percent confidence interval for p. Check that this interval satisfies the constraints of part (a).

14 How many times must you toss a coin so that the total length of the 95 percent confidence interval for the proportion of heads is no greater than 0.002?

3 • HYPOTHESIS TESTING

In the preceding section we used confidence intervals to estimate unknown means and proportions. We shall now consider a related technique with which we can decide the validity of a hypothesis.

Suppose, for example, that we wish to decide whether a certain pill is effective in preventing blistering from poison oak. Ordinarily, about 30 percent of those who get poison oak suffer blistering. A random sample of 100 people with poison oak are given the pill and 24 experience blistering. Is this significant evidence that the pill has an effect?

To answer this question we identify two hypotheses:

The **null** or **chance hypothesis:** The pill has no effect. The observed result is due to chance alone.

The **research** or **alternative hypothesis:** The pill has an effect (either positive or negative).

We wish to decide whether or not the observation of 24 cases of blistering is sufficient evidence to reject the null hypothesis and accept the alternative hypothesis. To do this, we proceed in the courtroom manner and assume the null hypothesis to be "innocent until proven guilty." In particular, we assume that the null hypothesis is true and use this assumption to compute the probability of obtaining the observed result. If this probability is very low, we rule out chance and conclude that the null hypothesis is false. For example, we reject the null hypothesis that the pill has no effect on blistering if it turns out that under this assumption, the probability of observing as few as 24 cases of blistering in a random sample of 100 people with poison oak is very small.

It is up to the experimenter to decide how unlikely an observation must be before it can be used to conclude that the null hypothesis is false. The level below which the probability of the observed result must drop to cause rejection of the null hypothesis is called the **significance level** of the statistical test and is denoted by the Greek letter α (alpha).

Significance level
The *significance level* α of a statistical test specifies how unlikely an observation must be (under the assumption of the null hypothesis) to cause the experimenter to decide that the null hypothesis is false.

The most commonly used significance level is $\alpha = 0.05$. For normal populations, the empirical rule gives the following simple procedure you can use to test hypotheses at this significance level.

Testing hypotheses at the significance level $\alpha = 0.05$ using normal random variables
Step 1: Compute the mean and standard deviation of the random variable using the assumption that the null hypothesis is true.
Step 2: Use the mean and standard deviation in Step 1 to convert the observed value of the random variable to the standard Z scale.
Step 3: Reject the null hypothesis if the value of Z in Step 2 is greater than 2 or less than -2. Otherwise do not reject the null hypothesis.

Since the standard deviation of the standard normal distribution is 1 and its mean is zero, the procedure just outlined tells us to reject the null hypothesis if the observed value of Z is more than 2 standard deviations from the mean of Z, or, equivalently, if the observed value of X is more than two standard deviations from its mean. The empirical rule states that 95 percent of all the values of a normal random variable will be within two standard deviations of the mean. Hence, this procedure works because 0.05 is the probability that a value will be more than two standard deviations from the mean.

In the test to determine the effectiveness of the poison oak pill, we observed that 24 out of 100 sufferers experienced blistering after being given the pill, whereas ordinarily 30 percent experience blistering. If the null hypothesis is true and the pill has no effect, then X, the number of people in the sample who experience blistering, has a *binomial* distribution with $n = 100$ and $p = 0.3$. Since n is large, this binomial distribution is approximately normal with

$$\mu = np = 30 \qquad \text{and} \qquad \sigma = \sqrt{np(1 - p)} = 4.58$$

The Z value of the observation $X = 24$ is

$$Z = \frac{24 - 30}{4.58} = -1.31$$

Since Z is between -2 and 2, we cannot rule out chance and do not reject the null hypothesis. That is, we conclude that there is not significant evidence that the pill has an effect.

The following example further illustrates this technique.

EXAMPLE 3.1 ● In a recent court case[1] a school district in Missouri was brought to trial by the United States government for allegedly not hiring a representative number of black teachers, thus violating a provision of the 1964 Civil Rights Act. Part of the government's evidence was the fact that between 1972 and 1974, the school district hired 405 teachers, of whom 15 were black, while according to the 1970 census, 15.4 percent of the teachers in the surrounding area (called the "relevant labor market area" by the court) were black. Test at significance level $\alpha = 0.05$ the research hypothesis that the proportion of blacks hired by the district is different from the corresponding proportion in the relevant labor market area.

SOLUTION The null hypothesis is that the hiring by the district is random (with regard to race) from the relevant labor market area. If the null hypothesis is true, then X, the number of black teachers hired, has a binomial distribution with $n = 405$ and $p = 0.154$. Since n is large, this binomial distribution is approximately normal with

$$\mu = np = 62.37 \qquad \text{and} \qquad \sigma = \sqrt{np(1 - p)} = 7.26$$

The Z value of the observation $X = 15$ is

$$Z = \frac{15 - 62.37}{7.26} = -6.52$$

Since the observed Z value is less than -2, we reject the null hypothesis. Such an observation would almost certainly not occur if the null hypothesis were true.

We note that although there is still a dispute over what constitutes the relevant labor market area in this case (this decision is to be made by a District Court) the Supreme Court has ruled that the statistical test we have used is an appropriate way to analyze such data. (The Supreme Court stated that "where gross statistical disparities can be shown, they alone may in a proper case constitute *prima facie* proof of a pattern or practice of discrimination." The Court further stated that ". . . if the difference between the expected value and the observed number is greater than two or three standard deviations, then the hypothesis that teachers were hired without regard to race would be suspect.") ●

Problems

1 You suspect that a certain gambler's coin is biased, but the gambler claims it is evenly balanced. The coin is tossed 100 times and you observe 65 heads.

[1] *Hazelwood School District et al.* v. *United States.*

At the significance level $\alpha = 0.05$, what conclusion can be drawn from this data? (The null hypothesis is that the coin is evenly balanced.)

2 With a former judge, 20 percent of convicted first-offenders were sentenced to jail terms. During the first 6 months with the new judge, 45 out of 200 convicted first-offenders were sentenced to jail terms. Determine at the significance level $\alpha = 0.05$ if this is statistically significant evidence that the conviction rate of the new judge is different from that of the old judge. (The null hypothesis is that there is no difference.)

3 In an area free of radioactive fallout, the rate of babies born with birth defects is 1 percent per year. In a 1-year study of an area of heavy radioactive fallout from atomic testing, it was found that 150 babies out of 10,000 were born with birth defects. Using a significance level of $\alpha = 0.05$, would you conclude that radioactive fallout increased the incidence of birth defects?

4 When a powerful pesticide was used, 10 percent of a certain crop was destroyed by pests annually. When this pesticide was found to be harmful to the ecology and declared illegal, a new type of pesticide was introduced. During the first year the new pesticide was in use, 160 out of a sample of 1,500 plants were killed by pests. Is this significant evidence at the significance level $\alpha = 0.05$ that the new pesticide is not as effective as the old one?

5 In order to decide whether or not someone has psychic powers, a researcher conducts the following test. There are two cards, one with a red circle and one with a green circle. On each trial the researcher selects one of the cards without showing it to the subject. The subject then has to state the color of the selected card. Suppose this procedure is repeated 100 times and the subject makes 70 correct responses.

(a) Is this significant evidence at the significance level $\alpha = 0.05$ that the subject is not guessing at random?

(b) Suppose the subject makes 30 correct responses. Is this significant evidence that the subject is not guessing at random?

6 The Brighteye Coffee Company wants to determine if more people like the taste of Brighteye coffee than the taste of a rival brand. Ninety people are selected at random from shoppers at a large shopping mall. Each person is asked to taste a cup of Brighteye coffee and a cup of the rival brand and to decide which cup tastes better. Suppose 48 of the 90 people choose Brighteye as the better-tasting coffee. Testing at the significance level of $\alpha = 0.05$, what should the coffee company conclude?

7 Over a 20-year period, 28 out of 50,000 workers exposed to a certain chemical developed a type of liver cancer, while in the general populace a fraction of only 0.0001 get this type of cancer. Testing at the significance level $\alpha = 0.05$, use this data to decide whether the risk of this type of cancer is greater for workers exposed to the chemical than for the general populace.

8 In a certain company 44 out of 125 employees are women, while in the relevant labor market area, a fraction of 0.41 are women. Testing at the significance level $\alpha = 0.05$, decide whether or not the fraction of women employees in this company is significantly different than the fraction of women in the relevant labor market area.

9 If you are conducting a statistical test at the significance level $\alpha = 0.01$, how large does $|Z|$ have to be before you reject the null hypothesis?

The next two problems involve testing hypotheses about population means μ. In each case, a value of the sample mean $\bar{X}$ is given. Transform $\bar{X}$ to Z using the fact that for large n, the distribution of $\bar{X}$ is approximately normal with mean μ and standard deviation $s/\sqrt{n}$. For μ, use the population mean specified by the null hypothesis. Reject the null hypothesis if $|Z| > 2$.

10 The average thickness of egg shells from a certain species of bird is 0.52 millimeters. The shells of eggs from 60 of the birds who have ingested the pesticide DDT are measured and the average thickness is found to be $\bar{X} = 0.50$ millimeters with $s = 0.04$ millimeters. Testing at significance level $\alpha = 0.05$, decide whether the average thickness of these egg shells is different from 0.52 millimeters.

11 An advertisement states that a certain compact car gets an average of $\mu = 28$ miles per gallon in city driving. A sample of 40 cars yields the following data: $\bar{X} = 26.7$ miles per gallon and $s = 6$ miles per gallon. Testing at significance level $\alpha = 0.05$, decide whether or not the results of the sample contradict the advertisement or fall within the bounds that chance allows.

4 • LINEAR CORRELATION AND PREDICTION

An important problem encountered by researchers in many disciplines is to determine if there is a relationship, or **correlation,** between two or more random variables. For example, a doctor may be interested in determining if there is a relationship between air pollution and respiratory disease; a psychologist may want to know if there is a relationship between adult neurosis and childhood trauma; an educator may be interested in knowing whether there is a relationship between SAT scores and college GPA. And if a relationship exists, the researcher may want to use information about a known variable to predict the value of an unknown variable.

In this section, X and Y will denote the values of a pair of variables resulting from a random sample. Since we shall be interested in the way X and Y behave together, it will be convenient to list the paired values as they occur in the sample and to construct a **scatter diagram,** which graphically depicts the relationship between X and Y. The following example will show you how to do this.

EXAMPLE 4.1 ● A random sample of seven college graduates was selected to see if there is a relationship between high-school grades and college grades. The table below shows the high-school grade point average, X, and college grade point average, Y, for each selected student. Construct a scatter diagram for this information.

X	2	1	3	3	4	4	4
Y	1.5	2	2	3	3.5	4	4

SOLUTION To construct a scatter diagram, we plot each pair (X, Y) on a graph. If any point occurs more than once (as the point (4, 4) does in this example), we indicate this on the scatter diagram. In Figure 4.1 we have constructed the scatter diagram for the given data.

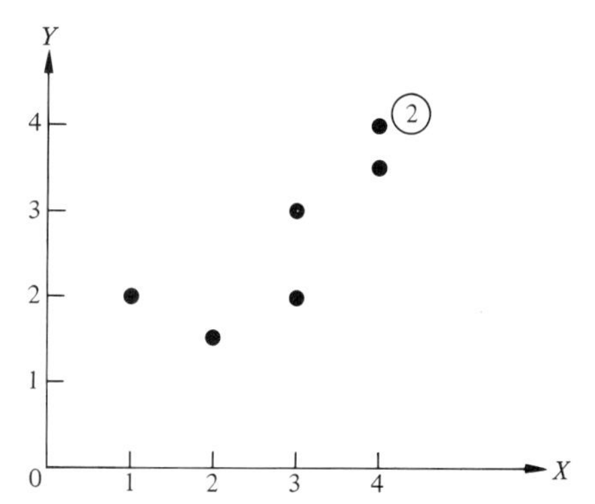

Figure 4.1 Scatter diagram of high school vs. college grades. ●

In the preceding example, the sample size of $n = 7$ was unrealistically small. In an actual research problem, a much larger sample would be used, and with the aid of computers the analyses that we shall be doing in this section can be easily carried out. For ease of presentation we shall continue to use small sample sizes.

In Figure 4.2 there are scatter diagrams illustrating three different types of possible relationships between X and Y. In Figure 4.2*a* the values of Y tend to increase linearly as the values of X increase. This type of relationship between X and Y is called **positive linear correlation.** In Figure 4.2*b* the values of Y tend to decrease linearly as the values of X increase. This is called **negative linear correlation.** In Figures 4.2*c* and 4.2*d* there is no clear linear relationship between X and Y and we say that X and Y are **linearly uncorrelated.**

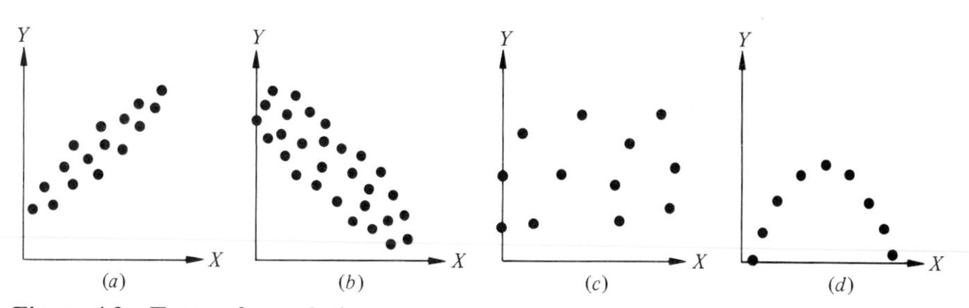

Figure 4.2 Types of correlation.

There are other types of correlation besides linear correlation, but linear correlation is the most important type and the easiest to analyze. Also, nonlinear correlation can often be changed into linear correlation by an appropriate transformation. In this section, we shall concern ourselves exclusively with linear correlation between X and Y and shall therefore often omit the adjective "linear."

Our first task is to define a measure of the amount of correlation between X and Y. We shall use a measure that is known as the **sample correlation coefficient,** and is denoted by the letter r. Here is the formula for computing r.

The sample correlation coefficient

The *sample correlation coefficient* r is a measure of the linear correlation between X and Y and is given by the formula

$$r = \frac{n\sum X_i Y_i - \left(\sum X_i\right)\left(\sum Y_i\right)}{\sqrt{n\sum X_i^2 - \left(\sum X_i\right)^2}\sqrt{n\sum Y_i^2 - \left(\sum Y_i\right)^2}}$$

where n is the size of the sample.

In the formula, ΣX_i tells us to add the X values, ΣY_i to add the Y values, $\Sigma X_i Y_i$ to add the products of the XY pairs, ΣX_i^2 to add the squared values of X, $(\Sigma X_i)^2$ to square the sum of the values of X, and so on. Here are some important properties of r.

Properties of the sample correlation coefficient

(a) The sample correlation coefficient r is always between -1 and 1.
(b) If r is close to 1, then X and Y are highly positively correlated.
(c) If r is close to -1, then X and Y are highly negatively correlated.
(d) If $r = 0$, X and Y are uncorrelated.

Properties (b) and (c) are vague because of the term "close to." In many situations, we use r as a comparative measure to see which of several variables has the highest degree of correlation with a particular variable. In other situations a confidence interval is constructed around r in order to account for its random variation, with correlation inferred if the confidence interval does not contain 0.

The next example illustrates how to calculate r.

EXAMPLE 4.2 ● For the data below, draw a scatter diagram and calculate the sample correlation coefficient r.

X	0	1	2	3	3
Y	1	2	4	3	4

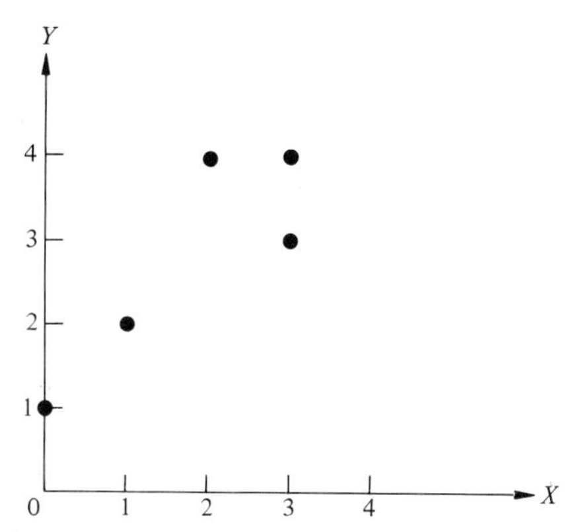

Figure 4.3 Scatter diagram for Example 4.2.

SOLUTION The scatter diagram is displayed in Figure 4.3. From the scatter diagram it appears that X and Y are positively correlated and we would expect r to be close to 1.

To calculate r, it is convenient to first construct a table as follows.

X	0	1	2	3	3	$\Sigma X_i = 9$
Y	1	2	4	3	4	$\Sigma Y_i = 14$
XY	0	2	8	9	12	$\Sigma X_i Y_i = 31$
X^2	0	1	4	9	9	$\Sigma X_i^2 = 23$
Y^2	1	4	16	9	16	$\Sigma Y_i^2 = 46$
						$n = 5$

We now use the formula for r and obtain

$$r = \frac{n \sum X_i Y_i - \left(\sum X_i\right)\left(\sum Y_i\right)}{\sqrt{n \sum X_i^2 - \left(\sum X_i\right)^2}\sqrt{n \sum Y_i^2 - \left(\sum Y_i\right)^2}}$$

$$= \frac{5(31) - 9(14)}{\sqrt{5(23) - 9^2}\sqrt{5(46) - 14^2}}$$

$$= 0.85$$

(Check these calculations using a calculator.) Since r is close to 1, we conclude as expected that X and Y are positively correlated. ●

We note that although the correlation coefficient measures the amount of linear relationship between X and Y, it does not by itself imply cause and effect. Other factors may be involved. For example, when the crippling disease polio was being studied in the early 1950s, researchers found a high positive correlation between soft-drink sales and the incidence of polio. This caused some researchers to suspect that an ingredient in soft drinks played a role in causing polio, until it was discovered that the polio virus was more active in warm weather, and hence polio incidence rose during that time. Of course, soft-drink sales also rose in warmer weather. Thus, there was a high positive correlation between polio incidence and soft-drink sales, yet the relationship was not causal.

Next, we shall introduce a method for predicting Y based on X, which is useful when there is a high correlation between X and Y.

Linear prediction

In many practical problems, a researcher may have information about a variable X and wish to predict the value of a variable Y. If there is high correlation between X and Y, a method known as **least squares prediction** can be used. Here is the basic idea. Suppose that we have a scatter diagram displaying the pairs (X, Y) and that we can draw a nonvertical straight line on the scatter diagram that "fits" the points reasonably well; that is, a line that comes close, on the average, to the points on the scatter diagram. Such a line is shown in Figure 4.4.

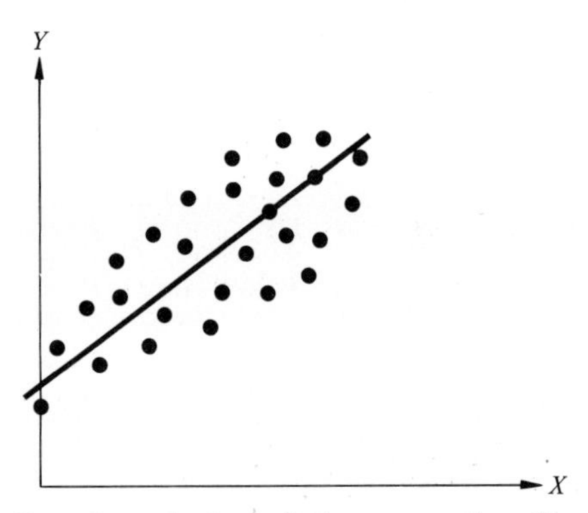

Figure 4.4 Fitting a line through the points on a scatter diagram.

We can use this line as a predictor in the following way: If we obtain a new value of X we predict Y to be the value that would make the point (X, Y) lie on the line. We shall call a line that is used in this way a **linear predictor.**

Any straight line can be used as a linear predictor, but of course, we would like to find the "best" line, that is, the one that is the closest on the average to the points on the scatter diagram. It is customary to define the best line to be the one for which the average squared vertical distance from the points on the scatter diagram to the line is minimized. This line is called the **least-squares line** or **regression line.**

Recall that the equation of a nonvertical straight line can be written in the form $Y = aX + b$, where a is the slope of the line and $(0, b)$ is the point at which the line intersects the Y axis. In order to find the equation of a particular line, we need to know the numbers a and b. Here are formulas you can use to compute a and b for a least-squares line.

The least-squares line

The best-fitting line to predict Y from X for a given set of paired data (X, Y) is

$$Y = aX + b$$

with

$$a = \frac{n \sum X_i Y_i - \left(\sum X_i\right)\left(\sum Y_i\right)}{n \sum X_i^2 - \left(\sum X_i\right)^2} \quad \text{and} \quad b = \overline{Y} - a\overline{X}$$

where $\overline{X}$ and $\overline{Y}$ are the sample means of the X and Y data, respectively.

The formula for the least-squares line is not difficult to derive, but we shall not do it here. The next example illustrates the use of the least-squares line.

EXAMPLE 4.3 ● Using the data of Example 4.1 showing high school versus college grade point averages, calculate the correlation coefficient r and find the equation for the least-squares line. Then draw the scatter diagram and graph the least-squares line on it. Finally, use the line to predict the college grade point average of a student whose high school grade point average is 3.3.

SOLUTION In the table below we have made the calculations necessary to obtain r and the equation of the least-squares line.

X	2	1	3	3	4	4	4	$\Sigma X_i = 21$
Y	1.5	2	2	3	3.5	4	4	$\Sigma Y_i = 20$
XY	3	2	6	9	14	16	16	$\Sigma X_i Y_i = 66$
X^2	4	1	9	9	16	16	16	$\Sigma X_i^2 = 71$
Y^2	2.25	4	4	9	12.25	16	16	$\Sigma Y_i^2 = 63.5$
								$n = 7$ $\overline{X} = \frac{21}{7} = 3$ $\overline{Y} = \frac{20}{7} = 2.86$

Using the formula for r we get

$$r = \frac{7(66) - 21(20)}{\sqrt{7(71) - 21^2}\,\sqrt{7(63.5) - 20^2}} = 0.84$$

Since r is close to 1, there is high positive correlation between X and Y.

Next we find the equation for the least-squares line as follows:

$$a = \frac{n\sum X_i Y_i - \left(\sum X_i\right)\left(\sum Y_i\right)}{n\sum X_i^2 - \left(\sum X_i\right)^2} = \frac{7(66) - 21(20)}{7(71) - 21^2} = 0.75$$

$$b = \overline{Y} - a\overline{X} = 2.86 - 0.75(3) = 0.61$$

and so

$$Y = 0.75X + 0.61$$

The graph of the least-squares line plotted on the scatter diagram is shown in Figure 4.5. If a student's high school grade point average is $X = 3.3$, we obtain a predicted college grade point average of $Y = 0.75(3.3) + 0.61 = 3.09$.

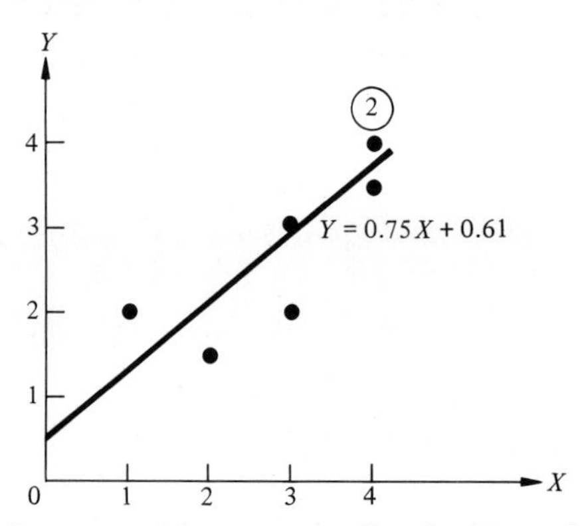

Figure 4.5 Scatter diagram and least-squares line for Example 4.3. ●

There are some important relationships between the least-squares line and the correlation coefficient. If there is positive correlation between X and Y, the slope of the line is positive. If there is negative correlation, the slope of the line is negative. If all the points of the scatter diagram lie on a (nonhorizontal or nonvertical) straight line, then there is "perfect" correlation and $r = 1$ or $r = -1$ depending upon whether the slope of the line is positive or negative. If $r = 0$, there is no linear relationship between X and Y and the least-squares line turns out to be the horizontal line $Y = \overline{Y}$. If you use this line as a predictor, you will predict $Y = \overline{Y}$ no matter what the value of X.

When using the least-squares line as a predictor, it is important to use it only in the region for which you have data. Many relationships are linear in a certain range but then change outside that range. For example, there is a positive correlation between blood pressure and dosage of certain medications, but the relationship must change at some point. (For example, if the dosage becomes toxic and the person dies, blood pressure drops to zero!) If there is ample data to demonstrate a correlation, the least-squares line can be an extremely powerful tool if used in the range of the data values.

Problems

1 For the following data, draw the scatter diagram, calculate the correlation coefficient, and find the equation of the least-squares line.

X	0	0	1	2.5	3
Y	0	1	3	3.5	4

2 For the following data, draw the scatter diagram, calculate the correlation coefficient, and find the equation of the least-squares line.

X	1	1	3	3	2
Y	0	3	0	3	1.5

3 Draw the scatter diagrams for the data below and, without doing any computation, determine the correlation coefficients r and the equations of the least-squares lines. Then, check your results by performing the calculations.

(a) X	1	2	3	4
Y	3	5	7	9

(b) X	0	2	3	4
Y	5	3	2	1

(c) X	3	4	6.7	8	8.1
Y	10	10	10	10	10

4 Draw scatter diagrams for which the following hold.

(a) $r = 0$
(b) $r = -1$
(c) The least-squares line is given by $Y = 4X - 3$.

5 Over the past four years, a college admissions officer has compiled the following data (measured in units of 1,000) relating the number X of college catalogs requested by high school students by December 1 to the number Y of completed applications received by March 1.

X	4.7	4.4	5.8	4.4
Y	1.2	1.1	1.5	1.0

(a) Draw a scatter diagram for this data.
(b) Find the correlation coefficient r.

(c) Find the equation of the least-squares line and graph the line on the scatter diagram.

(d) This year, the college received 4,800 requests for catalogs by December 1. Use the least-squares line to predict the number of completed applications that will be received by March 1.

6 A study is done with a group of mothers to determine whether or not there is a linear relationship between the number X of packs of cigarettes smoked per day and the weight Y of newborn babies. Here is the data.

X	0	0	0.5	1.1	1.5	2.3	2.5	3.0
Y	8.2	8.0	7.9	7.6	7.4	7.1	7.3	6.9

(a) Draw the scatter diagram.

(b) Find r.

(c) Find the equation of the least-squares line and graph the line on the scatter diagram.

7 A business manager has compiled the following data relating profit Y to advertising expenditure X (both measured in thousand-dollar units).

X	0.6	0.8	1.2	2.5	2.8	0.3	0.7
Y	3.1	4.1	2.1	5.7	6.2	4.0	5.1

(a) Draw the scatter diagram for this data.

(b) Find r.

(c) Find the equation of the least-squares line.

(d) Use the least-squares line to predict the profit if $2,000 is spent on advertising.

8 In 1965, a study was made of the effects on residents of northern Oregon of low-level radiation resulting from seepage of radioactive wastes from the AEC storage facility at Hanford, Washington, into the Columbia River.[2] For each of the 9 Oregon counties along the Columbia River or the Pacific Ocean, an "index of exposure" was computed and the cancer mortality rate (per 100,000 people per year from 1959 to 1964) was recorded. The data is given on the next page.

(a) Draw the scatter diagram for this data.

(b) Compute the correlation coefficient r.

(c) What do these findings suggest to you? (*Note:* To make your conclusion statistically meaningful, you would have to perform a hypothesis test for the random variable r. When the researcher doing this study

[2] Fadeley, Robert Cunningham, "Oregon Malignancy Pattern Physiographically Related to Hanford Washington Radioisotope Storage," *Journal of Environmental Health,* v. 27, 1965.

performed the appropriate test, he found that the probability of obtaining such a high correlation due to chance alone was less than 0.001.)

County	Index of exposure	Cancer mortality
Umatilla	2.49	147.1
Morrow	2.57	130.1
Gilliam	3.41	129.9
Sherman	1.25	113.5
Wasco	1.62	137.5
Hood River	3.83	162.3
Portland	11.64	207.5
Columbia	6.41	177.9
Clatsop	8.34	210.3

9 If a canned program that calculates r and the least-squares line is available at your campus computer center, use it to check the calculations you have done in this problem set.

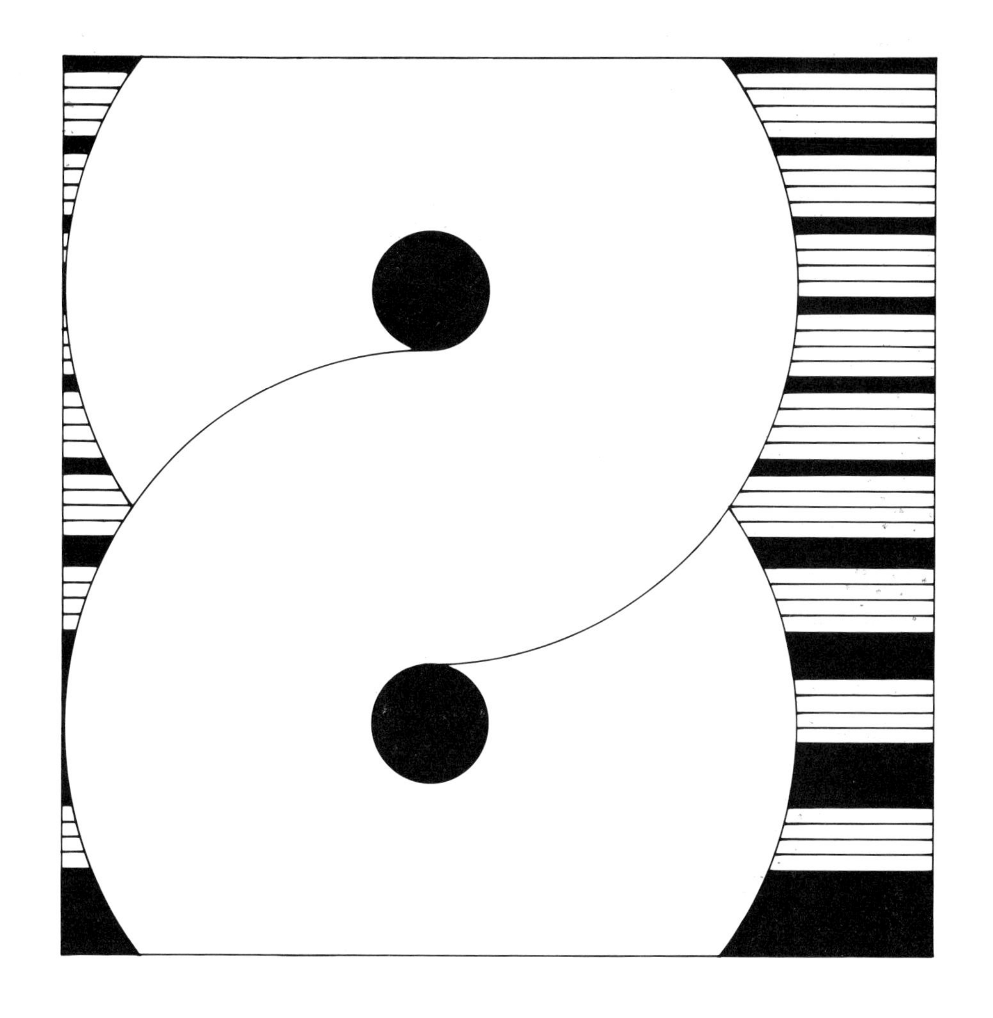

DECISION THEORY AND THE THEORY OF GAMES

1 • BAYESIAN METHODS

In this section, we shall analyze problems in which the goal is to maximize the expected value of a potential gain such as profit in a business venture or payoff in a gambling game. The methods we shall use are called **bayesian** because they involve Bayes' formula for conditional probability. For reference, we restate the definition of expected value from Chapter 6 and Bayes' formula from Chapter 5.

The expected value of a random variable
If X is a random variable that assumes values $x_1, x_2, \ldots, x_n$ with probabilities $p_1, p_2, \ldots, p_n$, respectively, the expected value of X is the sum

$$E(X) = x_1p_1 + x_2p_2 + \cdots + x_np_n$$

Bayes' formula
For any events A and B,

$$P(A \mid B) = \frac{P(B \mid A)P(A)}{P(B \mid A)P(A) + P(B \mid A')P(A')}$$

In the following three examples, different aspects of a decision-making problem will be analyzed. First, the criterion of maximizing expected gain will be introduced. Next, a procedure for incorporating additional information into the decision-making process will be described. And finally, the decision maker's overall expected gain when the additional information is used will be compared with the original expected gain in order to determine how much the additional information is worth.

EXAMPLE 1.1

● A chemical company must decide whether to begin immediate production of a powerful new pesticide that may be harmful to the ecology, or to wait a year until tests measuring the harmful effects of the pesticide have been completed. If the company begins production now, it can expect to make a profit of 5 million dollars during the year if the pesticide proves to be safe, but will suffer a net loss of 4 million dollars in fines and lawsuits if the pesticide is eventually found to be unsafe. If the company delays production of the pesticide and continues to produce a less-powerful pesticide already on the market, it can expect a profit of 2 million dollars. Based on preliminary tests, the company estimates that the probability is 0.8 that the pesticide will be declared safe and 0.2 that it will be declared unsafe. What action should the company take if it wants to maximize its expected gain?

SOLUTION

First, we display the company's possible gain (in millions of dollars) for each action and each "state of nature" in an array called a **payoff matrix.**

	Safe	Unsafe
Begin production now	5	−4
Delay production	2	2

Since the company estimates that

$$P(\text{safe}) = 0.8 \qquad \text{and} \qquad P(\text{unsafe}) = 0.2$$

we obtain the following expected gains E:

$$\text{If production begins now,} \qquad E = 5(0.8) + (-4)(0.2) = 3.2$$

$$\text{If production is delayed,} \qquad E = 2(0.8) + 2(0.2) = 2$$

Thus, in order to maximize its expected gain, the company should begin production now, for an expected gain of 3.2 million dollars. ●

In the preceding example, initial probability estimates were given for the safety of the pesticide. These initial probabilities are known as **prior probabilities.** In the next example, the company will have an opportunity to revise these probabilities by doing additional testing and to incorporate the revised probabilities into the decision-making process. These revised probabilities are called **posterior probabilities.**

EXAMPLE 1.2 ● Suppose that before deciding whether or not to begin production of the new pesticide, the company in Example 1.1 can conduct preliminary tests whose results will be either "positive" (implying safe) or "negative" (implying unsafe). The reliability of these tests is measured by the following conditional probabilities:

$$P(\text{positive}\,|\,\text{safe}) = 0.9 \qquad P(\text{positive}\,|\,\text{unsafe}) = 0.3$$

$$P(\text{negative}\,|\,\text{safe}) = 0.1 \qquad P(\text{negative}\,|\,\text{unsafe}) = 0.7$$

What actions should the company take after observing the results of the preliminary tests?

SOLUTION We shall use the new information to revise the prior probabilities, $P(\text{safe}) = 0.8$ and $P(\text{unsafe}) = 0.2$. The revised probabilities estimate the safety of the pesticide, given the results of the preliminary tests. Using Bayes' formula, we get the following posterior probabilities:

$$P(\text{safe}\,|\,\text{positive}) = \frac{P(\text{positive}\,|\,\text{safe})P(\text{safe})}{P(\text{positive}\,|\,\text{safe})P(\text{safe}) + P(\text{positive}\,|\,\text{unsafe})P(\text{unsafe})}$$

$$= \frac{0.9(0.8)}{0.9(0.8) + 0.3(0.2)} = 0.92$$

$$P(\text{unsafe}\,|\,\text{positive}) = 1 - 0.92 = 0.08$$

and

$$P(\text{safe} \mid \text{negative}) = \frac{P(\text{negative} \mid \text{safe})P(\text{safe})}{P(\text{negative} \mid \text{safe})P(\text{safe}) + P(\text{negative} \mid \text{unsafe})P(\text{unsafe})}$$

$$= \frac{0.1(0.8)}{0.1(0.8) + 0.7(0.2)} = 0.36$$

$$P(\text{unsafe} \mid \text{negative}) = 1 - 0.36 = 0.64$$

Using these posterior probabilities, we now find the company's expected gain for each possible action, given that the results of the preliminary tests are known.

If the results are positive and production begins now,

$$E = 5P(\text{safe} \mid \text{positive}) - 4P(\text{unsafe} \mid \text{positive})$$
$$= 5(0.92) - 4(0.08) = 4.28$$

If the results are negative and production begins now,

$$E = 5P(\text{safe} \mid \text{negative}) - 4P(\text{unsafe} \mid \text{negative})$$
$$= 5(0.36) - 4(0.64) = -0.76$$

If production is delayed, $E = 2$, regardless of the outcome of the tests.

Comparing these expected gains, we conclude that if the results of the preliminary tests are positive, the company should begin production now, for an expected gain of 4.28 million dollars, and if the results of the preliminary tests are negative, the company should delay production, for an expected gain of 2 million dollars. ●

In the next example, we shall determine how much the information obtained from the preliminary tests is worth to the company.

EXAMPLE 1.3 ● How much should the chemical company be willing to pay for the information obtained from the preliminary tests in Example 1.2?

SOLUTION In the preceding example we saw that if the results of the preliminary tests are positive, the company should begin production now for an expected gain of 4.28 million dollars, and if the results are negative, the company should delay production for an expected gain of 2 million dollars. In other words, the company's expected gain will be 4.28 million if the test results are positive and 2 million if they are negative. The overall expected gain in this case is therefore

$$E = 4.28P(\text{positive}) + 2P(\text{negative})$$

Using the intersection formula and the formula for the probability of the union of mutually exclusive events, we get

$$\begin{aligned} P(\text{positive}) &= P(\text{positive} \mid \text{safe})P(\text{safe}) + P(\text{positive} \mid \text{unsafe})P(\text{unsafe}) \\ &= 0.9(0.8) + 0.3(0.2) = 0.78 \\ P(\text{negative}) &= 1 - 0.78 = 0.22 \end{aligned}$$

Hence the overall expected gain when the information from the preliminary tests is used is

$$E = 4.28(0.78) + 2(0.22) = 3.78 \text{ million dollars}$$

Since the company's expected gain without the preliminary tests was found in Example 1.1 to be 3.2 million dollars, the tests are worth $3.78 - 3.2 = 0.58$ million dollars to the company. ●

The bayesian method of decision making illustrated in this section is useful in a variety of applications. The method depends on the decision maker's ability to obtain prior probabilities and to get additional information from which posterior probabilities can be calculated. In some situations, prior probabilities are either unavailable or unreliable. We shall take this up in the next section.

Problems

1 A toy manufacturer is planning to introduce a new toy for the coming Christmas season. The toy will be either a "space monster" or an "atomic laser cannon." The manufacturer estimates that the probability is 0.4 that by Christmas, public opposition to war toys will have created a market that is more favorable to the space monster than to the cannon. On the other hand, if the opponents of war toys are not successful, the market will be favorable to the cannon, which was featured in a popular movie. The manufacturer's sales estimates for the new toys (in millions of dollars) are summarized in the following payoff matrix.

	Market favorable to cannon	Market favorable to monster
Manufacture cannon	6	1
Manufacture monster	3	4

What action should the manufacturer take in order to maximize expected sales?

2 Suppose the manufacturer in Problem 1 decides to hire a market researcher to help predict the state of the market during the holiday season. The researcher is known to be "80 percent reliable" in situations such as this. That is, $P(\text{researcher recommends cannon} \mid \text{market favorable to cannon}) = 0.8$ and $P(\text{researcher recommends monster} \mid \text{market favorable to monster}) = 0.8$.

(a) Based on the researcher's advice, what actions should the manufacturer take?

(b) How much are the researcher's services worth to the manufacturer?

3 The government estimates that the probability is 0.3 that there will be a major flu epidemic next winter and must decide whether or not to initiate a nationwide immunization program. The total cost (in hundreds of millions of dollars) of each possible course of action, including the cost of developing the vaccine, the cost of administering the immunization program, and the cost of lost manpower if the epidemic occurs, is summarized in the following cost matrix.

	Epidemic occurs	Epidemic doesn't occur
Immunize	4	2.5
Don't immunize	9	0

Which action *minimizes* the government's expected cost?

4 Suppose that before making the final decision about the immunization program in Problem 3, the government decides to conduct further research to predict whether or not an epidemic will occur. The accuracy of these predictions is given by the following probabilities:

$$P(\text{epidemic predicted} \mid \text{epidemic occurs}) = 0.7$$

$$P(\text{no epidemic predicted} \mid \text{epidemic does not occur}) = 0.8$$

(a) Based on the results of this research, what actions should the government take?

(b) How much is the additional research worth to the government?

5 An oil prospector wishes to decide whether or not to drill for oil in a certain area. There are three possible results of the proposed drilling: no oil, moderate oil, and "bonanza." The net value of these results to the prospector, in thousands of dollars, is given in the following payoff matrix.

	No oil	Moderate oil	Bonanza
Drill	−50	50	100
Don't drill	0	0	0

Based on past experience, the prospector believes that the probability that there is no oil is 0.5, the probability that there is moderate oil is 0.3 and the probability that there is a bonanza is 0.2. Before drilling, the prospector can take a seismic sounding that gives two possible readings, + and −. The cost of the sounding is $5,000. Past experience has indicated the following relationship between the soundings and actual conditions.

$P(+ \mid \text{no oil}) = 0.2$ $\quad$ $P(+ \mid \text{moderate oil}) = 0.4$ $\quad$ $P(+ \mid \text{bonanza}) = 0.9$

(a) Find the prospector's best action and expected payoff if a seismic sounding is not taken.

(b) Determine the prospector's best strategy using the results of a sounding.

(c) How much is a seismic sounding worth to the prospector? Should the prospector take such a sounding?

6 There are two possible investments, A and B, and two possible states of the economy, "inflation" and "recession". The estimated percentage increase in value of each investment over the coming year for each state of the economy is given in the following payoff matrix.

$$\begin{array}{l} \\ \text{Invest in A} \\ \text{Invest in B} \end{array} \begin{array}{cc} \text{Inflation} & \text{Recession} \\ \left[\begin{array}{c} 10 \\ -5 \end{array}\right. & \left.\begin{array}{c} 5 \\ 20 \end{array}\right] \end{array}$$

Based on personal experience, the investor believes that the probability is 0.6 that the economy will be in a state of inflation next year and the probability is 0.4 that the economy will be in a state of recession. If more information is desired, the investor can consult a stock market indicator that will be either "↑" or "↓" and that is related to the state of the economy by the following conditional probabilities.

$$P(\uparrow|\text{inflation}) = 0.7 \qquad P(\uparrow|\text{recession}) = 0.2$$

$$P(\downarrow|\text{inflation}) = 0.3 \qquad P(\downarrow|\text{recession}) = 0.8$$

(a) Find the best investment and expected payoff if the market indicator is not consulted.

(b) Find the investor's best strategy using results of the market indicator.

(c) Find the investor's overall expected payoff when the market indicator is used and compute how much the information provided by the indicator is worth to the investor.

7 There are 10 coins. Eight of these coins are evenly balanced, while 2 of them are weighted in such a way that $P(\text{heads}) = 0.9$. One of these coins is selected at random and you must guess whether it is evenly balanced or weighted, being paid in dollars according to the following payoff matrix (in which negative entries denote losses).

$$\begin{array}{l} \\ \\ \text{Guess balanced} \\ \text{Guess weighted} \end{array} \begin{array}{cc} \text{Coin is} & \text{Coin is} \\ \text{balanced} & \text{weighted} \\ \left[\begin{array}{c} 3 \\ -1 \end{array}\right. & \left.\begin{array}{c} -8 \\ 8 \end{array}\right] \end{array}$$

(a) Find your best guess and expected payoff if no additional information is obtained.

(b) Suppose you are allowed to toss the selected coin and see how it lands. Find your best guessing strategy based on the results of the toss.

(c) Find your overall expected payoff for the strategy in part (b). What is the maximum amount you should pay to observe the result of one toss?

8 There are 20 boxes, 12 of type A, each containing 3 red balls and 1 green ball, and 8 of type B, each containing 2 red balls and 2 green balls. One of these boxes is chosen at random and you must guess what type of box it is. The payoff to you (in dollars) is determined by the following payoff matrix (in which negative entries denote losses).

	Box is type A	Box is type B
Guess type A	6	−5
Guess type B	−3	9

(a) Find your best guess and expected payoff if no additional information is obtained.

(b) Suppose that before making your guess you are allowed to select one ball at random from the chosen box. Find your best guessing strategy based on the results of this selection.

(c) How much is the information obtained from the selection of the ball in part (b) worth?

2 • MATRIX GAMES

In the preceding section we used expected value as a criterion for decision making. The situations we studied were not competitive, and relevant prior and posterior probabilities could be obtained by the decision maker. In this section, we shall discuss competitive situations in which one has to choose an action against an intelligent opponent who is competing for a reward. Reliable estimates of prior and posterior probabilities are not usually available in situations of this type.

Competitive encounters between two (or more) participants are analyzed using methods from a branch of mathematics known as the **theory of games.** Using these methods, competitive situations or **games** are represented in matrix form and optimal strategies for the "players" are obtained. The following two examples illustrate how you can represent a game using a payoff matrix.

EXAMPLE 2.1 ● Ruben and Catherine are going to play the game of "matching fingers." At the count of three, each will extend either 1 or 2 fingers. If they extend the same number of fingers, Ruben pays Catherine \$1. If they do not extend the same number of fingers, Catherine pays Ruben \$1. Represent this game in matrix form.

SOLUTION We shall construct a payoff matrix for which the columns represent the possible actions or "moves" for Catherine and the rows represent the possible moves for Ruben. The entries of the matrix are the amounts Ruben will win for the various combinations of moves. Negative entries are used to represent losses.

The resulting payoff matrix is displayed below.

		Catherine	
		1 finger	2 fingers
Ruben	1 finger	−1	1
	2 fingers	1	−1

●

For convenience, we shall denote the person whose possible actions are represented by the columns as the **column player** and the person whose possible actions are represented by the rows as the **row player.** It is customary to let the entries of the payoff matrix denote the row player's payoffs. Here is another example.

EXAMPLE 2.2 ● Two rival television networks, C and R, are competing for advertising revenue. Network C has two popular prime-time shows, a situation comedy and a police drama. Network R is planning to challenge one of these shows by scheduling an extravagant new science fiction adventure series at the same time. As a countermeasure, network C is planning to revamp one of its two shows by adding a famous rock star to the cast. Each network must make its decision well in advance of the new season, before knowing what the other is going to do. It is estimated that if network C revamps the comedy and network R challenges the comedy, C will lose 2 million dollars in advertising revenue to R. If C revamps the police drama and R challenges the comedy, C will lose 5 million to R. If C revamps the comedy and R challenges the police drama, C's loss will be 1 million dollars, and if C revamps the police drama and R challenges the police drama, neither network will gain or lose advertising revenue. Represent this competitive situation as a game in matrix form.

SOLUTION Letting network C be the column player and network R be the row player, and letting the entries denote the advertising revenue R gains from C, we get the following payoff matrix.

		Network C	
		Revamp comedy	Revamp drama
Network R	Challenge comedy	2	5
	Challenge drama	1	0

●

In the preceding two examples, the games involved two players and the payoff structure was such that for each pair of moves, one player's gain was the other's loss. Games with this property are called **two-person, zero-sum games** because the sum of the payoffs to each player for each combination of moves is always zero. In general, a game in which the sum of the payoffs to each player is the same for each combination of moves is called a **constant-sum game.** We now formulate strategies for playing such games.

Strategies

The entries of the payoff matrix are the payoffs to the row player. Hence, if the row player knew which column his opponent was going to choose, he would choose the row for which the corresponding entry was largest. Since he does not know what his opponent's move will be, he might decide to "play it safe" and adopt the following **conservative strategy.** He starts with the assumption that the worst will happen; namely, that no matter what row he chooses, his opponent will somehow manage to select the column that will result in the smallest payoff. To protect himself in this situation, the row player chooses the move whose smallest payoff is as large as possible. In particular, he examines each row, finds its minimum entry, and then chooses the row for which this minimum entry is maximum. For obvious reasons, this form of strategy is called the **maximin strategy.**

The column player, on the other hand, wants to maximize his payoff. For constant-sum games, this is the same as minimizing his opponent's payoff. Hence the column player's conservative strategy is to choose the column whose maximum entry is smallest. This is called the **minimax strategy.**

The maximin and minimax conservative strategies
For a game in matrix form, the row player's *maximin strategy* is to choose the row whose minimum entry is the maximum of the minimum row entries. The column player's *minimax strategy* is to choose the column whose maximum entry is the minimum of the maximum column entries.

The next example illustrates how to find these conservative strategies.

EXAMPLE 2.3

● For each of the following matrix games, find the maximin and minimax strategies for the row player and column player, respectively.

(a) $\begin{bmatrix} -1 & -3 & 5 \\ -2.5 & 1 & 3 \\ 4 & 0 & -2 \end{bmatrix}$ (b) $\begin{bmatrix} 8 & 4 \\ -4 & 0 \\ 5 & 3 \end{bmatrix}$

SOLUTION

(a) The minimum entry in the first row is -3, the minimum entry in the second row is -2.5 and the minimum entry in the third row is -2. Therefore, the maximin strategy for the row player is to choose the third row of the matrix.

The maximum entry in the first column is 4, the maximum entry in the second column is 1, and the maximum entry in the third column is 5. Therefore, the minimax strategy for the column player is to choose the second column. The procedure for selecting these strategies is illustrated in the following diagram.

				Row minima	
	−1	−3	5	−3	
	−2.5	1	3	−2.5	
	4	0	−2	(−2)	Maximum of the minima
Column maxima	4	(1)	5		
		Minimum of the maxima			

(b) Using reasoning similar to that in part (a), we find that the row player's maximin strategy is to select the first row while the column player's minimax strategy is to select the second column as illustrated in the following diagram.

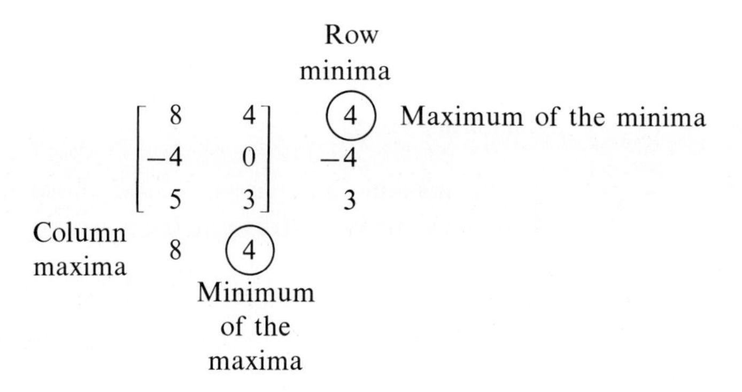

			Row minima	
	8	4	(4)	Maximum of the minima
	−4	0	−4	
	5	3	3	
Column maxima	8	(4)		
		Minimum of the maxima		

●

EXAMPLE 2.4 ● For the game of Example 2.2 in which two television networks were competing for advertising revenue, find the conservative strategy for each network.

SOLUTION From the payoff matrix

		Network C		
		Revamp comedy	Revamp drama	Row minima
Network R	Challenge comedy	2	5	(2)
	Challenge drama	1	0	0
	Column maxima	(2)	5	

we see that network C's conservative strategy is to revamp the comedy while network R's conservative strategy is to challenge the comedy. If each adopts its conservative strategy, R will gain 2 million dollars in advertising revenue from C. ●

Problems In each of the following problems, express the given competitive situation as a game in matrix form and determine each player's conservative strategy.

1 Miles and Tom play the following game. They simultaneously extend 1, 2, 3, or 4 fingers. If the sum of the number of fingers extended is odd, Miles receives that amount in dollars from Tom, and if the sum is even, Tom receives that amount from Miles.

2 Two competing sport shops, A and B, must decide how to price a popular brand of running shoe. Each of the shops will charge either \$25 or \$29 for the shoe. If A charges \$25 and B charges \$29, A will get 70 percent of the business for this item. If A charges \$29 and B charges \$25, A will get 40 percent of the business. If both charge the same amount, each will get 50 percent of the business.

3 Two candidates, R and C, are running for Congress in a district that includes a city and a suburban area. If both candidates campaign only in the suburbs, R will get 40 percent of the vote. If both campaign only in the city, R will get 45 percent of the vote. If R campaigns in the suburbs and C in the city, R will get 65 percent of the vote. And if R campaigns in the city and C in the suburbs, R will get 55 percent of the vote.

4 Two oil companies, Mainlander and Richmond, are planning to build gas stations at the intersection of State Street and University Avenue. Mainlander may build on the north side of University Avenue and Richmond on the

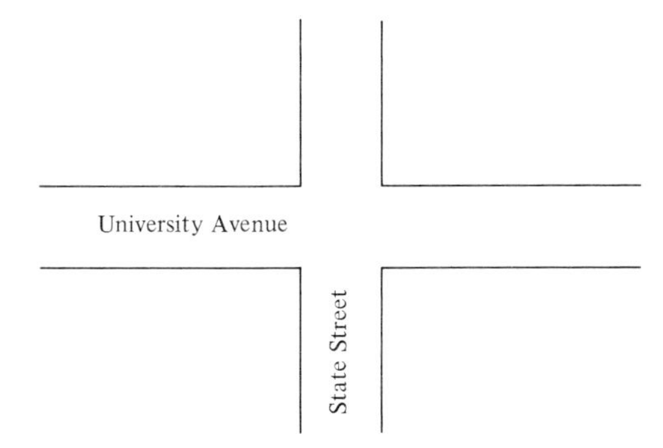

south side. It is estimated that if both build on the east side of State Street, Richmond will get 70 percent of the business and if both build on the west side, Richmond will get 40 percent of the business. If they build on opposite sides of State Street, each will get 50 percent of the business.

5 Player R wants to locate as far away from player C as possible, while player C wants to locate as close to player R as possible. C pays R \$1 for each meter between them. The five possible locations lie in a line as shown below, with the distance between adjacent locations equal to 1 meter.

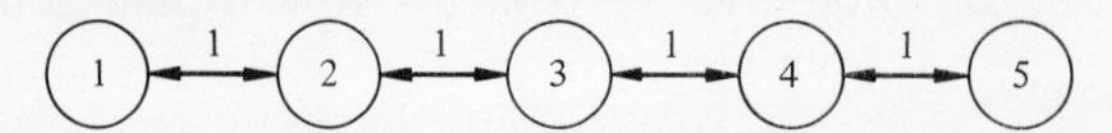

6 You are interested in two investments, A and B. If the economy goes into a state of inflation next year, investment A will go up in value by 10 percent and investment B will go down in value by 5 percent. If the economy is in a state of recession, investment A will go up in value by 5 percent and investment B will go up in value by 20 percent. (In this situation, you are the row player. Your opponent consists of the economic forces that determine the state of the economy and is not an active competitor.)

7 In a military tactics game, one player, in charge of 5 regiments, and his opponent, in charge of 4 regiments, are trying to gain control of 2 mountain outposts, A and B. Each must decide how many regiments to send to each outpost and neither knows in advance what the other is going to do. At each outpost, the player with the largest number of regiments captures the outpost as well as his opponent's outnumbered regiments. A player gets 1 point from his opponent for each captured outpost and 1 point for each captured regiment. If the players send the same number of regiments to an outpost, there is a draw and no points are awarded at that outpost.

8 Ruben and Catherine decide to modify the matching-fingers game of Example 2.1 by first writing down the total number of fingers they think will be shown and then extending 1 or 2 fingers. If both players guess the correct number of fingers, or if neither does, the game is a draw. If only one player guesses correctly, that player is the winner and receives $1 from the loser for each finger shown.

3 • SADDLE POINTS AND MIXED STRATEGIES

In some games, a player who knows that his opponent is using the conservative strategy can alter his strategy accordingly to increase his payoff. This is the case for the game in Example 2.3 (a) whose payoff matrix is reproduced below.

$$\begin{array}{r c c}
 & & \text{Row minima} \\
 & \begin{bmatrix} -1 & -3 & 5 \\ -2.5 & 1 & 3 \\ 4 & 0 & -2 \end{bmatrix} & \begin{array}{c} -3 \\ -2.5 \\ \boxed{-2} \end{array} \\
\text{Column maxima} & \begin{array}{ccc} 4 & \boxed{1} & 5 \end{array} &
\end{array}$$

In this game, the row player's conservative strategy is to choose row 3 and the column player's conservative strategy is to choose column 2. However, if the row player finds out that his opponent is playing conservatively, he can change his strategy and choose row 2, increasing his payoff from 0 to 1. Similarly, if the column player finds out that the row player is playing conservatively, he can change his strategy and choose column 3, decreasing his opponent's payoff from 0 to -2.

In other games, knowledge that one's opponent is playing conservatively cannot be used to advantage. This is the case for the game in Example 2.3 (b) whose payoff matrix is reproduced below.

$$\begin{array}{lcc} & \begin{bmatrix} 8 & 4 \\ -4 & 0 \\ 5 & 3 \end{bmatrix} & \begin{array}{c} \text{Row minima} \\ \textcircled{4} \\ -4 \\ 3 \end{array} \\ \text{Column maxima} & \begin{array}{cc} 8 & \textcircled{4} \end{array} & \end{array}$$

The row player's conservative strategy is to choose row 1 and the column player's conservative strategy is to choose column 2. In this case, knowledge that the column player will play conservatively and choose column 2 will not help the row player since his payoff of 4 if he plays conservatively cannot be increased. This is because the entry 4 in row 1, column 2 is the largest entry in its column. Similarly, knowledge that the row player will play conservatively will not help the column player. This is because the entry 4 is the smallest entry in its row.

Strictly determined games

Games in which neither player can benefit from the knowledge that his opponent is using the conservative strategy are said to be **strictly determined.** To tell if a game is strictly determined, look for an entry in the payoff matrix that is both the smallest entry in its row and the largest entry in its column. Such an entry is called a **saddle point.** The game is strictly determined if and only if its payoff matrix has a saddle point.

Saddle points and strictly determined games
An entry in a payoff matrix that is both the smallest entry in its row and the largest entry in its column is called a *saddle point.* A game whose payoff matrix has a saddle point is *strictly determined.* The conservative strategy in such a game is to choose the row or column containing the saddle point.

Here is an example.

EXAMPLE 3.1 ● Find the saddle points and conservative strategies for the following strictly determined games.

(a) $\begin{bmatrix} 8 & 0 & -2 \\ -6 & 4 & 1 \\ 4 & 3 & 2 \end{bmatrix}$ (b) $\begin{bmatrix} 3 & 5 & 4 & 3 \\ 2 & 4 & 0 & -2 \\ 1 & -6 & 5 & 0 \\ 0 & 2 & 6 & -1 \end{bmatrix}$

SOLUTION (a) The entry 2 in the third row and third column is the smallest entry in its row and the largest entry in its column and is, therefore, a saddle point. It follows that the row player's conservative strategy is to choose the third row, and the column player's conservative strategy is to choose the third column. (Verify this using the methods of the preceding section.)

(b) There are two saddle points, the two 3's in the first row. The row player's conservative strategy is to choose the first row, and the column player's is to choose either the first or fourth column.

Verify that in each of these games, knowledge that one player is playing conservatively will not affect the other player's decision to do the same. ●

Games that are not strictly determined

We have seen that for strictly determined games, each player can play conservatively without worrying that his opponent might detect his strategy. On the other hand, if a game that is not strictly determined is played over and over and one player plays conservatively every time, his opponent may notice this and alter his strategy to increase his payoff. In such games, therefore, it is a good idea to vary one's moves to avoid establishing a pattern that an opponent might detect. One way to do this is to choose moves in a random way, such as by tossing a coin. This is illustrated in the following example.

EXAMPLE 3.2 ● Consider the matching-fingers game of Example 2.1 which was represented by the following matrix:

	1 finger	2 fingers
1 finger	-1	1
2 fingers	1	-1

There is no saddle point, so this game is not strictly determined. Suppose the row player tosses a coin to decide which row to select. Compare the payoffs resulting from this strategy with those that would result if the row player used the conservative strategy of choosing row 1 each time and the column player found out.

SOLUTION If the row player chooses a row by tossing a coin, he will choose row 1 about half of the time and row 2 about half of the time, and the column player can't find out the move in advance. No matter which column is selected by the column player, the row player will win \$1 about half of the time and lose \$1 about half of the time, for an average payoff of zero. (Do you see why?)

On the other hand, if the row player chooses row 1 every time and the column player detects the pattern, he will choose column 1 every time and the row player will lose $1 on each play. ●

Strategies for repeated play

In the preceding example, the row player decided to play each row half of the time. This decision is an example of a **strategy** for repeated play of the game. In general, a **mixed strategy** is a decision to play more than one row (or column) and to play each a specified fraction of the time (or, equivalently, with a specified probability). The decision of the row player in Example 3.2 to play each row half of the time was a mixed strategy. A decision to play a particular row (or column) 100 percent of the time is called a **pure strategy.** For example, the decision always to use the conservative strategy when playing a strictly determined game is a pure strategy.

It is convenient to represent mixed strategies in matrix form. For example, in a 2×2 matrix game, the row player's decision to select row 1 with probability p_1 and row 2 with probability p_2 will be represented by the row vector

$$P = [p_1 \quad p_2]$$

and the column player's decision to select column 1 with probability q_1 and column 2 with probability q_2 will be represented by the column vector

$$Q = \begin{bmatrix} q_1 \\ q_2 \end{bmatrix}$$

Expected payoff

To decide which mixed strategy is best in a game that is not strictly determined, we will need to compute the **expected payoff** resulting from the use of a mixed strategy P by the row player and a mixed strategy Q by the column player. For example, suppose the payoff matrix is

$$A = \begin{bmatrix} a_{11} & a_{12} \\ a_{21} & a_{22} \end{bmatrix}$$

and the row and column players' mixed strategies are

$$P = [p_1 \quad p_2] \qquad \text{and} \qquad Q = \begin{bmatrix} q_1 \\ q_2 \end{bmatrix}$$

respectively. Each time the game is played, there are four possible outcomes that are summarized, along with their probabilities, in the accompanying table.

Row and column chosen	Probability	Payoff
1,1	p_1q_1	a_{11}
1,2	p_1q_2	a_{12}
2,1	p_2q_1	a_{21}
2,2	p_2q_2	a_{22}

The expected payoff E is simply the expected value of the payoffs listed in the third column of the table and is computed by multiplying each payoff by its probability and adding. Thus,

$$E = p_1 q_1 a_{11} + p_1 q_2 a_{12} + p_2 q_1 a_{21} + p_2 q_2 a_{22}$$

Notice that this sum is precisely the matrix product PAQ. (Check this.) This relationship between E and the product of P, A, and Q holds in general and can be stated as follows.

How to compute the expected payoff
The row player's expected payoff E is the matrix product

$$E = PAQ$$

where A is the payoff matrix for the game, P the row vector representing the row player's mixed strategy, and Q the column vector representing the column player's mixed strategy.

EXAMPLE 3.3 ● The payoff matrix for a certain game is

$$A = \begin{bmatrix} 4 & 2 \\ -1 & 4 \\ 5 & 3 \end{bmatrix}$$

Suppose the row player chooses row 1 with probability 0.25, row 2 with probability 0.5, and row 3 with probability 0.25, and the column player chooses column 1 with probability 0.2 and column 2 with probability 0.8. Using matrix multiplication, find the row player's expected payoff E.

SOLUTION

$$E = PAQ = [0.25 \quad 0.5 \quad 0.25] \begin{bmatrix} 4 & 2 \\ -1 & 4 \\ 5 & 3 \end{bmatrix} \begin{bmatrix} 0.2 \\ 0.8 \end{bmatrix}$$

$$= [1.75 \quad 3.25] \begin{bmatrix} 0.2 \\ 0.8 \end{bmatrix} = 2.95$$

●

Optimal strategies The concept of an **optimal strategy** can now be defined using the minimax and maximin criteria of Section 2. Consider the row player, for example. For each strategy, the row player seeks to determine the minimum expected payoff that will result if that strategy is used. The strategy whose minimum expected payoff is largest is said to be the optimal strategy. Similarly, the column player's optimal strategy is the one for which the row player's maximum expected payoff is smallest.

It can be shown that, like the conservative strategies for strictly determined games, optimal strategies in general have the property that knowledge that one's opponent is following an optimal strategy cannot be used to one's advantage. This is a consequence of a famous theorem known as the **minimax theorem** and accounts for the importance of optimal strategies in game theory.

For strictly determined games, the optimal strategies are the conservative pure strategies discussed in Section 2. For example, the row player's optimal strategy in such a game is to choose the row containing the saddle point with probability 1 and all the other rows with probability 0.

For games that are not strictly determined, the task of finding an optimal strategy is considerably more difficult. This is because there are infinitely many possible mixed strategies for each player, so one cannot methodically compare each strategy with all the opponent's possible counterstrategies and then choose the best one. The problem of finding optimal mixed strategies has been a major one for researchers ever since game theory was introduced. It turns out that finding optimal mixed strategies is equivalent to solving a related linear programming problem. Recent advances in computer technology have helped considerably in efforts to find efficient computational techniques.

Although finding optimal strategies for games with large payoff matrices is a difficult problem that requires use of a computer, there is a simple formula you can use for 2×2 games.

How to find optimal strategies for 2×2 games

If

$$A = \begin{bmatrix} a & b \\ c & d \end{bmatrix}$$

is the payoff matrix for a game that is not strictly determined, the optimal strategy for the row player is $P = [p_1 \quad p_2]$ where

$$p_1 = \frac{d - c}{a + d - b - c} \qquad \text{and} \qquad p_2 = 1 - p_1$$

and the optimal strategy for the column player is $Q = \begin{bmatrix} q_1 \\ q_2 \end{bmatrix}$ where

$$q_1 = \frac{d - b}{a + d - b - c} \qquad \text{and} \qquad q_2 = 1 - q_1$$

The value of a game

The row player's expected payoff if each player uses his optimal strategy is called the **value** of the game. If the value of a game is positive, the game is favorable to the row player, since the row player expects to receive at least this amount, even if the column player knows that the optimal strategy is being used. If the value of a game is negative, the game is favorable to the column player. If the value is zero, the game favors neither player and is said to be **fair.**

Here is the formula for the value of a 2×2 game.

How to compute the value of a 2×2 game
If

$$A = \begin{bmatrix} a & b \\ c & d \end{bmatrix}$$

is the payoff matrix for a game that is not strictly determined, the value of the game is

$$E = \frac{ad - bc}{a + d - b - c}$$

To derive this formula for the value of a 2×2 game, simply compute the matrix product PAQ where P and Q are the optimal strategies for the row and column player, respectively.

The next example illustrates the use of the formulas for the optimal strategies and for the value of a 2×2 game.

EXAMPLE 3.4 ● For the matching-fingers game of Example 2.1, find each player's optimal strategy and the value of the game.

SOLUTION The payoff matrix is

	1 finger	2 fingers
1 finger	−1	1
2 fingers	1	−1

Since there is no saddle point, the game is not strictly determined. Using the formulas for the optimal mixed strategies with $a = -1$, $b = 1$, $c = 1$, and $d = -1$, we find that the row player's optimal strategy is $P = [p_1 \quad p_2]$ where

$$p_1 = \frac{d - c}{a + d - b - c} = \frac{-1 - 1}{-1 - 1 - 1 - 1} = 0.5$$

and

$$p_2 = 1 - p_1 = 0.5$$

and the column player's optimal strategy is $Q = \begin{bmatrix} q_1 \\ q_2 \end{bmatrix}$ where

$$q_1 = \frac{d - b}{a + d - b - c} = \frac{-1 - 1}{-1 - 1 - 1 - 1} = 0.5$$

and

$$q_2 = 1 - q_1 = 0.5$$

These strategies are intuitively reasonable, since the payoffs in this game are symmetric. (The row player can implement his strategy by tossing an evenly balanced coin, choosing row 1 if the coin lands heads and row 2 if the coin lands tails. The column player can do the same thing in order to choose a column.)

The value of the game is

$$E = \frac{ad - bc}{a + d - b - c} = 0$$

This is also intuitively reasonable. Since both players have equal opportunities, the game should be fair and favor neither player. ●

The next example shows how a mixed strategy can be used to determine an investment allocation.

EXAMPLE 3.5 ● You have $500 to invest and can't make up your mind between two options, a real-estate investment and a stock-market investment. The real-estate investment will be profitable if the economy is in a state of inflation next year, while the stock-market investment will be profitable if the economy is in a state of recession. The following payoff matrix shows the estimated percentage increase or decrease in the value of each investment for each state of the economy.

	Inflation	Recession
Real-estate investment	25	0
Stock-market investment	−10	30

Think of this situation as a matrix game in which you are the row player, and find your optimal strategy and the value of the game. Interpret this strategy as an investment allocation.

SOLUTION The row player's optimal strategy is $P = [p_1 \quad p_2]$, where

$$p_1 = \frac{d - c}{a + d - b - c} = \frac{30 - (-10)}{25 + 30 - 0 - (-10)} = 0.62$$

$$p_2 = 1 - p_1 = 0.38$$

The value of this game is

$$E = \frac{ad - bc}{a + d - b - c} = 11.54$$

The probabilities p_1 and p_2 can be interpreted as fractions of the total investment money that the investor should invest in each option. If 62 percent of the investment money is invested in real estate and 38 percent is invested in the stock market, the investor can expect a profit of $E = 11.54$ percent, no matter which state the economy will be in. ●

Problems

In Problems 1 through 4, decide if the matrix game is strictly determined and, if it is, find the saddle points.

1 $\begin{bmatrix} 1 & 2 \\ 0 & 1 \end{bmatrix}$

2 $\begin{bmatrix} 4 & 3 & 0 \\ 5 & 7 & 8 \\ 3 & -1 & 9 \end{bmatrix}$

3 $\begin{bmatrix} 1 & 0 \\ 0 & 1 \end{bmatrix}$

4 $\begin{bmatrix} -1 & 0 & 0 & -1 \\ -2 & 4 & 5 & -2 \\ -3 & 5 & 4 & -3 \\ -1 & 0 & 0 & -1 \end{bmatrix}$

5 For each of the following games, fill in a number for x that will be a saddle point.

(a) $\begin{bmatrix} 5 & 1 & 15 & -2 \\ 3 & 8 & 7 & 6 \\ x & 10 & 7 & 8 \end{bmatrix}$ (b) $\begin{bmatrix} 14 & x \\ 6 & 12 \end{bmatrix}$

6 Determine which of the games in the problems for Section 2 are strictly determined.

In Problems 7 through 9, find the expected payoff when the players use the indicated mixed strategies.

7 $A = \begin{bmatrix} 2 & 0 \\ -1 & 1 \end{bmatrix}$ $P = [\frac{1}{2} \;\; \frac{1}{2}]$ $Q = \begin{bmatrix} \frac{1}{3} \\ \frac{2}{3} \end{bmatrix}$

8 $A = \begin{bmatrix} 1 & 0 & 6 \\ 0 & 2 & 0 \\ 5 & 0 & 3 \end{bmatrix}$ $P = [\frac{2}{5} \;\; \frac{1}{5} \;\; \frac{2}{5}]$ $Q = \begin{bmatrix} \frac{1}{3} \\ \frac{1}{3} \\ \frac{1}{3} \end{bmatrix}$

9 $A = \begin{bmatrix} 5 & 1 & 5 \\ 4 & 2 & 4 \\ 8 & 3 & 7 \end{bmatrix}$ $P = [0.3 \;\; 0.4 \;\; 0.3]$ $Q = \begin{bmatrix} 0.25 \\ 0.5 \\ 0.25 \end{bmatrix}$

10 In the game with payoff matrix

$$A = \begin{bmatrix} 4 & -3 \\ 0 & 5 \end{bmatrix}$$

the row player will use the strategy $P = [0.25 \;\; 0.75]$. Which is the better counterstrategy for the column player?

$$Q_1 = \begin{bmatrix} 0.8 \\ 0.2 \end{bmatrix} \quad \text{or} \quad Q_2 = \begin{bmatrix} 0.2 \\ 0.8 \end{bmatrix}$$

In Problems 11 through 14, find the optimal strategies for each player and the value E of the given matrix game.

11 $\begin{bmatrix} 5 & 0 \\ 0 & 3 \end{bmatrix}$

12 $\begin{bmatrix} -4 & -3 \\ -2 & -8 \end{bmatrix}$

13 $\begin{bmatrix} 21 & 8 \\ 4 & 16 \end{bmatrix}$

14 $\begin{bmatrix} 1 & 4 \\ 4 & 1 \end{bmatrix}$

15 Using the formulas for the optimal strategies, verify the formula for the value of a 2×2 game that is not strictly determined.

16 You are interested in two investments, A and B. If the economy goes into a state of inflation next year, investment A will go up in value by 10 percent and investment B will go down in value by 5 percent. If the economy is in a state of recession, investment A will go up in value by 5 percent and investment B will go up in value by 20 percent.

(a) Think of this situation as a game in which you are the row player and find your optimal strategy $P = [p_1 \quad p_2]$ and the value E of the game.

(b) Suppose you have a certain amount of money and you invest p_1 of it in investment A and p_2 of it in investment B, where p_1 and p_2 are the fractions in your optimal strategy in part (a). What is your expected percentage profit in this case?

17 The Warriors are going to play the Lakers in a basketball game. If you place your bet in Oakland you can get 3 to 2 odds for a bet on the Lakers and if you call your bookie in Los Angeles you can get 1 to 1 odds for a bet on the Warriors. This information is displayed in the following payoff matrix in which the entries represent the payoffs in dollars for each \$1 bet.

	Lakers win	Warriors win
Bet on Lakers	1.5	−1
Bet on Warriors	−1	1

(a) Think of this situation as a game in which you are the row player and find your optimal mixed strategy $P = [p_1 \quad p_2]$ and the value E of the game.

(b) Suppose you have k dollars to bet and you bet p_1 of it on the Lakers and p_2 of it on the Warriors, where p_1 and p_2 are the fractions in your optimal strategy in part (a). By at least how much can you expect to come out ahead, no matter who wins the game?

4 • INVESTMENT MODELS

In many decision problems, there is more than one method that can be used to reach an optimal decision. Since different methods can lead to different decisions, the decision maker must determine which method is the most suitable for a given problem. In this section we shall see how two of the techniques we have just studied can be applied to the same investment problem. In this case the choice of the "correct" technique depends in part on the type of information available to the decision maker.

In the first example we shall analyze an investment problem as if it were a matrix game. (This was Problem 16 in the preceding section.)

EXAMPLE 4.1 ● There are two possible investments, A and B, and two possible states of the economy, "inflation" and "recession." The estimated percentage increases in the value of the investments over the coming year for each possible state of the economy are shown in the following payoff matrix.

$$\begin{array}{lcc} & \text{Inflation} & \text{Recession} \\ \text{Invest in A} & \left[\begin{array}{c} 10 \\ -5 \end{array}\right. & \left.\begin{array}{c} 5 \\ 20 \end{array}\right] \\ \text{Invest in B} & & \end{array}$$

(a) Find the investor's optimal strategy $P = [p_1 \quad p_2]$.
(b) Find the value E of this game.
(c) What return can the investor expect if the optimal strategy is followed?

SOLUTION (a) To find the optimal strategy $P = [p_1 \quad p_2]$ of the investor (or row player) we use the formulas

$$p_1 = \frac{d - c}{a + d - b - c} \qquad \text{and} \qquad p_2 = 1 - p_1$$

with $a = 10$, $b = 5$, $c = -5$, and $d = 20$ and get

$$p_1 = 0.833 \qquad \text{and} \qquad p_2 = 0.167$$

(b) The value of the game is

$$E = \frac{ad - bc}{a + d - b - c} = 7.5$$

(c) If the investor follows the optimal strategy and invests 83.3 percent of his money in A and 16.7 percent in B, he can expect the value of his investment to increase by $E = 7.5$ percent, no matter what state the economy is in. ●

In the preceding example, the investor chose the strategy for which the minimum expected payoff is greatest. This maximin strategy is a conservative strategy designed to protect the investor if the state of the economy turns out to be the one less favorable to his investment. It is a reasonable strategy if the investor has no knowledge about the future state of the economy. Since the economy is not really a game player trying to defeat the investor, the investor may be able to obtain information about the future of the economy. If so, other strategies are usually preferable. In the next example we shall examine one such strategy.

EXAMPLE 4.2 ● We shall consider the same investment problem as in the preceding example. Suppose, however, that the investor has studied the economy and believes that the probability is 0.6 that the economy will be in a state of inflation next year, and that the probability is 0.4 that the economy will be in a state of recession. Based on these probabilities, which investment allocation gives the investor the highest expected gain?

SOLUTION We wish to find the fractions p_1 and $p_2 = 1 - p_1$ that represent the fractions of money that should be invested in A and B, respectively, in order to maximize the investor's expected gain. If the economy goes into a state of inflation, the investor's percentage gain will be

$$10p_1 - 5p_2$$

and if the economy goes into a state of recession, the investor's percentage gain will be

$$5p_1 + 20p_2$$

Since $P(\text{inflation}) = 0.6$ and $P(\text{recession}) = 0.4$ we find that the investor's overall expected gain will be

$$E = 0.6(10p_1 - 5p_2) + 0.4(5p_1 + 20p_2)$$

Since $p_2 = 1 - p_1$, we can rewrite this expression as

$$E = 0.6[10p_1 - 5(1 - p_1)] + 0.4[5p_1 + 20(1 - p_1)] = 3p_1 + 5$$

We wish to choose the value of p_1 between 0 and 1 that maximizes this expression. Clearly, this value is $p_1 = 1$. In other words, the investor should invest all his money in investment A, for an expected percentage gain of

$$E = 3(1) + 5 = 8$$

Notice that this is greater than the expected percentage gain of 7.5 generated by the maximin strategy in Example 4.1. ●

In the preceding example, the investor's decision depended on the probabilities assigned to the states of the economy. The probability of inflation was high and so investment A was chosen. If the probability of recession had been high, the investor would have chosen investment B. The wisdom of these decisions depends upon the accuracy of the investor's probabilities.

Probabilities initially assigned to the possible "states of nature" (in this case, the states of the economy), are called *prior probabilities.* They can be obtained from past data or may merely represent the opinions of the decision maker. If the decision maker obtains additional information, it can be used to update the prior probabilities by means of Bayes' formula. The resulting probabilities, called *posterior probabilities,* can then be used by the decision maker to maximize the expected gain. This bayesian approach was studied in Section 1. In the next example we shall apply bayesian methods to our investment problem.

EXAMPLE 4.3 ● Suppose that the prior probabilities of the states of the economy are the same as in Example 4.2; that is,

$$P(\text{inflation}) = 0.6 \qquad P(\text{recession}) = 0.4$$

Suppose that in order to gain more information, the investor looks at a stock market indicator that will either be "↑" or "↓". Finally, suppose that the state of the economy for the next year is related to the market indicator by the following conditional probabilities:

$$P(\uparrow|\text{inflation}) = 0.7 \qquad P(\uparrow|\text{recession}) = 0.2$$

$$P(\downarrow|\text{inflation}) = 0.3 \qquad P(\downarrow|\text{recession}) = 0.8$$

(a) Find the posterior probabilities of the states of the economy for each possible position of the market indicator. That is, find

$$P(\text{inflation}|\uparrow) \qquad P(\text{inflation}|\downarrow)$$

$$P(\text{recession}|\uparrow) \qquad P(\text{recession}|\downarrow)$$

(b) Using the posterior probabilities, find the investor's best decision for each possible position of the market indicator.

SOLUTION (a) We shall use Bayes' formula to compute the posterior probabilities as follows:

$$P(\text{inflation}|\uparrow) = \frac{P(\uparrow|\text{inflation})P(\text{inflation})}{P(\uparrow|\text{inflation})P(\text{inflation}) + P(\uparrow|\text{recession})P(\text{recession})}$$

$$= \frac{0.7(0.6)}{0.7(0.6) + 0.2(0.4)} = 0.84$$

$$P(\text{recession}|\uparrow) = 1 - P(\text{inflation}|\uparrow) = 0.16$$

$$P(\text{inflation}|\downarrow) = \frac{P(\downarrow|\text{inflation})P(\text{inflation})}{P(\downarrow|\text{inflation})P(\text{inflation}) + P(\downarrow|\text{recession})P(\text{recession})}$$

$$= \frac{0.3(0.6)}{0.3(0.6) + 0.8(0.4)} = 0.36$$

$$P(\text{recession} \mid \downarrow) = 1 - P(\text{inflation} \mid \downarrow) = 0.64$$

(b) Referring to the payoff matrix

$$\begin{array}{l} \\ \text{Invest in A} \\ \text{Invest in B} \end{array} \begin{array}{c} \text{Inflation} \quad \text{Recession} \\ \begin{bmatrix} 10 & 5 \\ -5 & 20 \end{bmatrix} \end{array}$$

and using the posterior probabilities, we find that if the market indicator is $\uparrow$, the investor's expected gain if he invests in A is

$$E = 10P(\text{inflation} \mid \uparrow) + 5P(\text{recession} \mid \uparrow) = 10(0.84) + 5(0.16) = 9.2$$

and his expected gain if he invests in B is

$$E = -5P(\text{inflation} \mid \uparrow) + 20P(\text{recession} \mid \uparrow) = -5(0.84) + 20(0.16) = -1$$

If the indicator is $\downarrow$, the investor's expected gain if he invests in A is

$$E = 10P(\text{inflation} \mid \downarrow) + 5P(\text{recession} \mid \downarrow) = 10(0.36) + 5(0.64) = 6.8$$

and his expected gain if he invests in B is

$$E = -5P(\text{inflation} \mid \downarrow) + 20P(\text{recession} \mid \downarrow) = -5(0.36) + 20(0.64) = 11$$

Therefore, if the indicator is $\uparrow$, the investor should invest in A, for an expected gain of 9.2 percent, and if the indicator is $\downarrow$, the investor should invest in B for an expected gain of 11 percent.

Using reasoning similar to the reasoning in the preceding example, it can be shown that with this decision procedure, the investor cannot increase his expected gain by allocating a fraction of his money to each possible investment. ●

We have studied two investment models, one based on the game theory or maximin method and the other on the bayesian method. These methods led to different decisions for the same problem. The maximin method gave rise to a mixed strategy in which the investor allocated a fraction of his money to each investment, while the bayesian method gave rise to a strategy in which the investor allocated all of his money to one investment. Each method has advantages and disadvantages. For example, the maximin method protects the investor from unfavorable states of the economy which could produce large losses, but does not make use of information the investor might have about future states of the economy. On the other hand, although the bayesian method provides an efficient procedure for incorporating such information into the decision-making structure, if the information is unreliable or if the investor is unlucky, a substantial loss can occur. There are more complicated procedures that incorporate the best features of each of these methods, some of which introduce nonmonetary criteria into the decision-making structure.

Problems

1 An investor has a choice of two investments, A and B. The percentage gain of each investment over the next year depends on whether the economy is "up" or "down." This investment information is displayed in the following payoff matrix.

$$\begin{array}{lcc} & \text{Economy up} & \text{Economy down} \\ \text{Invest in A} & -5 & 20 \\ \text{Invest in B} & 18 & 0 \end{array}$$

(a) Find the best investment allocation if the investor wishes to maximize the minimum expected gain. That is, find the row player's optimal strategy in the corresponding matrix game.

(b) Find the percentage gain the investor is assured of when using this optimal strategy. That is, find the value of the corresponding matrix game.

2 Suppose the investor in Problem 1 thinks that the (prior) probability is 0.4 that the economy will be "up" in the coming year and the probablity is 0.6 that the economy will be "down." Based on these probabilities, find the investment that maximizes the investor's expected gain.

3 Suppose the investor in Problems 1 and 2 does more research on the economy and finds that the state of the economy is related to this year's weather in the following way:

$$P(\text{rainy}\,|\,\text{up}) = 0.3 \qquad P(\text{rainy}\,|\,\text{down}) = 0.5$$

$$P(\text{dry}\,|\,\text{up}) = 0.7 \qquad P(\text{dry}\,|\,\text{down}) = 0.5$$

Based on these probabilities and the prior probabilities given in Problem 2, find the investor's best investment for each possible state of the weather. That is, find the bayesian solution to the investment problem.

4 There are 12 boxes of type A, each containing 3 red balls and 1 green ball, and 8 boxes of type B, each containing 2 red balls and 2 green balls. One of these boxes is selected at random and you must guess which type of box is selected. Your payoff is determined by the matrix

	Type A is selected	Type B is selected
Guess type A	6	−5
Guess type B	−3	9

(a) In Problem 8 of Section 1 of this chapter, you were asked to use bayesian methods to find your best guessing strategy assuming you are allowed to select one ball at random from the chosen box before making your guess. Do this problem if you have not already done so.
(b) Now think of this situation as a matrix game in which you are the row player and find your optimal mixed strategy. What is your expected payoff if you use this strategy?
(c) Which decision method is more appropriate in this situation, the bayesian method of part (a) or the game-theoretic method of part (b)? Explain.

5 In Example 3.4 of this chapter, we found optimal mixed strategies for each player in the matching-fingers game with payoff matrix

	1 finger	2 fingers
1 finger	−1	1
2 fingers	1	−1

(a) Suppose the row player thinks the (prior) probability is $\frac{2}{3}$ that the column player will select column 1 and the probability is $\frac{1}{3}$ that the column player will select column 2. Based on these probabilities, find the move that maximizes the row player's expected payoff.
(b) After playing this game a number of times, the row player observes that the column player has a nervous habit of touching his chin before making some of his moves, and that this is related to the subsequent move by the following conditional probabilities.

$$P(\text{touch} \mid \text{column 1}) = \tfrac{1}{5} \qquad \text{and} \qquad P(\text{touch} \mid \text{column 2}) = \tfrac{3}{4}$$

Using these probabilities and the prior probabilities given in part (a), find the row player's best strategy.
(c) Which strategy, the game-theoretic strategy used in Example 3.4 or the bayesian strategy in part (b), do you think is more appropriate for the row player in repeated play of this game? Explain.

6 In Problem 17 of the preceding section you were asked to find your optimal betting strategy if you could bet on the Lakers in Los Angeles and the Warriors in Oakland at the odds represented in the following payoff matrix.

	Lakers win	Warriors win
Bet on Lakers	1.5	−1
Bet on Warriors	−1	1

(a) Do Problem 17 if you have not already done so.

(b) Suppose you think the (prior) probability is $\frac{1}{4}$ that the Lakers will win and the (prior) probability is $\frac{3}{4}$ that the Warriors will win. Based on these probabilities, find the bet that maximizes your expected gain.

(c) Which betting method involves more risk, the one in part (a) or the one in part (b)? Which method yields the larger expected gain? Which method would you use?

THE MATHEMATICS OF FINANCE

1 • SIMPLE AND COMPOUND INTEREST

Interest is the fee charged for the privilege of borrowing money. If you get a loan from a bank or other financial institution, you are charged interest for the privilege of using the institution's money. If you invest money in a savings account, the bank pays you interest for the privilege of using *your* money. If you buy an item like a car or stereo "on time," you must pay interest for the privilege of postponing full payment until after you have taken possession of the item.

Simple interest

The total amount of money borrowed (or invested) is called the **principal.** Interest that is computed on the principal alone is called **simple interest.** For example, suppose you have to pay simple interest at a rate of 5 percent per year on a 2-year loan of \$300. The interest charged will be $300(0.05)(2) = \$30$, and the total amount owed at the end of the 2-year period will be $300 + 30 = \$330$.

Here is the general formula for computing simple interest.

Simple interest
If P denotes the principal, r the yearly interest rate expressed in decimal form, and n the number of years for which the principal is borrowed, then the total amount A owed at the end of n years is given by the formula

$$A = P + Prn = P(1 + rn)$$

The following example illustrates the use of the simple-interest formula.

EXAMPLE 1.1 ● Suppose you borrow \$2,000 from a bank at a simple-interest rate of 10 percent per year.

(a) How much will you owe at the end of 3 years?
(b) At the end of how many years will you owe \$2,100?

SOLUTION (a) Applying the simple interest formula, with $P = 2{,}000$, $r = 0.1$, and $n = 3$, we get

$$A = P(1 + rn) = 2{,}000[1 + 0.1(3)] = 2{,}000(1.3) = \$2{,}600$$

(b) We wish to find the value of n for which $A = 2{,}100$ when $P = 2{,}000$ and $r = 0.1$. Substituting these values into the simple-interest formula $A = P(1 + rn)$ we get

$$2{,}100 = 2{,}000(1 + 0.1n) \quad \text{or} \quad 2{,}100 = 2{,}000 + 200n$$

which we solve for n to get

$$n = \frac{100}{200} = 0.5$$

Since n represents the number of years, we conclude that you will owe \$2,100 after half a year or 6 months. ●

Compound interest

Interest that is computed on the previously accumulated interest as well as on the principal is called **compound interest.**

For example, suppose you invest \$500 in a savings account at an annual interest rate of 8 percent compounded quarterly (4 times a year). This means that $\frac{8}{4} = 2$ percent of the existing balance is added to your account every 3 months. Hence, at the end of 3 months, your balance will be

$$500 + 500(0.02) = 500(1.02) = \$510$$

At the end of 6 months your balance will be

$$510 + 510(0.02) = 510(1.02) = \$520.20$$

and so on.

In general, if the annual interest rate is r and interest is compounded k times per year, the year is divided into k equal interest periods and the interest rate during each is $i = r/k$. Thus, after the first interest period, the new amount A in the account will equal the principal plus the first interest payment:

$$A = P + Pi = P(1 + i)$$

The second interest payment will be made on the new balance, so the balance after the second interest period will be

$$A = P(1 + i) + P(1 + i)(i) = P(1 + i)^2$$

Continuing in this way, we find the balance after n interest periods to be

$$A = P(1 + i)^n$$

Compound interest

If interest is compounded k times per year at an annual rate r, and if P is the principal, then the amount A owed (or accumulated) after n interest periods is given by the formula

$$A = P(1 + i)^n$$

where $i = r/k$.

The following example illustrates the use of the compound-interest formula.

EXAMPLE 1.2 ● Suppose you invest in a savings account that pays interest at an annual rate of 8 percent compounded quarterly.

(a) If you start with a principal of $1,200, what will your balance be at the end of 3 years?
(b) If your balance at the end of 1 year is $800, what was your original investment?

SOLUTION (a) Applying the formula $A = P(1 + i)^n$ with $P = 1{,}200$, $n = 3(4) = 12$, and $i = 0.08/4 = 0.02$, we get

$$A = 1{,}200(1.02)^{12} = \$1{,}521.89$$

(b) At the end of 1 year there will have been 4 interest periods. If your balance at that time is $800, the compound interest formula yields

$$800 = P(1.02)^4$$

Solving for P we get

$$P = \frac{800}{(1.02)^4} = \$739.08$$

●

The more often interest is compounded, the faster your balance will grow. Banks sometimes compete on the basis of how frequently they compound interest. Several cases are compared in the next example.

EXAMPLE 1.3 ● Suppose you invest $2,000 at an annual interest rate of 6 percent. Find your balance at the end of 1 year if interest is compounded

(a) yearly
(b) semiannually
(c) quarterly
(d) monthly

SOLUTION In each case we use the formula $A = P(1 + i)^n$ with $P = 2{,}000$ and appropriate values for i and n.

(a) In this case, $i = 0.06$ and $n = 1$. Hence,

$$A = 2{,}000(1.06) = \$2{,}120.00$$

(b) In this case, $i = 0.06/2 = 0.03$ and $n = 2$. Hence,

$$A = 2{,}000(1.03)^2 = \$2{,}121.80$$

(c) This time, $i = 0.06/4 = 0.015$ and $n = 4$, so

$$A = 2{,}000(1.015)^4 = \$2{,}122.73$$

(d) This time, $i = 0.06/12 = 0.005$ and $n = 12$, so

$$A = 2{,}000(1.005)^{12} = \$2{,}123.36$$

Effective interest rate When interest is compounded with a certain frequency, it is often important to know the equivalent simple annual interest rate. This is known as the **effective interest rate** and can be easily obtained from the compound-interest formula.

Suppose interest is compounded k times per year at the annual rate r. Then the balance at the end of one year is

$$A = P\left(1 + \frac{r}{k}\right)^k$$

On the other hand, if x is the effective interest rate, the corresponding balance at the end of one year is

$$A = P(1 + x)$$

Equating the two expressions for A we get

$$P\left(1 + \frac{r}{k}\right)^k = P(1 + x) \qquad \text{or} \qquad x = \left(1 + \frac{r}{k}\right)^k - 1$$

Effective interest rate
If interest is compounded k times per year at an annual rate r, the effective interest rate is the simple annual interest rate that yields the same interest after one year. It is given by the formula

$$\text{Effective rate} = \left(1 + \frac{r}{k}\right)^k - 1$$

The next example illustrates the use of the effective-interest-rate formula.

EXAMPLE 1.4 • Suppose a certain investment yields interest at the annual rate of 10 percent, compounded quarterly. What is the effective interest rate?

SOLUTION We use the formula for the effective interest rate with $k = 4$ and $r = 0.10$ to get

$$\text{Effective rate} = \left(1 + \frac{0.10}{4}\right)^4 - 1 = 0.104$$

Hence the effective interest rate is 10.4 percent.

Present value In many situations we are interested in knowing the amount of money we must invest now at a fixed compound interest rate in order to obtain a desired balance at some time in the future. This amount is called the **present value** of the desired balance.

To obtain a formula for the present value, we start with the compound-interest formula

$$A = P(1 + i)^n$$

and solve for P, getting

$$P = \frac{A}{(1 + i)^n}$$

Present value

The *present value* of A dollars payable n interest periods from now is the amount P that must be invested today so that it will be worth A dollars at the end of the n periods. If interest is compounded k times per year at an annual rate r, the present value is given by the formula

$$P = \frac{A}{(1 + i)^n}$$

where $i = r/k$.

The following example illustrates the use of the present-value formula.

EXAMPLE 1.5 ● How much should be invested at 5 percent interest, compounded quarterly, if the desired balance after 10 years is $5,000?

SOLUTION We apply the formula for present value with $A = 5{,}000$, $i = 0.05/4 = 0.0125$, and $n = 40$ (4 periods a year for 10 years) to get

$$P = \frac{5{,}000}{(1.0125)^{40}} = \$3{,}042.07$$

That is, $3,042.07 invested today at 5 percent interest, compounded quarterly, will yield a balance of $5,000 after 10 years. ●

Problems

1 Suppose you obtain a loan of $500 for 1 year at a simple interest rate of 10 percent. How much will you have to pay back at the end of the year?

2 A certain investment yields simple interest at an annual rate of 15 percent. If the investment (principal plus interest) is worth $1,544 after 4 years, how much was invested originally?

3 The simple annual interest on a \$3,000 loan is 9 percent. Find the amount of interest charged if the loan is for a period of
(a) 6 months (b) 1 year (c) 3 years

4 If an investment that pays 6 percent annual simple interest is worth \$817.50 after 1.5 years, what was the principal?

5 A principal of \$900 was invested in an account that pays simple interest at an annual rate of 4 percent. If the accumulated interest is now \$180, how long ago was the money invested?

6 Find the annual rate of simple interest for which \$900 will grow to \$1,044 in 2 years.

7 Mr. Jones borrowed \$500 for 1 month from a loan company. A month later he paid off the loan with a check for \$600. What was the annual simple interest rate charged by the company?

8 Herman borrowed \$100 from Jane and paid off the loan 2 weeks later with a check for \$110. Find the annual simple interest rate charged by Jane. (Assume there are exactly 52 weeks in a year.)

9 Find a general formula for the simple interest rate r in terms of the amount of the loan, the total interest, and the length of the loan.

10 Leon invested \$500 for 6 months at an annual simple interest rate of 10 percent.
(a) How much interest did Leon earn?
(b) By what percentage did Leon's investment grow?

11 (a) How long will it take a \$100 investment to double at an annual simple interest rate of 0.05?
(b) How long will it take an investment of P dollars to double at an annual simple interest rate of 0.05?
(c) How long will it take an investment of P dollars to double at an annual simple interest rate of r?
(d) How long will it take an investment of P dollars to grow by a factor of q at an annual simple interest rate of r?

12 A certain bank offers interest at an annual rate of 6 percent compounded quarterly. If you invest \$700, find your accumulated savings after
(a) 6 months (b) 1 year (c) 5 years (d) 20 years

13 Find the resulting balance for each of the following investments.
(a) \$1,000 invested for 5 years at an annual rate of 6 percent compounded quarterly.
(b) \$2,000 invested for 3 years at an annual rate of 6 percent compounded quarterly.
(c) \$3,000 invested for 1 year at an annual rate of 4 percent compounded semiannually.

14 Find the balance at the end of 1 year if \$600 is invested at an annual rate of 8 percent compounded
(a) yearly (b) semiannually
(c) quarterly (d) monthly (12 times per year)

15 In 6 years, an investment of \$200 has grown to \$321.69. If interest is compounded quarterly, find the annual interest rate.

16 Find the amount of interest earned on the following investments.
- (a) \$750 invested for 3 years at an annual rate of 6 percent compounded quarterly.
- (b) \$800 invested for 3.5 years at an annual rate of 5 percent compounded monthly.
- (c) \$50,000 invested for 5 years at an annual rate of 10 percent compounded quarterly.

17 Find the effective interest rate equivalent to an annual rate of 6 percent compounded
(a) monthly (b) quarterly (c) semiannually

18 In terms of effective interest rate, order the following investments from lowest to highest: An annual rate of
- (a) 8 percent compounded annually.
- (b) 7.5 percent compounded semiannually.
- (c) 7.4 percent compounded quarterly.
- (d) 7.3 percent compounded monthly.

19 Find the present value of
- (a) \$4,000 payable in 3 years, if the annual interest rate is 6 percent, compounded quarterly.
- (b) \$10,000 payable in 5 years, if the annual interest rate is 10 percent, compounded semiannually.
- (c) \$50,000 payable in 10 years, if the annual interest rate is 9 percent, compounded monthly.

20 How much should you invest at an annual rate of 4 percent compounded quarterly if you want your investment to be worth \$6,000 in 5 years?

21 The Smiths wish to save for their daughter's college education. How much should they invest now at an annual rate of 5 percent compounded semiannually if they wish to have \$30,000 at the end of 10 years?

22 You owe \$500 which is to be paid at the end of 1 year. If you want to pay off your debt now, how much should your creditor demand if the money can be invested at an annual rate of 5 percent compounded quarterly?

2 • ANNUITIES

An **annuity** is a sequence of equal payments made at regular intervals over a period of time called the **term** of the annuity. We assume that the first payment is made at the end of the first interest period. After each payment is made, it earns interest at a fixed rate until the end of the term of the annuity. Annuities are a popular way to save money because most people don't have large amounts to invest all at once and prefer to save gradually. We shall study annuities in which the payments are made at the same time that interest is compounded. These are known as **ordinary annuities.** The **amount,** A, of an annuity is its total value after all payments are made and all interest is creditied. In the following example we shall calculate the amount of an annuity.

EXAMPLE 2.1 ● An annuity is set up in which there are yearly payments of \$200 for 5 years with interest being paid at the annual rate of 6 percent compounded annually. Find the amount of the annuity.

SOLUTION The first payment of \$200 is received at the end of the first year and earns interest at a rate of 6 percent per year for each of the remaining 4 years. To compute the value, A_1 (principal plus interest), of this payment at the expiration of the annuity, we use the compound interest formula $A = P(1 + i)^n$ with $P = 200$, $i = 0.06$, and $n = 4$. We get

$$A_1 = 200(1.06)^4 = \$252.50$$

The second payment of \$200 will earn interest at the rate of 6 percent per year for 3 years. Hence its value at the expiration of the annuity will be

$$A_2 = 200(1.06)^3 = \$238.20$$

The third payment will earn interest for 2 years, the fourth payment will earn interest for 1 year, and the final payment (made at the expiration of the annuity) earns no interest. The corresponding values are

$$A_3 = 200(1.06)^2 = \$224.72$$

$$A_4 = 200(1.06) = \$212.00$$

$$A_5 = \$200.00$$

The amount of the annuity is the sum

$$\begin{aligned} A &= A_1 + A_2 + A_3 + A_4 + A_5 \\ &= \$252.50 + \$238.20 + \$224.72 + \$212.00 + \$200.00 \\ &= \$1{,}127.42 \end{aligned}$$

●

We shall now derive a general formula for the amount of an annuity. Let R denote the size of each payment, i the interest rate per period, and n the number of interest periods. Using the compound interest formula as in the preceding example we find that the individual payments are worth the following amounts at the expiration of the annuity:

$$A_1 = R(1 + i)^{n-1} \qquad A_2 = R(1 + i)^{n-2} \ldots A_{n-1} = R(1 + i) \qquad A_n = R$$

The total amount of the annuity is thus

$$\begin{aligned} A &= A_1 + A_2 + \cdots + A_n \\ &= R[(1 + i)^{n-1} + (1 + i)^{n-2} + \cdots + (1 + i) + 1] \end{aligned}$$

The sum inside the brackets, $(1 + i)^{n-1} + (1 + i)^{n-2} + \cdots + (1 + i) + 1$, is commonly denoted by $s_{\overline{n}|i}$, read "s angle n at i," and can be rewritten more conveniently as

$$s_{\overline{n}|i} = \frac{(1 + i)^n - 1}{i}$$

(See Problem 9 for a derivation of this.) It follows that the accumulated amount of an annuity is given by

$$A = Rs_{\overline{n}|i} = R\frac{(1+i)^n - 1}{i}$$

The amount of annuity
If an annuity consists of n payments of R dollars, each of which earns interest at a rate i per period, then the amount A of the annuity is given by the formula

$$A = Rs_{\overline{n}|i}$$

where

$$s_{\overline{n}|i} = \frac{(1+i)^n - 1}{i}$$

The next two examples illustrate the use of this formula.

EXAMPLE 2.2 ● In order to save for a new car, you decide to put \$150 dollars every 3 months into a savings account that pays interest at an annual rate of 6 percent compounded quarterly. How much will you have at the end of

(a) 1 year (b) 5 years

SOLUTION We use the formula for the amount of an annuity, with $R = 150$ and $i = 0.06/4 = 0.015$.

(a) At the end of 1 year you will have made 4 payments, so $n = 4$ and the formula yields

$$A = Rs_{\overline{4}|0.015} = 150\frac{(1.015)^4 - 1}{0.015} = \$613.64$$

(b) After 5 years, there will have been $n = 20$ payments, and you will have accumulated

$$A = 150\frac{(1.015)^{20} - 1}{0.015} = \$3{,}468.55$$ ●

EXAMPLE 2.3 ● In order to save \$10,000 for his retirement, Bob wants to put a fixed amount each month into a savings account that pays interest at an annual rate of 6 percent compounded monthly. If he plans to save in this way for 20 years, how much should Bob's monthly payments be?

SOLUTION We wish to find the payment R for which the amount A of the annuity is \$10,000. Substituting $A = 10{,}000$, $i = 0.06/12 = 0.005$, and $n = 12(20) = 240$

into the formula for A, we get

$$10{,}000 = R\frac{(1.005)^{240} - 1}{0.005} = 462.04R$$

Solving this equation for R we find that

$$R = \frac{10{,}000}{462.04} = \$21.64$$ ●

Present value of an annuity

In Section 1 we defined the present value of an amount of money to be the principal that must be invested today to yield the desired amount at some specified time in the future. The **present value of an annuity** that consists of n payments of R dollars is the amount of money that must be invested today to generate the same sequence of payments. This means that withdrawals of R dollars can be made from the investment for each of the next n periods, after which the investment will be used up. To compute the present value of an annuity, we simply add the present values of the payments. Here is an example using the present-value formula.

EXAMPLE 2.4 ● Find the present value of an annuity that pays $50 semiannually for 2 years if the annual interest rate is 6 percent compounded semiannually.

SOLUTION Since the payments are made twice a year, the number of payments during the 2-year period is $n = 4$ and the interest rate per period is $i = 0.06/2 = 0.03$. Using the present-value formula $P = A/(1 + i)^n$ with $A = 50$ and $n = 1$, we find the present value of the first payment to be

$$P_1 = \frac{50}{(1.03)} = \$48.54$$

The present value of the second payment will be based on 2 interest periods and thus will be

$$P_2 = \frac{50}{(1.03)^2} = \$47.13$$

Similarly,

$$P_3 = \frac{50}{(1.03)^3} = \$45.76$$

and

$$P_4 = \frac{50}{(1.03)^4} = \$44.42$$

The present value of the entire annuity is

$$\begin{aligned} P &= P_1 + P_2 + P_3 + P_4 \\ &= 48.54 + 47.13 + 45.76 + 44.42 = \$185.85 \end{aligned}$$

In other words, \$185.85 invested today in a savings account at an annual interest rate of 6 percent compounded semiannually would enable you to make exactly 4 semiannual withdrawals of \$50 before the money is used up, the first withdrawal taking place at the end of the first interest period. ●

The calculations in Example 2.4 were simplified by the fact that the interest was compounded with exactly the same frequency with which the withdrawals were made, and that the first withdrawal was made at the end of the first interest period. We shall make these assumptions throughout the discussion of present value. In general, if an annuity consists of n payments of R dollars, and the interest rate per period is i, then the present value of the annuity is given by

$$\begin{aligned} P &= \frac{R}{(1+i)} + \frac{R}{(1+i)^2} + \cdots + \frac{R}{(1+i)^n} \\ &= R\left[\frac{1}{(1+i)} + \frac{1}{(1+i)^2} + \cdots + \frac{1}{(1+i)^n}\right] \end{aligned}$$

The sum in the brackets is denoted by $a_{\overline{n}|i}$ and can be rewritten (see Problem 10) as

$$a_{\overline{n}|i} = \frac{1 - 1/(1+i)^n}{i}$$

Present value of an annuity
The present value P of an annuity that consists of n payments of R dollars, each of which earns interest at a rate i per payment period, is given by the formula

$$P = Ra_{\overline{n}|i}$$

where

$$a_{\overline{n}|i} = \frac{1 - 1/(1+i)^n}{i}$$

The following example illustrates the use of the present-value formula.

EXAMPLE 2.5 ● How much money should be invested now at an annual interest rate of 4 percent compounded quarterly so that withdrawals of \$100 can be made every 3 months for the next 5 years, with the first withdrawal being made at the time of the first interest payment?

SOLUTION We wish to find the present value of an annuity with $R = 100$, $n = 4(5) = 20$, and $i = 0.04/4 = 0.01$. The present-value formula yields

$$P = 100a_{\overline{20}|0.01} = 100\frac{1 - 1/(1.01)^{20}}{0.01} = \$1{,}804.56$$ ●

Problems

1 Find the amount A of each of the following ordinary annuities.
   (a) \$100 per month for 5 years with interest paid monthly at the annual rate of 6 percent.
   (b) \$500 per quarter for 3 years with interest paid quarterly at the annual rate of 10 percent.
   (c) \$10,000 semiannually for 8 years with interest paid semiannually at the annual rate of 20 percent.

2 Find the payments for annuities with the following amounts, rates, and terms:
   (a) $A = \$8{,}000$, $i = 0.02$, $n = 48$
   (b) $A = \$10{,}000$, $i = 0.01$, $n = 5$
   (c) $A = \$6{,}000$, $i = 0.06$, $n = 10$

3 Luis invests \$100 per month for 2 years in an account that pays interest at an annual rate of 6 percent compounded monthly, and then, without making any withdrawals, invests \$200 per month in the same account for the next 2 years. How much will Luis have when the 4 years are up?

4 Which annuity accumulates more, \$100 per month for 2 years at an annual interest rate of 6 percent compounded monthly, or \$300 per quarter for 2 years at an annual interest rate of 6.5 percent compounded quarterly?

5 James decides to put enough money aside each month so that at the end of 2 years he will have enough money for a vacation in France. If James estimates the vacation cost to be \$4,000, what should his payments be if he can invest at an annual interest rate of 9 percent compounded monthly?

6 Find the present values of the following annuities.
   (a) \$60 per month for 5 years at an annual interest rate of 10 percent compounded monthly.
   (b) \$5,000 per quarter for 7 years at an annual interest rate of 20 percent compounded quarterly.
   (c) \$10,000 per year for 10 years at an annual interest rate of 40 percent compounded yearly.

7 How much should Karen invest now at an annual interest rate of 6 percent compounded quarterly to enable her to make quarterly withdrawals of \$500 for the next 4 years?

8 Ed invests \$4,000 in a savings account that pays interest at an annual rate of 4 percent compounded quarterly. Find the amount he can withdraw each quarter so that at the end of 5 years the balance in the account will be zero.

9 Verify the steps of the following proof of the formula for $s_{\overline{n}|i}$.

$$
\begin{aligned}
s_{\overline{n}|i} &= (1+i)^{n-1} + (1+i)^{n-2} + \cdots + (1+i) + 1 \\
&= \frac{[(1+i)^{n-1} + (1+i)^{n-2} + \cdots + (1+i) + 1][1-(1+i)]}{1-(1+i)} \\
&= \frac{1-(1+i)^n}{-i} \\
&= \frac{(1+i)^n - 1}{i}
\end{aligned}
$$

10 Verify the steps of the following proof of the formula for $a_{\overline{n}|i}$.

$$\begin{aligned} a_{\overline{n}|i} &= \frac{1}{1+i} + \frac{1}{(1+i)^2} + \cdots + \frac{1}{(1+i)^n} \\ &= \frac{1 + (1+i) + \cdots + (1+i)^{n-1}}{(1+i)^n} \\ &= \frac{s_{\overline{n}|i}}{(1+i)^n} \\ &= \frac{1 - 1/(1+i)^n}{i} \end{aligned}$$

3 • AMORTIZATION AND SINKING FUNDS

If a debt is paid off by periodic equal payments that include both principal and interest, then the debt is said to be **amortized.** (Sometimes this term is used more generally to mean any method of discharging a debt.) When the purchase of expensive items such as cars, stereos, and jewelry is made on time, an initial down payment is made and the rest of the purchase price and interest is amortized by a series of equal periodic payments. Another example of an amortized loan is the mortgage on a house, for which equal monthly payments, including principal and interest, are made over a period of many years.

If we think of a loan carrying an interest charge as being an investment by the lender, then we can think of an amortized loan as an annuity with the amount of the loan being the present value of the annuity and the payments made by the borrower being "withdrawals" by the lender. In fact, one of the main applications of the present-value formula for annuities is to make calculations for amortized loans. The next two examples will illustrate this technique.

EXAMPLE 3.1 ● Jane purchased a new car for $300 down plus payments of $100 per month for 3 years.

(a) Find the cash purchase price of the car (excluding interest) if the monthly payments include 18 percent annual interest, compounded monthly.
(b) Find the total amount of interest Jane has to pay.

SOLUTION (a) The cash purchase price C of the car can be written as

$$C = 300 + P$$

where 300 is the down payment and P is the present value of an annuity with $R = 100$, $n = 12(3) = 36$, and $i = 0.18/12 = 0.015$. Applying the present-value formula, we get

$$P = 100a_{\overline{36}|0.015} = 100\frac{1 - 1/(1.015)^{36}}{0.015} = \$2{,}766.07$$

Thus the cash price of the car is $300 + 2{,}766.07 = \$3{,}066.07$.

(b) Jane has to make 36 payments of \$100 each plus a down payment of \$300, for a total payment of \$3,900. Since the cash price of the car is only \$3,066.07, the amount of interest Jane will pay is

$$\$3{,}900 - \$3{,}066.07 = \$833.93$$

●

EXAMPLE 3.2 ● Mr. and Mrs. Smith buy a house for \$65,000. Suppose they make a 20 percent down payment, with the balance to be amortized in a 30-year mortgage at an annual interest rate of 9 percent compounded monthly.

(a) Find the monthly payments.
(b) Find the total amount of interest that will be paid.
(c) Find the **equity** that the Smiths will have in their house after 5 years. That is, find the amount of the mortgage they will have already paid.

SOLUTION (a) The mortgage will be for the amount owed after the 20 percent down payment, namely for $65{,}000(0.8) = \$52{,}000$. This is the present value P of an annuity with $n = 12(30) = 360$ monthly payments of R dollars with an interest rate of $i = 0.09/12 = 0.0075$. The present-value formula yields

$$52{,}000 = R\frac{1 - 1/(1.0075)^{360}}{0.0075} = 124.28R$$

Solving for R we find that the size of the monthly payments is

$$R = \$418.41$$

(b) Monthly payments of \$418.41 over a 30-year period yield a total payment of $418.41(360) = \$150{,}627.60$ to pay off a \$52,000 mortgage. Thus, the amount of interest is

$$\$150{,}627.60 - \$52{,}000 = \$98{,}627.60$$

(c) After 5 years there will still be $25(12) = 300$ mortgage payments of \$418.41 to make. The amount of the mortgage that is still unpaid is the present value of this annuity, which is

$$P = 418.41\frac{1 - 1/(1.0075)^{300}}{0.0075} = \$49{,}858.41$$

The equity is the difference

$$\$52{,}000 - \$49{,}858.41 = \$2{,}141.59$$

●

Sinking funds

A **sinking fund** is a fund set up to pay off a debt or in anticipation of a future debt. At regular intervals, equal payments are made into the fund while interest is compounded on the previous payments. The size of the payments is chosen so that after a specified period of time the fund will contain the amount of money needed to pay off the debt in a single payment. Sinking funds of this type are ordinary annuities, and therefore the formula for the amount of an annuity can be used to determine the payments needed to pay off a specified debt over a specified period of time. The next two examples illustrate this use of the formula.

EXAMPLE 3.3 ● A company wishes to spend \$40,000 for new equipment and decides to set up a sinking fund to accumulate this money over a 3-year period. If payments are to be made to the fund quarterly, with interest compounded quarterly at an annual rate of 5 percent, how large should the payments be?

SOLUTION The sinking fund is an ordinary annuity with amount $A = 40{,}000$, interest rate $i = 0.05/4 = 0.0125$, and $n = 4(3) = 12$ payments. We use the formula $A = Rs_{\overline{n}|i}$ to obtain

$$40{,}000 = Rs_{\overline{12}|0.0125} = R\frac{(1.0125)^{12} - 1}{0.0125} = 12.86R$$

Solving for R we get

$$R = \$3{,}110.42$$

and conclude that each payment should be \$3,110.42. ●

EXAMPLE 3.4 ● Harry has accumulated a \$7,000 gambling debt. The casino allows him to set up a sinking fund with equal monthly payments and interest compounded monthly at the annual rate of 4 percent, so that at the end of 2 years he will have paid off the debt. The casino also requires Harry to make separate semiannual interest payments of 20 percent of the total debt.

(a) Find the size of Harry's monthly payments into the sinking fund.
(b) How much interest will Harry have to pay altogether?

SOLUTION (a) Harry's sinking fund is an ordinary annuity with amount $A = 7{,}000$, interest rate $i = 0.04/12 = 0.0033$, and $n = 12(2) = 24$ payments. Substituting into the formula $A = Rs_{\overline{n}|i}$ we get

$$7{,}000 = Rs_{\overline{24}|0.0033} = R\frac{(1.0033)^{24} - 1}{0.0033} = 24.93R$$

Solving for R we find that the size of the monthly payments is

$$R = \$280.79$$

(b) No portion of the debt will be paid for 2 years, so the 4 semiannual interest payments will each be based on the total debt of \$7,000. Thus, Harry will have to pay a total interest of 4(7,000)(0.20) = \$5,600. ●

Problems

1 Ann bought a new car for \$800 down plus \$93 per month for 3 years. If the monthly payments include 18 percent annual interest compounded monthly, what was the purchase price of the car (excluding interest) and how much interest did Ann have to pay?

2 Joey bought a diamond engagement ring for Myrna. He had to pay \$5 down plus \$15 per month for 30 months. If the monthly payments included 20 percent annual interest compounded monthly, what was the purchase price of the ring (excluding interest) and how much interest did Joey have to pay?

3 Mr. and Mrs. Spiff are going to borrow \$7,000 to take a luxury 'round-the-world cruise. To pay off this debt they wish to make equal monthly payments for 2 years. If the lender charges 18 percent annual interest compounded monthly, find the Spiffs' monthly payments and the total interest they will have to pay.

4 The newlyweds Joey and Myrna decide to buy a new house whose cash selling price is \$40,000. Joey and Myrna make a 20 percent down payment and obtain a 30-year mortgage at $9\frac{1}{2}$ percent annual interest compounded monthly.

(a) Find Joey and Myrna's monthly payments.

(b) Find the total amount of interest they will have to pay if they keep the house for the full 30 years.

(c) Find the amount of equity they will have in their house after 10 years.

(d) Find the amount of interest they will have paid after 10 years.

5 A house is purchased for \$6,000 down and a mortgage with monthly payments of \$175 for 30 years. If the annual interest rate on the mortgage is $8\frac{1}{2}$ percent, compounded monthly, find the cash price of the house.

6 You inherit \$15,000 and invest the money in a bank that pays interest at an annual rate of 6 percent compounded quarterly. If you use this money by making equal quarterly withdrawals over a 5-year period, how much should each withdrawal be?

7 Find the present value of an annuity that consists of annual payments of \$5,000 for 6 years followed by annual payments of \$8,000 for 4 years, if the prevailing interest rate is 10 percent compounded annually.

8 A city has \$500,000 worth of school bonds due in 15 years. If a sinking fund can be established that pays 8 percent interest per annum compounded annually, how much should be deposited each year in order to pay off the debt?

9 A company borrows $60,000 to purchase a computer system. At the time of purchase the company agrees to pay interest on the loan as it arises and to set up a sinking fund to pay off the principal in 5 years. If the sinking fund is to have quarterly payments and the annual interest rate is 5 percent compounded quarterly, find the size of the payments.

10 A certain sinking fund requires monthly payments of $150 for 6 years, with interest compounded monthly at a 6 percent annual rate. Find the amount of the debt that the sinking fund is to pay off.

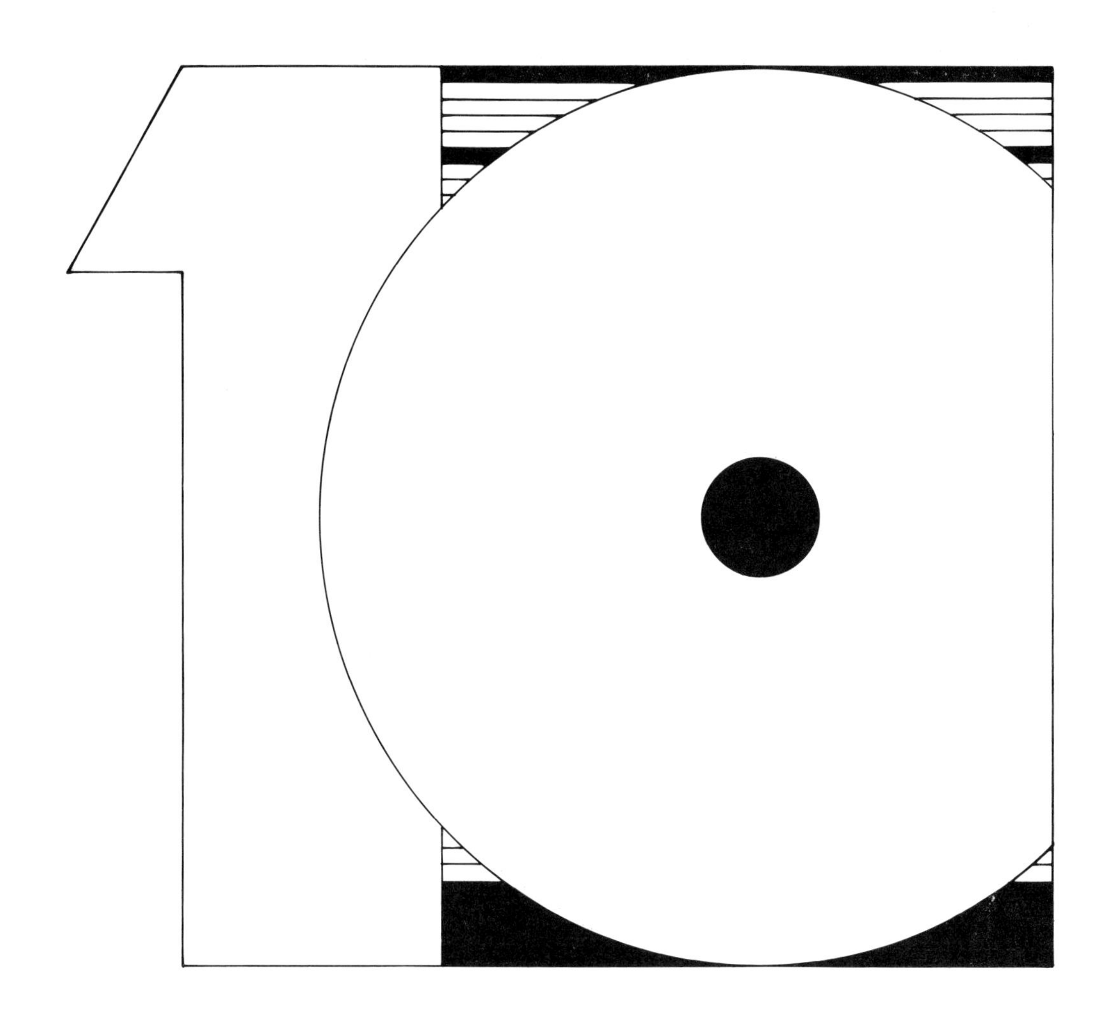

FUNCTIONS AND GRAPHS

1 • ALGEBRAIC FUNCTIONS

In Chapter 1, you were introduced to linear functions and their graphs. You learned that a linear function is a function of the form $f(x) = mx + b$, that the graph of a linear function is a straight line, and that the rate of change of a linear function is constant. In this chapter, you will see some useful *nonlinear* functions. The graphs of most nonlinear functions are curves. Functions of this type are used to describe situations in which one quantity changes at a variable rate with respect to another. Here is an example.

EXAMPLE 1.1 ● An environmental study of a certain suburban community suggests that x years from now, the average daily level of carbon monoxide in the air will be $f(x) = 3 + 0.1x^2$ parts per million.

(a) What will the level of carbon monoxide be 2 years from now?
(b) By how much will the level of carbon monoxide change during the 2d year?
(c) By how much will the level change during the 3d year?
(d) Sketch the relevant portion of the graph of the function $f(x)$.

SOLUTION (a) The level of carbon monoxide 2 years from now will be

$$f(2) = 3 + 0.1(4) = 3.4 \text{ parts per million}$$

(b) The change in the level of carbon monoxide during the 2d year will be

$$f(2) - f(1) = 3.4 - 3.1 = 0.3 \text{ parts per million}$$

(c) The change in the level of carbon monoxide during the 3d year will be

$$f(3) - f(2) = 3.9 - 3.4 = 0.5 \text{ parts per million}$$

Notice that this is different from the change in the carbon monoxide level during the 2d year.

(d) The function $f(x)$ has a practical interpretation in this context for $x \geq 0$. To sketch the corresponding portion of the graph, we first compute $f(x)$ for several convenient nonnegative values of x. The results of these computations are summarized in the following table.

x	0	1	2	3	4	5
$f(x)$	3	3.1	3.4	3.9	4.6	5.5

We then plot the corresponding points $(x, f(x))$ and connect them by a smooth curve. The resulting graph is shown in Figure 1.1. ●

The method of curve sketching illustrated in part (d) of Example 1.1 has obvious shortcomings. For many functions, the computation of a table of values is tedious and time-consuming. Moreover, no matter how many points you plot,

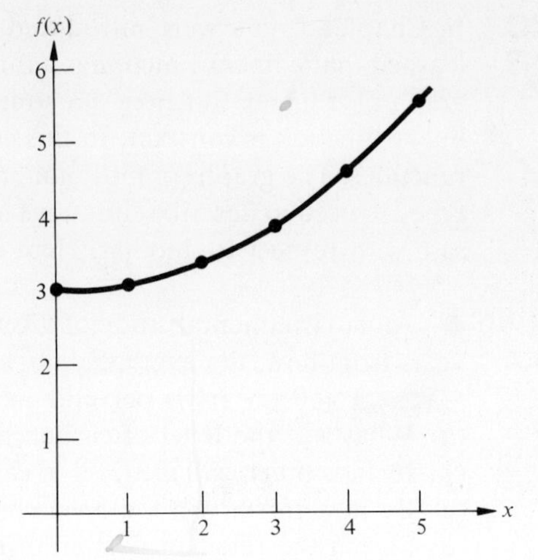

Figure 1.1 The graph of $f(x) = 3 + 0.1x^2$ for $x \geq 0$.

you cannot be sure of the behavior of the graph between these points. These difficulties will be overcome by use of techniques involving differential calculus that will be introduced in Chapter 11.

Quadratic functions

A **quadratic function** is a function of the form

$$f(x) = ax^2 + bx + c$$

where a, b, and c are constants and $a \neq 0$. The function $f(x) = 3 + 0.1x^2$ in Example 1.1 is a quadratic function with $a = 0.1$, $b = 0$, and $c = 3$.

The graph of a quadratic function is a **parabola.** A parabola is a curve formed by the intersection of a circular cone and a plane as shown in Figure 1.2.

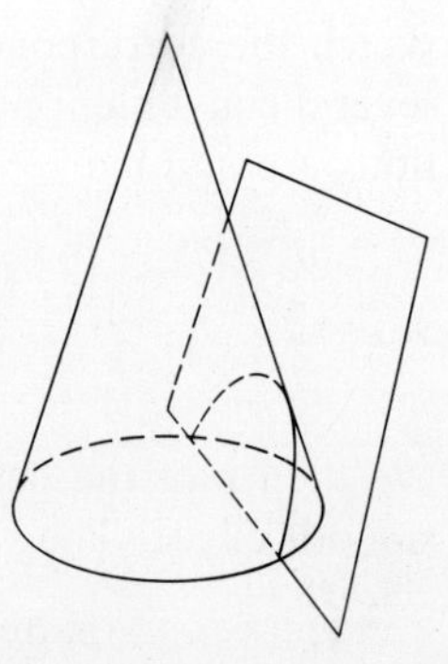

Figure 1.2 A parabola: the intersection of a cone and a plane.

The graph of a quadratic function may be **concave upward** as in Figure 1.3*a* or **concave downward** as in Figure 1.3*b*.

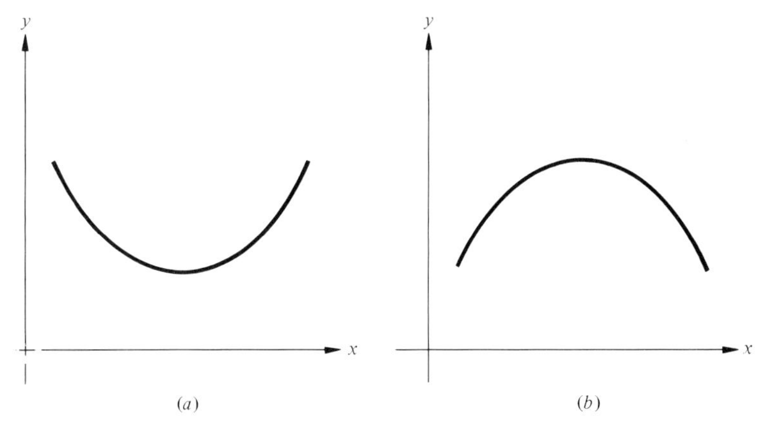

Figure 1.3 The graph of a quadratic function.

Polynomials

A **polynomial** is a function of the form

$$f(x) = a_0 + a_1x + a_2x^2 + \cdots + a_nx^n$$

where n is a nonnegative integer, and $a_0, a_1, \ldots, a_n$ are constants. If $a_n \neq 0$, the integer n is said to be the **degree** of the polynomial. For example, the function $f(x) = 3x^5 - 6x^2 + 7$ is a polynomial of degree 5. Quadratic functions are polynomials of degree 2. Linear functions are also polynomials. Their degree is 1. The factored function $f(x) = (x - 1)(x + 1)(x - 3)$ is a polynomial since it can be rewritten as $f(x) = x^3 - 3x^2 - x + 3$. Factored polynomials are particularly easy to sketch because their x intercepts are known. A simple procedure for obtaining a rough sketch of such a polynomial is illustrated in the next example. It is based on the fact that polynomials are **continuous** functions, that is, functions whose graphs are unbroken curves.

EXAMPLE 1.2 ● Sketch the graph of the factored polynomial $f(x) = (x - 1)(x + 1)(x - 3)$.

SOLUTION From the factorization we see that the values of x for which $f(x) = 0$ are $x = 1$, $x = -1$, and $x = 3$. The corresponding points $(1, 0)$, $(-1, 0)$, and $(3, 0)$ are the x intercepts of the graph.

Since f is continuous and its graph an unbroken curve, there is no way that $f(x)$ can change sign between two adjacent x intercepts. Hence the portion of the graph between the intercepts $(1, 0)$ and $(3, 0)$, for example, must be a curve that joins these intercepts and that lies either entirely above the x axis as in Figure 1.4*a*, or entirely below it as in Figure 1.4*b*.

We can settle the matter by simply plotting on the graph one point whose x coordinate lies between 1 and 3. Any such point will do, so we make the computation as simple as possible by taking $x = 2$. Since $f(2) = -3$, the point $(2, -3)$ is on the graph and we conclude that the curve lies below the x axis as in Figure 1.4*b*.

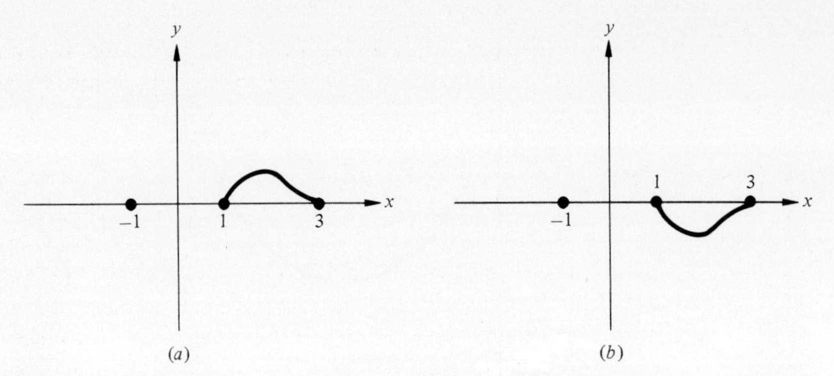

Figure 1.4 The behavior of a continuous function between adjacent x intercepts.

Similarly, since $f(0) = 3$, the point $(0, 3)$ is on the graph and we conclude that the curve lies above the x axis between $x = -1$ and $x = 1$.

Finally, we compute $f(x)$ for some values of x to the right of $x = 3$ and to the left of $x = -1$ to determine the behavior of the graph beyond the x intercepts.

x	-2	4
$f(x)$	-15	15

The resulting sketch is shown in Figure 1.5. ●

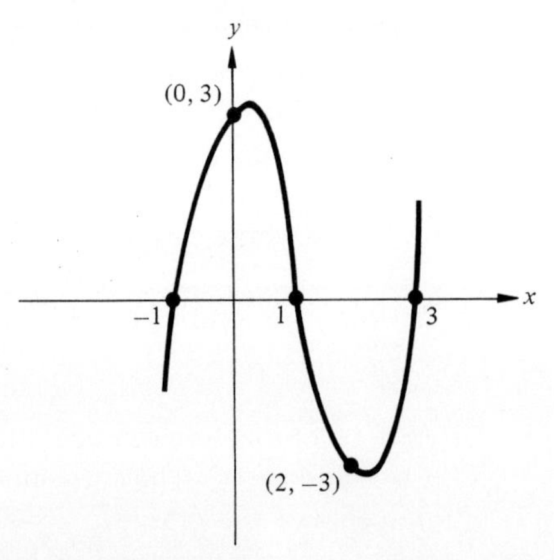

Figure 1.5 A sketch of the polynomial $y = (x - 1)(x + 1)(x - 3)$.

The sketch in Figure 1.5 is only a rough approximation of the actual graph of the polynomial. It does not show, for example, the precise location of the peak that occurs somewhere between $x = -1$ and $x = 1$. In Chapter 11, you will learn how to use differential calculus to obtain this information, as well as other details about the shape of the graph.

When the factors of a polynomial are not known, explicit computation of its

x intercepts may be impossible. In fact, some polynomials have no x intercepts at all. In such cases, you will have to plot additional points to obtain a sketch.

Rational functions

Many problems lead to functions that can be written as quotients of polynomials. A quotient of two polynomials is called a **rational function.** For example, the functions $f(x) = (x^2 + 1)/(x - 2)$ and $f(x) = 1/x$ are both rational. So is the function $f(x) = 1 + 1/x$ since it can be rewritten as $f(x) = (x + 1)/x$.

In general, rational functions are harder to graph than polynomials. One reason for this is that rational functions need not be continuous. In fact, there will be a break or **discontinuity** in the graph of a rational function at each value of x for which the denominator is equal to zero. Here is an example.

EXAMPLE 1.3 ● Graph the rational function $f(x) = 1/(x - 2)$.

SOLUTION We begin with the following table of values.

x	-2	-1	0	1	2	3	4	5	6
$f(x)$	$-\frac{1}{4}$	$-\frac{1}{3}$	$-\frac{1}{2}$	-1	$\times$	1	$\frac{1}{2}$	$\frac{1}{3}$	$\frac{1}{4}$

Since division by zero is impossible, we cannot calculate $f(2)$. It follows that there is no point on the graph whose x coordinate is 2, and consequently, there is a break in the graph when $x = 2$. This suggests that we look more closely at the behavior of f for values of x near 2. Continuing our table of values we get

x	$\frac{3}{2}$	$\frac{7}{4}$	2	$\frac{9}{4}$	$\frac{5}{2}$
$f(x)$	-2	-4	$\times$	4	2

When the corresponding points are plotted, we obtain the graph in Figure 1.6. ●

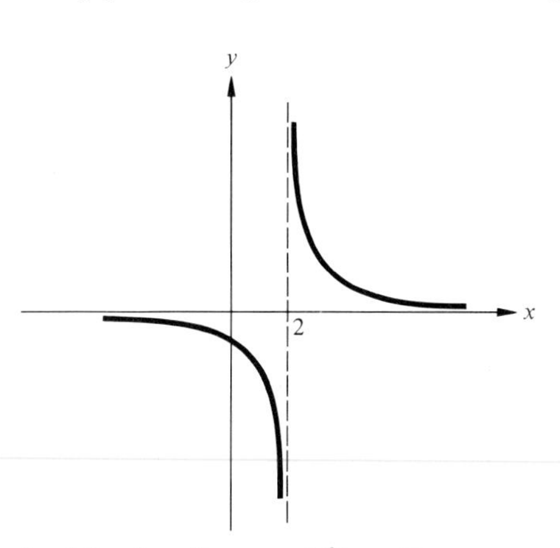

Figure 1.6 The graph of the function $y = 1/(x - 2)$.

Curve-fitting In many practical problems, the goal is to find a function of a particular type that agrees with a set of data. A problem of this type is illustrated next.

EXAMPLE 1.4 ● In 1971, the circulation of a local newspaper was 5,000; in 1975 the circulation was 8,200; and in 1979 the circulation was 14,600. The circulation manager believes that circulation is a quadratic function of time. Assuming the manager is right, express the circulation of the newspaper as a function of time. Use this function to predict the circulation of the newpaper in 1983.

SOLUTION We let x denote the number of years since 1971 and $C(x)$ the corresponding circulation of the newspaper. In geometric terms, the information about the circulation in 1971, 1975, and 1979 says that the points (0, 5,000), (4, 8,200) and (8, 14,600) all lie on the graph of the quadratic function we are seeking. We substitute the coordinates of these three points into the general quadratic function $C(x) = ax^2 + bx + c$ and find that

$$5{,}000 = c$$
$$8{,}200 = 16a + 4b + c$$
$$14{,}600 = 64a + 8b + c$$

This is a system of three linear equations in the variables a, b, and c. Using the matrix-reduction method from Chapter 2 to solve this system we find that

$$a = 100 \qquad b = 400 \qquad c = 5{,}000$$

and conclude that the quadratic function representing circulation as a function of time is

$$C(x) = 100x^2 + 400x + 5{,}000$$

To predict the circulation in 1983, we compute $C(12)$ and get

$$C(12) = 100(144) + 400(12) + 5{,}000 = 24{,}200$$

The graph of the circulation function is shown in Figure 1.7. ●

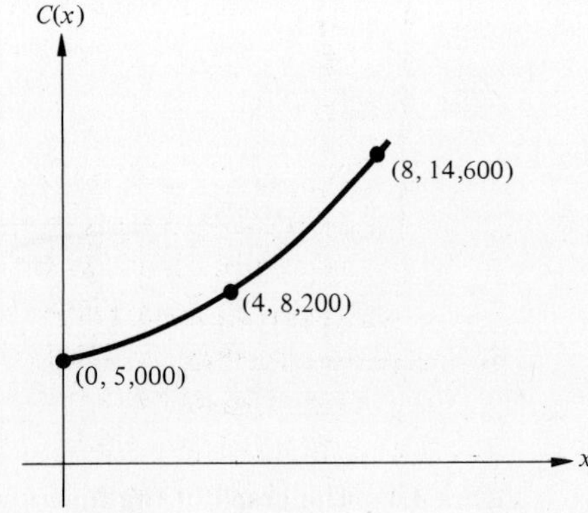

Figure 1.7 Increasing newspaper circulation: $C(x) = 100x^2 + 400x + 5{,}000$.

Problems

1 An efficiency study of the morning shift at a certain factory indicates that an average worker who arrives on the job at 8:00 A.M. will have assembled $f(x) = -x^3 + 6x^2 + 15x$ transistor radios x hours later.
- (a) How many radios will such a worker have assembled by 10:00 A.M.?
- (b) How many radios will such a worker assemble between 9:00 A.M. and 10:00 A.M.?

2 Suppose the total cost in dollars of manufacturing q units of a certain commodity is given by the function $C(q) = q^3 - 30q^2 + 400q + 500$.
- (a) Compute the cost of manufacturing 20 units.
- (b) Compute the cost of manufacturing the 20th unit.

3 Suppose that t hours past midnight, the temperature in Miami was $C(t) = -\frac{1}{6}t^2 + 4t + 10$ degrees Celsius.
- (a) What was the temperature at 2:00 P.M.?
- (b) By how much did the temperature increase or decrease between 6:00 P.M. and 9:00 P.M.?

4 It is estimated that t years from now, the population of a certain suburban community will be $P(t) = 20 - 6/(t + 1)$ thousand.
- (a) What will the population of the community be 9 years from now?
- (b) By how much will the population increase during the 9th year?
- (c) What will happen to the size of the population in the long run?

5 Biologists have found that the speed of blood in an artery is a function of the distance of the blood from the artery's central axis. According to **Poiseuille's law,** the speed (in centimeters per second) of blood that is r centimeters from the central axis of an artery of radius R is given by the function $S(r) = C(R^2 - r^2)$, where C is a constant. Suppose that for a certain artery, $C = 1.76 \times 10^5$ centimeters and $R = 1.2 \times 10^{-2}$ centimeters.
- (a) Compute the speed of the blood at the central axis of this artery.
- (b) Compute the speed of the blood midway between the artery's wall and central axis.

6 To study the rate at which animals learn, a group of psychology students performed an experiment in which a rat was sent repeatedly through a laboratory maze. The students found that the time required for the rat to traverse the maze on the nth trial was approximately $f(n) = 3 + 12/n$ minutes.
- (a) For what values of n does $f(n)$ have meaning in this context?
- (b) How long did it take the rat to traverse the maze on the 3d trial?
- (c) How many trials were required before the rat traversed the maze in 4 minutes or less?
- (d) According to the function f, what will happen to the time required for the rat to traverse the maze as the number of trials increases? Will the rat ever be able to traverse the maze in less than 3 minutes?

7 It is estimated that the number of worker-hours required to distribute new telephone books to x percent of the households in a certain rural community is given by the function $f(x) = 600x/(300 - x)$.
- (a) For what values of x does $f(x)$ have a practical interpretation in this context?

(b) How many worker-hours were required to distribute new telephone books to the first 50 percent of the households?
(c) How many worker-hours were required to distribute new telephone books to the entire community?
(d) What percentage of the households in the community had received new telephone books by the time 150 worker-hours had been expended?

8 During a nationwide program to immunize the population against a virulent form of influenza, public health officials found that the cost of inoculating x percent of the population was approximately $f(x) = 150x/(200 - x)$ million dollars.
(a) For what values of x does $f(x)$ have a practical interpretation in this context?
(b) What was the cost of inoculating the first 50 percent of the population?
(c) What was the cost of inoculating the second 50 percent of the population?
(d) What percentage of the population had been inoculated by the time 37.5 million dollars had been spent?

In Problems 9 through 14, sketch the graph of the given function by plotting a representative collection of points.

9 $f(x) = x$
10 $f(x) = x^2$
11 $f(x) = x^3$
12 $f(x) = x^4$
13 $f(x) = x^2 - 1$
14 $f(x) = 2 - x^3$

In Problems 15 through 22, sketch the graph of the given polynomial.

15 $f(x) = (x + 1)(x - 2)$
16 $f(x) = (x - 2)(x + 3)(x - 1)$
17 $f(x) = (x - 2)^2(x + 3)(x - 1)$
18 $f(x) = -x^2 + 2x + 3$
19 $f(x) = x^2 - 4x + 4$
20 $f(x) = x^3 + 4x^2 - x - 4$
21 $f(x) = -2x^2 + 4x$
22 $f(x) = x^3 + 3x^2$

In Problems 23 through 30, sketch the graph of the given rational function.

23 $f(x) = \dfrac{1}{x}$
24 $f(x) = \dfrac{1}{x^2}$
25 $f(x) = \dfrac{1}{x - 3}$
26 $f(x) = \dfrac{1}{(x - 3)^2}$
27 $f(x) = \dfrac{x + 2}{x - 3}$
28 $f(x) = \dfrac{x - 2}{x + 3}$
29 $f(x) = 1 + \dfrac{1}{x}$
30 $f(x) = 1 - \dfrac{1}{x}$

31 The consumer demand for a certain commodity is $D(p) = -200p + 12{,}000$ units per month when the market price is p dollars per unit.
(a) Graph this demand function.

(b) Where does the graph of the demand function cross the p axis?
(c) Express consumers' **total monthly expenditure** for the commodity as a function of p. (The total monthly expenditure is the total amount of money consumers spend each month on the commodity.)
(d) Sketch the graph of the total monthly expenditure function.
(e) Discuss the economic significance of the p intercepts of the expenditure curve.

32 Suppose the total cost of manufacturing x units of a certain commodity is $C(x) = x^2 + 6x + 19$ dollars. Express the average cost per unit as a function of the number of units produced and, on the same set of axes, sketch the total cost and average cost curves.

33 The president of a local civic association believes that since 1975, the membership of the association has been growing quadratically. In 1975, the membership of the association was 600; in 1977 the membership was 708; and in 1979 it was 832.
(a) Find a quadratic function that is consistent with these figures that expresses the membership of the association as a function of time.
(b) Use the function in part (a) to predict what the membership of the association will be in 1985.

34 The following data were compiled by a medical student during the first 30 minutes of an experiment designed to study the growth of bacteria in a certain culture.

Number of minutes	0	10	30
Number of bacteria	5,000	8,000	20,480

(a) The student guesses that the number of bacteria grows quadratically. Find a quadratic function that is consistent with the student's data that expresses the number of bacteria as a function of time.
(b) At the end of one hour, the student finds that 83,886 bacteria are present in the culture. On the basis of this additional data, what conclusion should the student draw about the conjecture that the number of bacteria grows quadratically? Explain.

35 An importer of Norwegian cheese estimates that retailers will buy 240 pounds of the cheese per week when the wholesale price is \$1 per pound; they will buy 210 pounds per week when the wholesale price is \$2 per pound; and they will buy 160 pounds per week when the wholesale price is \$3 per pound.
(a) Using the importer's estimates, find a quadratic function expressing retailers' weekly demand for the cheese as a function of its wholesale price.
(b) Use the function in part (a) to estimate what the demand for the cheese would be if the wholesale price were \$4 per pound.

2 • FUNCTIONAL MODELS

In the next chapter you will encounter techniques that can be used to visualize and analyze functions. Before these techniques can be applied to the solution of practical problems, practical relationships must be represented mathematically as functions. That is, mathematical models must be built that correspond to the situations under consideration. In this section, you will see some of the techniques you can use to build such models.

EXAMPLE 2.1 ● A manufacturer can produce radios at a cost of \$2 apiece. The radios have been selling for \$5 apiece and, at this price, consumers have been buying 4,000 radios a month. The manufacturer is planning to raise the price of the radios and estimates that for each \$1 increase in the price, 400 fewer radios will be sold each month. Express the manufacturer's monthly profit as a function of the price at which the radios are sold.

SOLUTION We begin by stating the desired relationship in words.

$$\text{Profit} = (\text{number of radios})(\text{profit per radio})$$

Since our goal is to express profit as a function of price, we introduce the variable x to represent the price in dollars at which the radios will be sold, and let $P(x)$ be the corresponding profit.

Next, we express the number of radios sold in terms of the variable x. We know that 4,000 radios were sold each month when the price was \$5, and that 400 fewer radios will be sold each month for each \$1 increase in the price. Thus,

$$\text{Number of radios} = 4{,}000 - 400(\text{number of \$1 increases})$$

The number of \$1 increases in the price is the difference, $x - 5$, between the new and old selling prices. Hence,

$$\begin{aligned}\text{Number of radios} &= 4{,}000 - 400(x - 5)\\ &= 6{,}000 - 400x\\ &= 400(15 - x)\end{aligned}$$

Substituting

$$\text{Profit per radio} = x - 2$$

and

$$\text{Number of radios} = 400(15 - x)$$

into the verbal equation with which we began, we conclude that

$$P(x) = 400(15 - x)(x - 2)$$

The graph of this factored polynomial is sketched in Figure 2.1. (Actually, only the portion of the graph for $x \geq 5$ is relevant to the original problem as stated. Can you give a practical interpretation of the portion between $x = 2$ and $x = 5$?) Notice that the profit function reaches a maximum for some value of x between $x = 2$ and $x = 15$. In Chapter 11 you will learn how to calculate this optimal selling price. ●

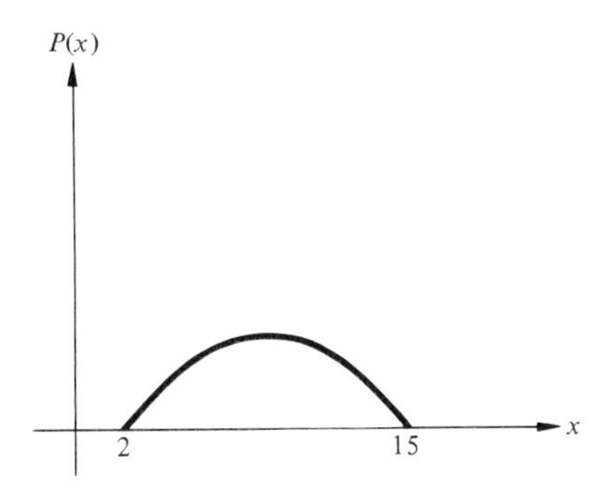

Figure 2.1 A profit function: $P(x) = 400(15 - x)(x - 2)$.

In the next example, the quantity we are seeking is expressed most naturally in terms of two variables. We will have to eliminate one of these before we can write the quantity as a function of a single variable.

EXAMPLE 2.2 ● The highway department is planning to build a picnic area for motorists along a major highway. It is to be rectangular with an area of 5,000 square meters and is to be fenced off on the three sides not adjacent to the highway. Express the number of meters of fencing required as a function of the length of the unfenced side.

SOLUTION It is natural to start by introducing two variables, say x and y, to denote the lengths of the sides of the picnic area (Figure 2.2) and to express the number of meters of fencing required in terms of these two variables.

$$\text{Number of meters of fencing} = x + 2y$$

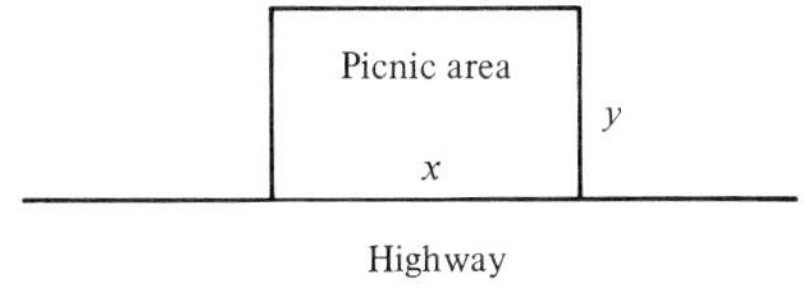

Figure 2.2 Rectangular picnic area.

Since we want to express the number of meters of fencing as a function of x alone, we must find a way to express y in terms of x. That is, we must find an equation relating x and y. The fact that the area is to be 5,000 square meters gives us this equation. Specifically,

$$xy = 5{,}000$$

Solving this equation for y we get

$$y = \frac{5{,}000}{x}$$

Substituting this into the formula for the number of meters of fencing and letting F denote the corresponding function, we conclude that

$$F(x) = x + \frac{10{,}000}{x}$$

A graph of this rational function is sketched in Figure 2.3. Of course, only when x is positive does this graph represent the number of meters of required fencing. For this reason, the portion of the graph corresponding to negative values of x is drawn as a broken curve.

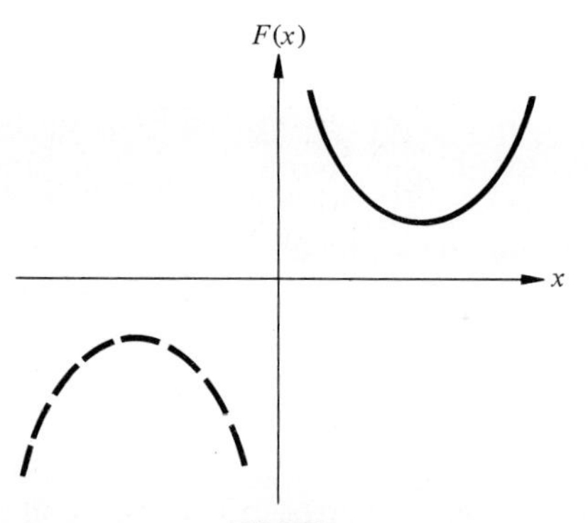

Figure 2.3 **The graph of** $F(x) = x + \frac{10{,}000}{x}$.

Notice that the height of the graph increases without bound as x increases without bound and also as positive values of x approach zero. Can you explain why in terms of the original problem? Notice also that there is some length x for which the amount of required fencing is minimal. In Chapter 11, differential calculus will be used to compute this optimal value of x. ●

Problems

1 A college bookstore can obtain the book *Social Groupings of the American Dragonfly* from the publisher at a cost of \$3 per book. The bookstore has been offering the book at a price of \$15 per copy and, at this price, has been selling 200 copies a month. The bookstore is planning to lower its price to stimulate sales and estimates that for each \$1 reduction in the price, 20 more books will be sold each month.

(a) Express the bookstore's monthly profit from the sale of this book as a function of the selling price and draw the graph.
(b) Compute the monthly profit if the selling price of the book is lowered to $10.

2 A manufacturer has been selling lamps at a price of $2 per lamp and, at this price, consumers have been buying 2,000 lamps per month. The manufacturer wishes to raise the price and estimates that for each 1-cent increase in the price, 10 fewer lamps will be sold each month. The manufacturer can produce the lamps at a cost of 40 cents per lamp.
(a) Express the manufacturer's monthly profit as a function of the price at which the lamps are sold and draw the graph.
(b) Compute the profit if the lamps are sold for $2.50 apiece.

3 A bus company has adopted the following pricing policy for groups wishing to charter its buses: Groups containing no more than 40 people will be charged a fixed amount of $2,400 (40 times $60). In groups containing between 40 and 80 people, everyone will pay $60 minus 50 cents for each person in excess of 40. The company's lowest fare of $40 per person will be offered to groups that have 80 members or more. Express the bus company's revenue as a function of the size of the group.

4 A city recreation department plans to build a rectangular playground 3,600 square meters in area. The playground is to be surrounded by a fence.
(a) Express the length of the fencing as a function of the length of one of the sides of the playground.
(b) How much fencing is needed if the dimensions of the playground are 100 by 36 meters?

5 A closed box with a square base is to have a volume of 250 cubic meters. The material for the top and bottom of the box costs $2 per square meter, and the material for the sides costs $1 per square meter. Express the construction cost of the box as a function of the length of its base.

6 An open box with a square base is to be built for $48. The sides of the box will cost $3 per square meter, and the base will cost $4 per square meter. Express the volume of the box as a function of the length of its base.

7 A truck is 975 kilometers due east of a car and is traveling west at a constant speed of 60 kilometers per hour. Meanwhile, the car is going north at a constant speed of 90 kilometers per hour. Express the distance between the car and the truck as a function of time.

8 A cable is to be run from a power plant on one side of a river 900 meters wide to a factory on the other side, 3,000 meters downstream. The cable will be run in a straight line from the power plant to some point P on the opposite bank, and then along the bank to the factory. The cost of running the cable across the water is $5 per meter, while the cost over land is $4 per meter. Let x be the distance from P to the point directly across the river from the power plant, and express the cost of installing the cable as a function of x.

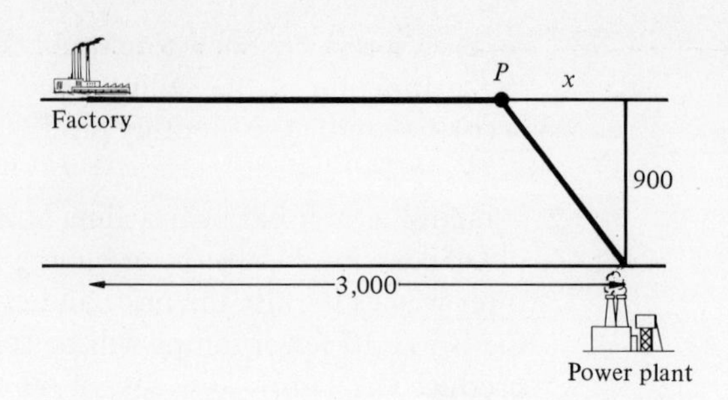

9 A rectangular poster contains 25 square centimeters of print surrounded by margins of 2 centimeters on each side and 4 centimeters on the top and bottom. Express the total area of the poster (printing plus margins) as a function of the width of the printed portion.

10 A plastics firm has received an order from the city recreation department to manufacture 8,000 special Styrofoam kickboards for its summer swimming program. The firm owns several machines, each of which can produce 30 kickboards an hour. The cost of setting up the machines to produce these particular kickboards is \$20 per machine. Once the machines have been set up, the operation is fully automated and can be supervised by a single foreman earning \$4.80 per hour. Express the cost of producing the 8,000 kickboards as a function of the number of machines used, and sketch the relevant portion of the graph.

11 A Florida citrus grower estimates that if 60 orange trees are planted in a field, the average yield per tree will be 400 oranges. The average yield will decrease by 4 oranges per tree for each additional tree planted. Express the grower's total yield as a function of the number of additional trees planted, draw the graph, and estimate the total number of trees the grower should plant to maximize yield.

12 Farmers can get \$2 per bushel for their potatoes on July first, and after that, the price drops by 2 cents per bushel per day. On July first, a farmer has 80 bushels of potatoes in the field and estimates that the crop is increasing at a rate of one bushel per day. Express the farmer's revenue from the sale of the potatoes as a function of the time at which the crop is harvested, draw the graph, and estimate when the farmer should harvest the potatoes to maximize revenue.

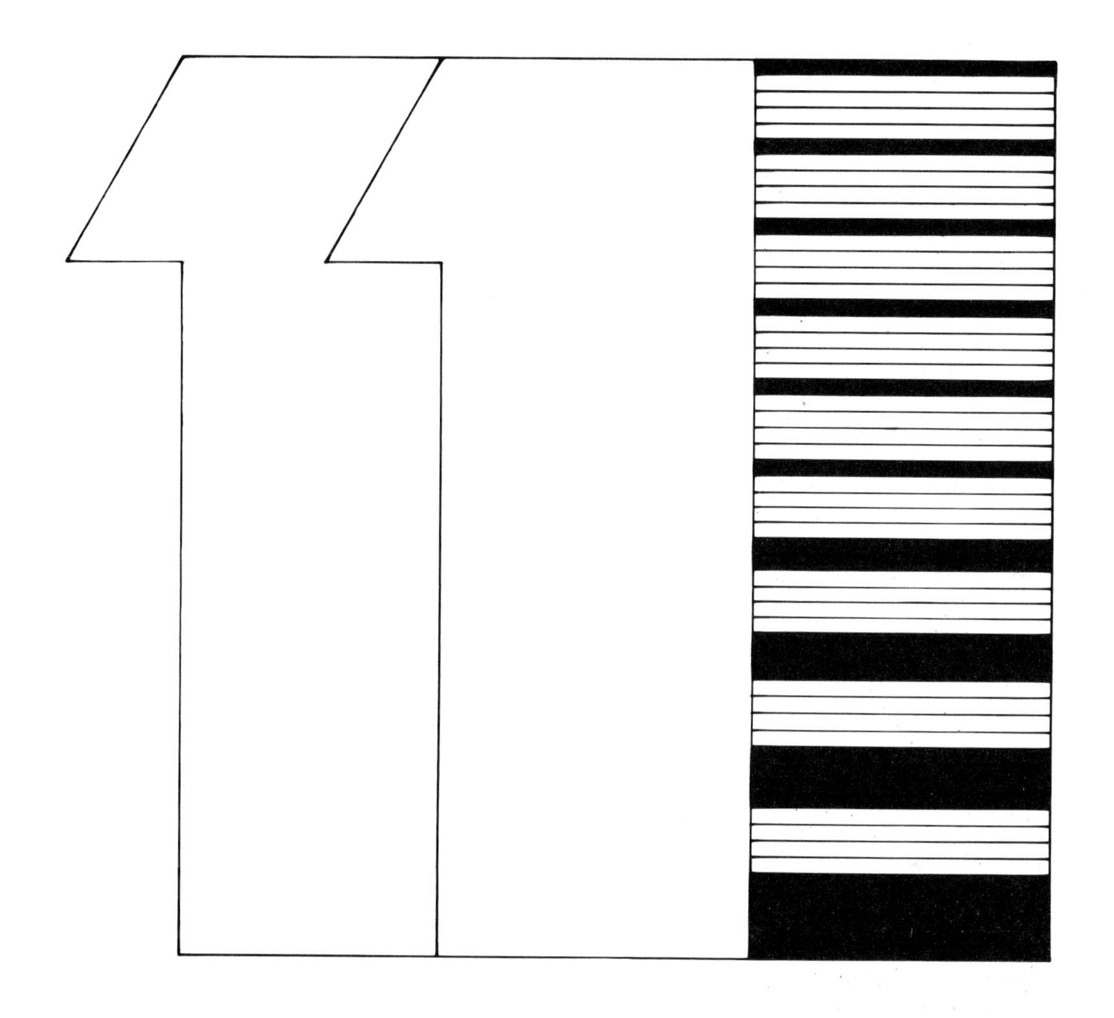

DIFFERENTIAL CALCULUS

1 • THE DERIVATIVE

Differentiation is a mathematical technique of exceptional power and versatility. It is one of the two central concepts in the branch of mathematics called **calculus** and has a variety of applications including curve sketching, the optimization of functions, and the analysis of rates of change. We begin our discussion with a practical example involving profit maximization.

A practical optimization problem

In Example 2.1 of Chapter 10, we considered a problem in which a manufacturer's monthly profit from the sale of radios was $P(x) = 400(15 - x)(x - 2)$ dollars when the radios were sold for x dollars apiece. The graph of this profit function, which is reproduced in Figure 1.1, suggests that there is an optimal selling price x at which the manufacturer's profit will be greatest. In geometric terms, the optimal price is the x coordinate of the peak of the graph.

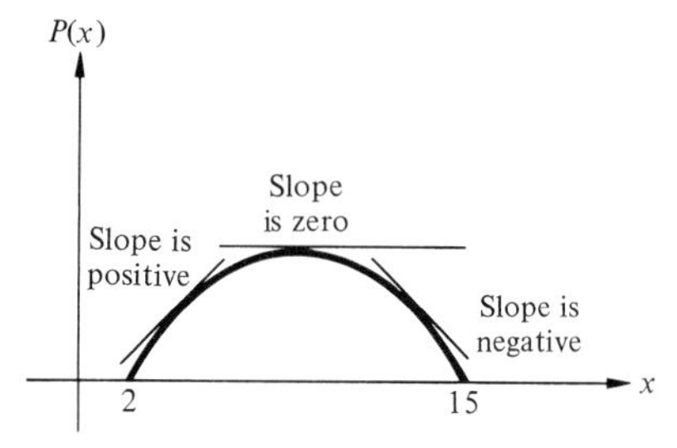

Figure 1.1 The profit function $P(x) = 400(15 - x)(x - 2)$.

In this relatively simple example, the peak can be characterized in terms of lines that are tangent to the graph. In particular, the peak is the only point on the graph at which the tangent line is horizontal; that is, at which the slope of the tangent is zero. To the left of the peak, the slope of the tangent is positive. To the right of the peak, the slope is negative. But just at the peak itself, the curve "levels off" and the slope of its tangent is zero.

These observations suggest that we could solve the optimization problem if we had a procedure for computing slopes of tangent lines. We shall now develop such a procedure. Throughout the development, we shall rely on our intuitive understanding that the tangent to a curve at a point is the line that indicates the direction of the curve at that point.

The slope of a tangent

The goal is to solve the following general problem: Given a point $(x, f(x))$ on the graph of a function f, find the slope of the line that is tangent to the graph at this point. The situation is illustrated in Figure 1.2.

In Chapter 1, Section 1 you learned that the slope of the line passing through two points (x_1, y_1) and (x_2, y_2) is given by the formula

$$\text{Slope} = \frac{\Delta y}{\Delta x} = \frac{y_2 - y_1}{x_2 - x_1}$$

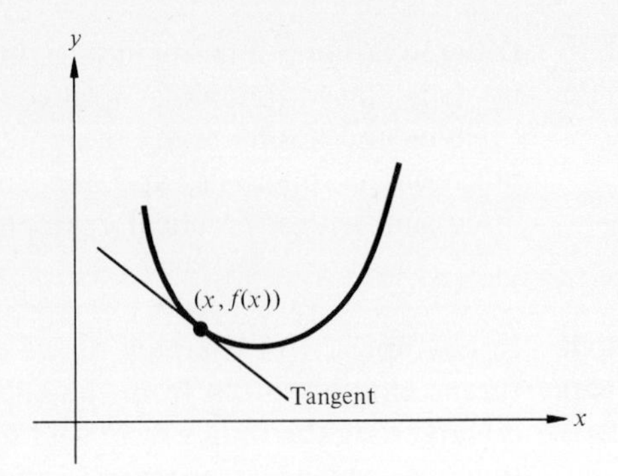

Figure 1.2 A tangent to the curve $y = f(x)$.

Unfortunately, in the present situation we know only one point on the tangent line, namely, the point of tangency $(x, f(x))$. Hence, direct computation of the slope is impossible and we are forced to adopt an indirect approach.

The strategy is to approximate the tangent by other lines whose slopes can be computed directly. In particular, we consider lines joining the given point $(x, f(x))$ to neighboring points on the graph of f. These lines, shown in Figure 1.3, are called **secants** and are good approximations to the tangent provided the neighboring point is close to the given point $(x, f(x))$.

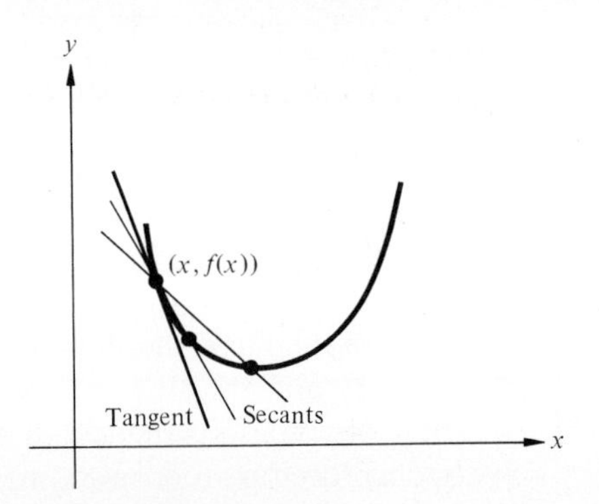

Figure 1.3 Secants approximating a tangent.

We can make the slope of the secant as close as we like to the slope of the tangent by choosing the neighboring point sufficiently close to the given point $(x, f(x))$. This suggests that we should be able to determine the slope of the tangent by first computing the slope of a related secant and then studying the behavior of this slope as the neighboring point gets closer and closer to the given point.

To compute the slope of a secant, we first label the coordinates of the neighboring point as indicated in Figure 1.4. In particular, we let Δx denote the change in the x coordinate between the given point $(x, f(x))$ and the neighboring point. The x coordinate of the neighboring point is $x + \Delta x$ and, since the point lies on the graph of f, its y coordinate is $f(x + \Delta x)$.

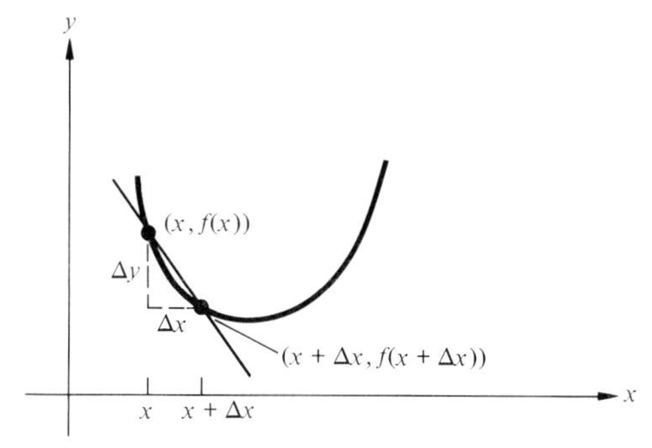

Figure 1.4 A secant through the curve $y = f(x)$.

Since the change in the y coordinate is $\Delta y = f(x + \Delta x) - f(x)$ it follows that

$$\text{Slope of secant} = \frac{\Delta y}{\Delta x} = \frac{f(x + \Delta x) - f(x)}{\Delta x}$$

This quotient is not the slope of the tangent, but only an approximation to it. If Δx is small, however, the neighboring point $(x + \Delta x, f(x + \Delta x))$ is close to the given point $(x, f(x))$, and the approximation is a good one. In fact, the slope of the actual tangent is the number that this quotient approaches as Δx approaches zero.

A calculation based on these observations is performed in the following example.

EXAMPLE 1.1 ● Find the slope of the line that is tangent to the graph of the function $f(x) = x^2$ at the point (2, 4).

SOLUTION A sketch of f showing the given point (2, 4) and a related secant is drawn in Figure 1.5.

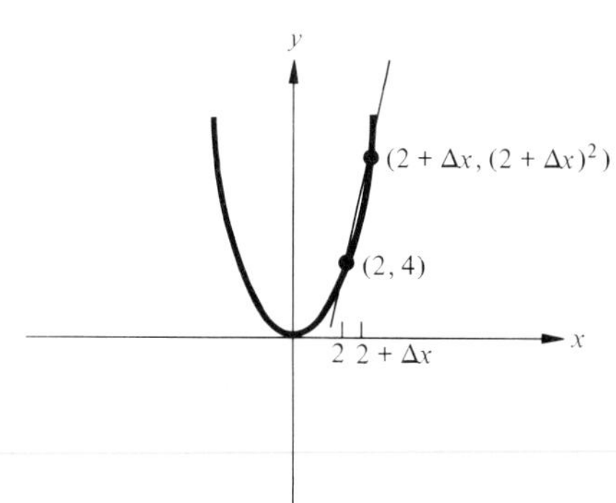

Figure 1.5 The curve $y = x^2$ and a secant through (2, 4).

Since the x coordinate of the given point is 2, it follows that the x coordinate of the neighboring point is $2 + \Delta x$, and the y coordinate of this point is $(2 + \Delta x)^2$. Hence,

$$\text{Slope of secant} = \frac{(2 + \Delta x)^2 - 4}{\Delta x}$$

Our goal is to find the number that this quotient approaches as Δx approaches zero. Before we can do this, we must rewrite the quotient in a simplier form. (Do you see what would happen if we let Δx approach zero in the unsimplified quotient?) To simplify the quotient, we expand the term $(2 + \Delta x)^2$, rewrite the numerator, and then divide numerator and denominator by Δx. This gives us

$$\begin{aligned} \frac{(2 + \Delta x)^2 - 4}{\Delta x} &= \frac{4 + 4\Delta x + (\Delta x)^2 - 4}{\Delta x} \\ &= \frac{4\Delta x + (\Delta x)^2}{\Delta x} \\ &= 4 + \Delta x \end{aligned}$$

Hence,

$$\text{Slope of secant} = 4 + \Delta x$$

Since $4 + \Delta x$ approaches 4 as Δx approaches zero, we conclude that at the given point (2, 4), the slope of the tangent is 4. ●

The derivative In Example 1.1, we found the slope of the tangent to the curve $y = x^2$ at a particular point (2, 4). In the next example, we perform the same calculation again, this time representing the given point algebraically as (x, x^2). The result is a formula into which we can substitute any value of x to calculate the slope of the tangent to the curve at any point (x, x^2).

EXAMPLE 1.2 ● Derive a formula expressing the slope of the tangent to the curve $y = x^2$ as a function of the x coordinate of the point of tangency.

SOLUTION We represent the point of tangency as (x, x^2) and the neighboring point as $(x + \Delta x, (x + \Delta x)^2)$. The slope of the corresponding secant is

$$\begin{aligned} \frac{(x + \Delta x)^2 - x^2}{\Delta x} &= \frac{x^2 + 2x\Delta x + (\Delta x)^2 - x^2}{\Delta x} \\ &= 2x + \Delta x \end{aligned}$$

Since $2x + \Delta x$ approaches $2x$ as Δx approaches 0, we conclude that at the point (x, x^2), the slope of the tangent is $2x$.

For example, at the point (2, 4), $x = 2$ and so the slope of the tangent is $2(2) = 4$. ●

In Example 1.2, we started with a function f and derived a related function that expressed the slope of its tangent in terms of the x coordinate of the point of tangency. This derived function is known as the **derivative** of f and is frequently denoted by the symbol f'. In Example 1.2, we discovered that the derivative of x^2 is $2x$. That is, we found that if $f(x) = x^2$, then $f'(x) = 2x$.

The derivative f' of a function f

Algebraic Definition:

$$\frac{f(x + \Delta x) - f(x)}{\Delta x} \text{ approaches } f'(x) \text{ as } \Delta x \text{ approaches zero}$$

Geometric Interpretation:

The derivative expresses the slope of the tangent to the curve $y = f(x)$ as a function of the x coordinate of the point of tangency.

Notation

Symbols other than f' are sometimes used to denote the derivative. For example, if y rather than $f(x)$ is used to denote the function itself, the symbol dy/dx (suggesting slope $\Delta y/\Delta x$) is frequently used instead of $f'(x)$. Hence, instead of the statement

$$\text{If } f(x) = x^2, \text{ then } f'(x) = 2x$$

we could write

$$\text{If } y = x^2, \text{ then } \frac{dy}{dx} = 2x$$

Sometimes the two notations are combined as in the statement

$$\text{If } f(x) = x^2, \text{ then } \frac{df}{dx} = 2x$$

By omitting reference to y and f altogether, we can condense these statements and write

$$\frac{d}{dx}(x^2) = 2x$$

to indicate that the derivative of x^2 is $2x$.

The maximization profit

To conclude this section, we solve the optimization problem with which we began the discussion.

EXAMPLE 1.3 ● A manufacturer can produce radios at a cost of \$2 apiece. The radios have been selling for \$5 apiece and, at this price, consumers have been buying 4,000 radios a month. The manufacturer is planning to raise the price of the radios and estimates that for each \$1 increase in the price, 400 fewer radios will be sold each month. At what price should the manufacturer sell the radios to generate the largest possible profit?

SOLUTION As we saw in Example 2.1 of Chapter 10, the profit P is given by the function

$$P(x) = 400(15 - x)(x - 2)$$

where x is the selling price of the radios. The graph of this factored polynomial is sketched once again in Figure 1.6.

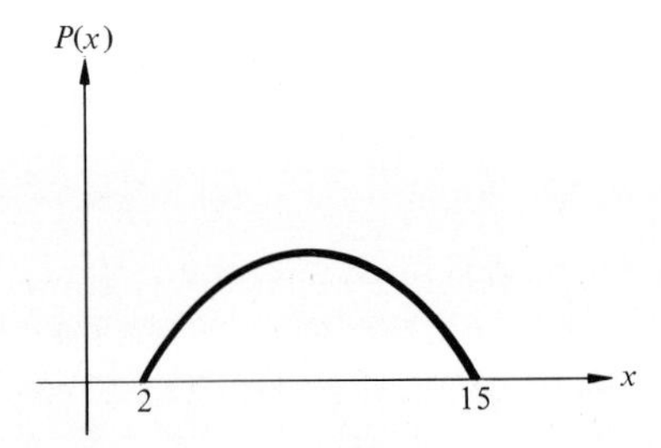

Figure 1.6 The profit function $P(x) = 400(15 - x)(x - 2)$.

The goal is to find the value of x for which the profit $P(x)$ is maximal. This is the value of x for which the slope of the tangent is zero. Since this slope is given by the derivative, we begin by computing $P'(x)$. For convenience, we work with the unfactored form of the profit function

$$P(x) = -400x^2 + 6{,}800x - 12{,}000$$

First we form the quotient

$$\frac{P(x + \Delta x) - P(x)}{\Delta x}$$

$$= \frac{[-400(x + \Delta x)^2 + 6{,}800(x + \Delta x) - 12{,}000] - [-400x^2 + 6{,}800x - 12{,}000]}{\Delta x}$$

$$= \frac{-400(\Delta x)^2 - 800x\,\Delta x + 6{,}800\,\Delta x}{\Delta x}$$

$$= -400\,\Delta x - 800x + 6{,}800$$

Since $-400\,\Delta x - 800x + 6{,}800$ approaches $-800x + 6{,}800$ as Δx approaches zero, it follows that

$$P'(x) = -800x + 6{,}800$$

To find the value of x for which the slope of the tangent is zero, we solve the equation $P'(x) = 0$ and get

$$-800x + 6{,}800 = 0 \qquad \text{or} \qquad x = 8.5$$

We conclude that to generate the largest possible profit, the manufacturer should sell the radios for \$8.50 apiece. ●

Problems

In Problems 1 through 6, compute the derivative of the given function and find the slope of the line that is tangent to the graph of the function for the given value of x.

1 $f(x) = 5x - 3$; $x = 2$

2 $y = x^2 - 1$; $x = -1$

3 $y = 2x^2 - 3x + 5$; $x = 0$

4 $f(x) = x^3 - 1$; $x = 2$

5 $y = 1/x$; $x = \frac{1}{2}$

6 $f(x) = \sqrt{x}$; $x = 9$

7 Let $f(x) = x^2$.

(a) Compute the slope of the secant joining the points on the graph of f whose x coordinates are $x = -2$ and $x = -1.9$.

(b) Use calculus to compute the slope of the line that is tangent to the graph when $x = -2$, and compare this slope with your answer in part (a).

8 Let $f(x) = x^3$.

(a) Compute the slope of the secant joining the points on the graph of f whose x coordinates are $x = 1$ and $x = 1.1$.

(b) Use calculus to compute the slope of the line that is tangent to the graph when $x = 1$, and compare this slope with your answer in part (a).

9 (a) Find the derivative of the linear function $y = 3x - 2$.

(b) Write an equation of the tangent to the graph of this function at the point $(-1, -5)$.

(c) Explain how the answers to parts (a) and (b) could have been obtained from geometric considerations, with no calculation whatsoever.

10 Sketch the graph of the function $y = x^2 - 3x$ and use calculus to find its lowest point.

11 Sketch the graph of the function $y = 1 - x^2$ and use calculus to find its highest point.

12 Sketch the graph of the function $y = x^3 - x^2$. Determine the values of x for which the derivative dy/dx is zero. What happens to the graph at the corresponding points?

13 What can you conclude about the graph of a function between $x = a$ and $x = b$ if its derivative is positive whenever $a \leq x \leq b$?

14 Sketch the graph of a function f whose derivative has all of the following properties.

(a) $f'(x) > 0$ when $x < 1$ and when $x > 5$

(b) $f'(x) < 0$ when $1 < x < 5$

(c) $f'(1) = 0$ and $f'(5) = 0$

15 A manufacturer can produce cassette tape recorders at a cost of $20 apiece. It is estimated that if the tape recorders are sold for x dollars apiece, consumers will buy $120 - x$ tape recorders a month. Use calculus to determine the price at which the manufacturer should sell the tape recorders to generate the largest possible profit.

16 (a) Find the derivatives of the functions $y = x^2$ and $y = x^2 - 3$, and account geometrically for their similarity.

(b) Without further computation, find the derivative of the function $y = x^2 + 5$.

17 (a) Find the derivative of the function $y = x^2 + 3x$.

(b) Find the derivatives of the functions $y = x^2$ and $y = 3x$ separately.

(c) How is the answer in part (a) related to the answers in part (b)?

(d) In general, if $f(x) = g(x) + h(x)$, what would you guess is the relationship between the derivative of f and the derivatives of g and h?

18 (a) Compute the derivatives of the functions $y = x^2$ and $y = x^3$.

(b) Examine your answers in part (a). Can you detect a pattern? What do you think is the derivative of $y = x^4$? And the derivative of $y = x^{27}$?

2 • TECHNIQUES OF DIFFERENTIATION

In Section 1, you learned how to find the derivative of a function by letting Δx approach zero in the expression for the slope of a secant. For all but the simplest functions, this process is tedious and time consuming. In this section, you will see some short-cuts. Proofs of the validity of these short-cuts can be found in most calculus texts.

The derivative of a power function

A **power function** is a function of the form $f(x) = x^n$, where n is a real number. For example, $f(x) = x^2$, $f(x) = x^{-3}$, and $f(x) = x^{1/2}$ are all power functions. So are $f(x) = 1/x^2$ and $f(x) = \sqrt[3]{x}$ since they can be rewritten as $f(x) = x^{-2}$ and $f(x) = x^{1/3}$, respectively. Here is a simple rule you can use to find the derivative of any power function.

The derivative of a power function

For any number n,

$$\frac{d}{dx}(x^n) = nx^{n-1}$$

That is, to find the derivative of x^n, reduce the power of x by 1 and multiply by the original power.

According to this rule, the derivative of x^2 is $2x^1$ or $2x$, which agrees with the result we obtained in Example 1.2. Here are a few more calculations.

EXAMPLE 2.1 ● Differentiate (find the derivative of) each of the following functions.

(a) $y = x^{27}$ (b) $y = 1/x^{27}$
(c) $y = \sqrt{x}$ (d) $y = 1/\sqrt[3]{x}$

SOLUTION In each case, we first write the function as a power function and then apply the general rule.

(a) $\frac{d}{dx}(x^{27}) = 27x^{26}$

(b) $\frac{d}{dx}\left(\frac{1}{x^{27}}\right) = \frac{d}{dx}(x^{-27}) = -27x^{-28}$

(c) $\frac{d}{dx}(\sqrt{x}) = \frac{d}{dx}(x^{1/2}) = \frac{1}{2}x^{-1/2}$

(d) $\frac{d}{dx}\left(\frac{1}{\sqrt[3]{x}}\right) = \frac{d}{dx}(x^{-1/3}) = -\frac{1}{3}x^{-4/3}$ ●

The derivative of a constant

The derivative of any constant function is zero. This is because the graph of a constant function $y = c$ is a horizontal line and its slope is zero.

The derivative of a constant
For any constant c,

$$\frac{d}{dx}(c) = 0$$

That is, the derivative of a constant function is zero.

The derivative of a constant times a function

The next rule expresses the fact that the curve $y = cf(x)$ is c times as steep as the curve $y = f(x)$.

The constant-multiple rule
For any constant c,

$$\frac{d}{dx}(cf) = c\frac{df}{dx}$$

That is, the derivative of a constant times a function is equal to the constant times the derivative of the function.

The use of this rule is illustrated in the next example.

EXAMPLE 2.2 ● Differentiate the function $y = 5x^2$.

SOLUTION We already know that $d/dx\,(x^2) = 2x$. Combining this with the constant-multiple rule we get

$$\frac{d}{dx}(5x^2) = 5\frac{d}{dx}(x^2) = 5(2x) = 10x$$ ●

The derivative of a sum The next rule states that a sum can be differentiated term by term.

> **The sum rule**
>
> $$\frac{d}{dx}(f + g) = \frac{df}{dx} + \frac{dg}{dx}$$
>
> That is, the derivative of a sum is the sum of the individual derivatives.

EXAMPLE 2.3 ● Differentiate the function $y = x^2 + 3x$.

SOLUTION We know that $d/dx\,(x^2) = 2x$ and that $d/dx\,(3x) = 3$. According to the sum rule, it follows that

$$\frac{d}{dx}(x^2 + 3x) = \frac{d}{dx}(x^2) + \frac{d}{dx}(3x) = 2x + 3$$ ●

By combining the sum rule with the power and constant-multiple rules, you can differentiate any polynomial. Here is an example.

EXAMPLE 2.4 ● Differentiate the polynomial $y = 5x^3 - 4x^2 + 12x - \frac{1}{2}$.

SOLUTION According to the sum rule, y can be differentiated term by term. Thus,

$$\frac{dy}{dx} = \frac{d}{dx}(5x^3) + \frac{d}{dx}(-4x^2) + \frac{d}{dx}(12x) + \frac{d}{dx}(-\tfrac{1}{2})$$
$$= 15x^2 - 8x + 12$$ ●

The derivative of a product Suppose you wanted to differentiate the product $y = x^2(3x + 1)$. You might be tempted to differentiate the factors x^2 and $3x + 1$ separately, and then multiply your answers. That is, since $d/dx\,(x^2) = 2x$ and $d/dx\,(3x + 1) = 3$, you might conclude that $dy/dx = 6x$. However, this answer is wrong. To see this, rewrite the function as $y = 3x^3 + x^2$ and observe that the derivative is $9x^2 + 2x$ and not $6x$. The derivative of a product is *not* the product of the individual derivatives. Here is the correct formula for the derivative of a product.

The product rule

$$\frac{d}{dx}(fg) = f\frac{dg}{dx} + g\frac{df}{dx}$$

That is, the derivative of a product is the sum of two terms, one of which is the first factor times the derivative of the second factor, and the other is the second factor times the derivative of the first factor.

The use of this rule is illustrated in the next example.

EXAMPLE 2.5 ● Differentiate the function $y = x^2(3x + 1)$.

SOLUTION According to the product rule,

$$\begin{aligned}\frac{d}{dx}[x^2(3x + 1)] &= x^2\frac{d}{dx}(3x + 1) + (3x + 1)\frac{d}{dx}(x^2)\\ &= x^2(3) + (3x + 1)(2x)\\ &= 9x^2 + 2x\end{aligned}$$

●

The derivative of a quotient

The derivative of a quotient is not the quotient of the individual derivatives. Here is the correct rule.

The quotient rule

$$\frac{d}{dx}\left(\frac{f}{g}\right) = \frac{g\,df/dx - f\,dg/dx}{g^2}$$

Using the quotient rule, you can now differentiate any rational function. Here is an example.

EXAMPLE 2.6 ● Differentiate the rational function $y = \dfrac{x^2 + 2x - 21}{x - 3}$.

SOLUTION According to the quotient rule,

$$\begin{aligned}\frac{dy}{dx} &= \frac{(x - 3)\,d/dx\,(x^2 + 2x - 21) - (x^2 + 2x - 21)\,d/dx\,(x - 3)}{(x - 3)^2}\\ &= \frac{(x - 3)(2x + 2) - (x^2 + 2x - 21)(1)}{(x - 3)^2}\\ &= \frac{x^2 - 6x + 15}{(x - 3)^2}\end{aligned}$$

●

Problems In Problems 1 through 16, differentiate the given function. In each case, do as much of the computation as possible in your head.

1 $y = 3x^5 - 4x^3 + 9x - 6$
2 $f(x) = \frac{1}{4}x^8 - 5x^4 - x + 2$
3 $g(u) = 2u - 3/u^2 + 5u^{1/3}$
4 $y = 3\sqrt{x} + 2/5x^3 - \sqrt{2}$
5 $h(t) = -16t + 1/\sqrt{t} - t^{3/2} + 1/3t + t/3$
6 $f(x) = (5x^4 - 3x^2 + 12x - 6)(1 - 6x)$
7 $g(x) = \frac{1}{3}(x^6 - 2x^3 + 1)$
8 $y = (t^3 - t^2 + t - 1)(t^2 + 1)$
9 $g(x) = x/(x^2 - 2)$
10 $y = (x + 2)/(3x - 1)$
11 $h(u) = (u^2 + 1)/(3u - u^3)$
12 $f(t) = 1/(t - 2)$
13 $y = (x^2 + 2x + 1)/3$
14 $h(x) = x^3/2 - 1/2x$
15 $f(x) = (2x + 1)(x^2 + 6x - 3)(1 - x^5)$
16 $g(u) = (u^2 + 2u + 1)(u + 5)/(2u - 1)$

In Problems 17 through 20, find the equation of the line that is tangent to the graph of the given function at the specified point.

17 $y = x^5 - 3x^3 - 5x + 2$; $(1, -5)$
18 $y = (x^2 + 1)(1 - x^3)$; $(0, 1)$
19 $f(x) = (x + 1)/(x - 1)$; $(0, -1)$
20 $f(x) = 1 - 1/x + 2/\sqrt{x}$; $(4, \frac{7}{4})$

In Problems 21 through 24, find the equation of the line that is tangent to the graph of the given function at the point $(x, f(x))$ for the specified value of x.

21 $f(x) = x^4 - 3x^3 + 2x^2 - 6$; $x = 2$
22 $f(x) = x - 1/x^2$; $x = 1$
23 $f(x) = (x^2 + 2)/(x^2 - 2)$; $x = -1$
24 $f(x) = (x^3 - 2x^2 + 3x - 1)(x^5 - 4x^2 + 2)$; $x = 0$

25 (a) Differentiate the function $y = 2x^2 - 5x - 3$.
(b) Now factor the function as $y = (2x + 1)(x - 3)$ and differentiate using the product rule. Compare your answers.

26 (a) Use the quotient rule to differentiate the function $y = (2x - 3)/x^3$.
(b) Now rewrite this function as $y = x^{-3}(2x - 3)$ and differentiate using the product rule.
(c) Show that your answers to parts (a) and (b) are the same.

27 The product rule tells us how to differentiate the product of *any* two functions, while the constant-multiple rule tells us how to differentiate products in which one of the factors is constant. Show that the two rules are consistent. That is, use the product rule to show that $d/dx\,(cf) = c\,df/dx$ if c is a constant.

28 Sketch the graph of the function $f(x) = x^2 - 4x - 5$ and use calculus to determine its lowest point.

29 Sketch the graph of the function $f(x) = 3 - 2x - x^2$ and use calculus to determine its highest point.

30 (a) Use calculus to find all the points on the graph of the function $f(x) = x^3 + x^2$ at which the tangent line is horizontal.
(b) Sketch a graph showing f and its horizontal tangents.

31 The consumer demand for a certain commodity is $D(p) = -200p + 12{,}000$ units per month when the market price is p dollars per unit.
(a) Express consumers' total monthly expenditure for the commodity as a function of p and graph this expenditure function.
(b) Use calculus to determine the market price that will generate the largest possible consumer expenditure.

3 • RATE OF CHANGE

In this section, you will see how the derivative of a function can be interpreted as its rate of change. Viewed in this way, a derivative may represent such quantities as the rate at which population grows, a manufacturer's marginal cost, the speed of a moving object, the rate of inflation, or the rate at which natural resources are being depleted.

You may have already sensed the connection between derivatives and rate of change. The derivative of a function is the slope of its tangent line, and the slope of any line is the rate at which it is rising or falling. The purpose of this section is to make this connection more precise. Here is a familiar practical situation that will serve as a model for the general discussion.

Average and instantaneous speed

Imagine that a car is moving along a straight road, and that $D(t)$ is its distance from its starting point after t hours. Suppose we want to determine the speed of the car at a particular time t. Without access to the car's speedometer, we must resort to indirect methods.

Suppose we record the position of the car at time t and then do so again at some later time $t + \Delta t$. That is, suppose we know $D(t)$ and $D(t + \Delta t)$. We can then compute the **average speed** of the car between times t and $t + \Delta t$ as follows:

$$\text{Average speed} = \frac{\text{change in distance}}{\text{change in time}} = \frac{D(t + \Delta t) - D(t)}{\Delta t}$$

Since the speed of the car may fluctuate during the time interval from t to $t + \Delta t$, it is unlikely that this average speed is equal to the **instantaneous speed** (the speed shown on the speedometer) at time t. However, if Δt is small, the possibility of drastic changes in speed is small, and the average speed may be a fairly good approximation to the instantaneous speed. Indeed, we can find the instantaneous speed at time t by letting Δt approach zero in the expression for the average speed. But as Δt approaches zero, the quotient $[D(t + \Delta t) - D(t)]/\Delta t$ approaches the derivative of D. We conclude, therefore, that the instantaneous speed at time t is just the derivative $D'(t)$ of the distance function.

Average and instantaneous rate of change

These ideas can be extended to more general situations. Suppose that y is a function of x, say $y = f(x)$. Corresponding to a change from x to $x + \Delta x$, the variable y changes by an amount $\Delta y = f(x + \Delta x) - f(x)$. Thus, the quotient

$$\frac{\text{Change in } y}{\text{Change in } x} = \frac{\Delta y}{\Delta x} = \frac{f(x + \Delta x) - f(x)}{\Delta x}$$

represents the resulting **average rate of change** of y with respect to x. As the interval over which we are averaging becomes shorter (that is, as Δx approaches zero), this quotient approaches a quantity that we would intuitively call the **instantaneous rate of change** of y with respect to x. But this quantity is precisely the derivative of f. Hence the instantaneous rate of change of y with respect to x is just the derivative dy/dx.

Instantaneous rate of change
If $y = f(x)$, the instantaneous rate of change of y with respect to x is given by the derivative of f. That is

$$\text{Rate of change} = \frac{dy}{dx} = f'(x)$$

Here is an example.

EXAMPLE 3.1 ● It is estimated that x months from now, the population of a certain community will be $P(x) = x^2 + 20x + 8{,}000$.

(a) At what rate will the population be changing 15 months from now?
(b) By how much will the population actually change during the 16th month?

SOLUTION (a) The rate of change of the population is the derivative $P'(x) = 2x + 20$. Since $P'(15) = 30 + 20 = 50$, we conclude that 15 months from now, the population will be growing at a rate of 50 people per month.
(b) The actual change in the population during the 16th month is the difference between the population at the end of 16 months and the population at the end of 15 months. That is,

$$P(16) - P(15) = 8{,}576 - 8{,}525 = 51 \text{ people}$$

The reason for the difference between the actual change in population during the 16th month in part (b) and the monthly rate of change at the beginning of that month in part (a) is that the rate of change of the population varied during the month. The instantaneous rate in part (a) can be thought of as the change in population that would occur during the 16th month if the rate of change of the population remained constant throughout that month. ●

Marginal analysis in economics

In economics, the rate of change of the total production cost with respect to the number of units produced is called the **marginal cost.** It is measured in dollars per unit and is often a good approximation to the cost of producing one additional unit.

EXAMPLE 3.2 ● Suppose the total cost in dollars of manufacturing q units of a certain commodity is $C(q) = 3q^2 + 5q + 10$.

(a) Derive a formula for the marginal cost.
(b) What is the marginal cost when 50 units have been produced?
(c) What is the actual cost of producing the 51st unit?

SOLUTION (a) The marginal cost is the derivative $C'(q) = 6q + 5$.
(b) When 50 units have been produced, $q = 50$ and the marginal cost is $C'(50) = \$305$ per unit.
(c) The actual cost of producing the 51st unit is the difference between the cost of producing 51 units and the cost of producing 50 units. That is,

$$\text{Cost of 51st unit} = C(51) - C(50) = 8{,}068 - 7{,}760 = \$308 \quad ●$$

Notice that in the preceding example, the marginal cost in part (b) was close to, but not equal to the actual cost in part (c) of producing one additional unit. In geometric terms, the difference between these two quantities is the difference between the slope of a tangent and the slope of a nearby secant. The marginal cost $C'(50)$ in part (b) is the slope of the line that is tangent to the cost curve when $q = 50$ while the difference $C(51) - C(50)$ in part (c) is the slope

$$\frac{\text{Change in } C(q)}{\text{Change in } q} = \frac{C(50 + 1) - C(50)}{1}$$

of the secant joining the two points on the cost curve whose q coordinates are 50 and 51. The situation is illustrated in Figure 3.1.

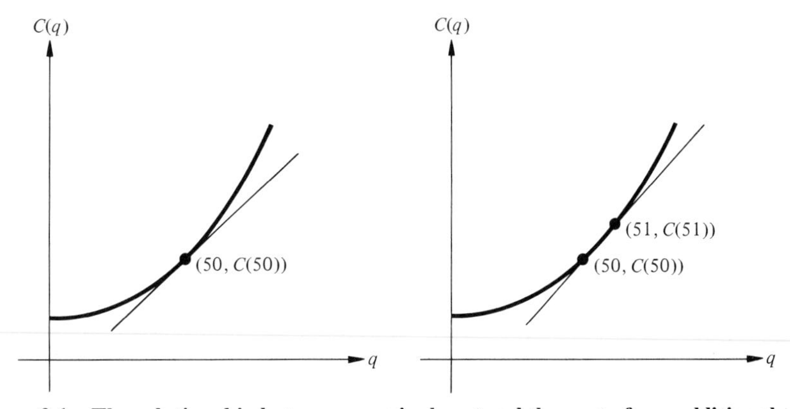

Figure 3.1 The relationship between marginal cost and the cost of one additional unit.

The answers to parts (b) and (c) are almost equal because the points $(50, C(50))$ and $(51, C(51))$ are close together and lie on a portion of the cost curve that is practically linear. For such points, the slope of the secant is a good approximation to the slope of the tangent. Because the similarity of the answers in parts (b) and (c) is typical, and because it is usually easier to compute the marginal cost for one value of q than the total cost for two values of q, economists often used the marginal cost to approximate the actual cost of producing one addtional unit.

In general, the term **marginal analysis** in economics refers to the practice of using the derivative of a function to estimate the change in the dependent variable produced by a unit increase (an increase of 1) in the size of the independent variable. In the next example, marginal analysis is used to estimate the effect of a unit increase in the size of the labor force on the output of a factory.

EXAMPLE 3.3 ● It is estimated that the weekly output at a certain plant is $f(x) = -x^3 + 60x^2 + 1{,}200x + 26{,}400$ units (for $0 \leq x \leq 50$), where x is the number of workers employed at the plant. Currently there are 30 workers employed at the plant. Use marginal analysis to estimate the change in the weekly output that will result from the addition of one more worker to the work force.

SOLUTION The derivative

$$f'(x) = -3x^2 + 120x + 1{,}200$$

is the rate of change of the output f with respect to the number x of workers. For any value of x, this derivative is an approximation to the number of additional units that will be produced each week due to the hiring of the $(x + 1)$st worker. Hence, the change in the weekly output that will result if the number of workers is increased from 30 to 31 is approximately

$$f'(30) = -3(30)^2 + 120(30) + 1{,}200 = 2{,}100 \text{ units}$$

For practice, compute the change in output exactly and compare your answer to the approximation. Is the approximation a good one? ●

Problems

1 An object moves along a straight line so that after t minutes its distance from a fixed reference point is $D(t) = 10t + 5/(t + 1)$ meters.
   (a) At what speed is the object moving at the end of 4 minutes?
   (b) How far does the object actually travel during the 5th minute?

2 It is estimated that t years from now the population of a certain suburban community will be $P(t) = 20 - 6/(t + 1)$ thousand.
   (a) Derive a formula for the rate at which the population will be changing with respect to time.

(b) At what rate will the community's population be growing 1 year from now?
(c) By how much will the population actually increase during the 2d year?
(d) At what rate will the population be growing 9 years from now?
(e) What will happen to the rate of population growth in the long run?

3 It is estimated that t years from now, the circulation of a local newspaper will be $C(t) = 100t^2 + 400t + 5{,}000$.
(a) Derive an expression for the rate at which the circulation will be changing t years from now.
(b) At what rate will the circulation be changing 5 years from now? Will it be increasing or decreasing?
(c) By how much will the circulation actually change during the 5th year?

4 It is estimated that x years from now, the average SAT score of the incoming freshmen at an eastern liberal arts college will be $f(x) = -6x + 582$.
(a) Derive an expression for the rate at which the average SAT score will be changing with respect to time.
(b) What is the significance of the fact that the expression in part (a) is a constant? What is the significance of the fact that the constant in part (a) is negative?

5 Use calculus to prove that if y is a linear function of x, the rate of change of y with respect to x is constant.

6 Two cars leave an intersection at the same time. One travels east at a constant speed of 60 kilometers per hour, while the other goes north at a constant speed of 80 kilometers per hour. Find an expression for the rate at which the distance between the cars is changing with respect to time.

7 An efficiency study of the morning shift at a certain factory indicates that an average worker who arrives on the job at 8:00 A.M. will have assembled $f(x) = -x^3 + 6x^2 + 15x$ transistor radios x hours later.
(a) Derive a formula for the rate at which the worker will be assembling radios after x hours.
(b) At what rate will the worker be assembling radios at 9:00 A.M.?
(c) How many radios will the worker actually assemble between 9:00 A.M. and 10:00 A.M.?

8 Suppose the total cost in dollars of manufacturing q units is $C(q) = 3q^2 + q + 500$.
(a) Use marginal analysis to estimate the cost of manufacturing the 41st unit.
(b) Compute the actual cost of manufacturing the 41st unit.

9 A manufacturer's total monthly revenue is $R(q) = 240q + 0.05q^2$ dollars when q units are produced during the month. Currently, the manufacturer is producing 80 units a month and is planning to increase the monthly output by 1 unit.
(a) Use marginal analysis to estimate the additional revenue that will be generated by the production of the 81st unit.

(b) Use the revenue function to compute the actual additional revenue that will be generated by the production of the 81st unit.

10 It is estimated that the weekly output at a certain plant is $f(x) = -x^2 + 2{,}100x + 86{,}000$ units, where x is the number of workers employed at the plant. Currently there are 60 workers employed at the plant.

(a) Use marginal analysis to estimate the effect that one additional worker will have on the weekly output.

(b) Compute the actual change in the weekly output that will result if one additional worker is hired.

11 At a certain factory, the daily output is $600K^{1/2}$ units, where K denotes the capital investment measured in units of \$1,000. The current capital investment is \$900,000. Use marginal analysis to estimate the effect that an additional capital investment of \$1,000 will have on the daily output.

12 At a certain factory, the daily output is $3{,}000K^{1/2}L^{1/3}$ units, where K denotes the firm's capital investment measured in units of \$1,000, and L the size of the labor force measured in worker-hours. Suppose that the current capital investment is \$400,000 and that 1,331 worker-hours of labor are used each day. Use marginal analysis to estimate the effect that an additional capital investment of \$1,000 will have on the daily output if the size of the labor force is not changed.

13 Suppose the total manufacturing cost C at a certain factory is a function of the number q of units produced which, in turn, is a function of the number t of hours during which the factory has been operating.

(a) What quantity is represented by the derivative dC/dq? In what units is this quantity measured?

(b) What quantity is represented by the derivative dq/dt? In what units is this quantity measured?

(c) What quantity is represented by the product $(dC/dq)(dq/dt)$? In what units is this quantity measured?

4 • THE CHAIN RULE

Suppose the total manufacturing cost at a certain factory is a function of the number of units produced which, in turn, is a function of the number of hours during which the factory has been operating. Let C, q and t denote the cost (in dollars), the number of units, and the number of hours, respectively. Then,

$$\frac{dC}{dq} = \begin{array}{l}\text{rate of change of cost}\\ \text{with respect to output}\end{array} \qquad \text{(dollars per unit)}$$

and

$$\frac{dq}{dt} = \begin{array}{l}\text{rate of change of output}\\ \text{with respect to time}\end{array} \qquad \text{(units per hour)}$$

The product of these two rates is the rate of change of cost with respect to time.

$$\frac{dC}{dq}\frac{dq}{dt} = \begin{array}{l}\text{rate of change of cost}\\ \text{with respect to time}\end{array} \qquad \text{(dollars per hour)}$$

Since the rate of change of cost with respect to time is also given by the derivative dC/dt, it follows that

$$\frac{dC}{dt} = \frac{dC}{dq}\frac{dq}{dt}$$

This formula is a special case of a powerful rule called the **chain rule.**

The chain rule
Suppose y is a function of u and u is a function of x. Then y can be regarded as a function of x and

$$\frac{dy}{dx} = \frac{dy}{du}\frac{du}{dx}$$

That is, the derivative of y with respect to x is the derivative of y with respect to u times the derivative of u with respect to x.

Notice that one way to remember the chain rule is to pretend that the derivatives dy/du and du/dx are quotients and cancel du, reducing the expression $(dy/du)(du/dx)$ on the right-hand side of the equation to the expression dy/dx on the left.

EXAMPLE 4.1 ● Find dy/dx if $y = u^3 - 3u^2 + 1$ and $u = x^2 + 2$.

SOLUTION Since

$$\frac{dy}{du} = 3u^2 - 6u \qquad \text{and} \qquad \frac{du}{dx} = 2x$$

it follows that

$$\frac{dy}{dx} = (3u^2 - 6u)(2x)$$

Notice that this derivative is expressed in terms of the variables x and u. Since we are thinking of y as a function of x, it is more natural to express dy/dx in terms of x alone. To do this, we use the fact that $u = x^2 + 2$ and we conclude that

$$\frac{dy}{dx} = [3(x^2 + 2)^2 - 6(x^2 + 2)](2x) = 6x^3(x^2 + 2)$$

For practice, check this answer by first substituting $u = x^2 + 2$ into the original expression for y and then differentiating with respect to x. ●

The chain rule for powers

In Section 2, you learned the rule

$$\frac{d}{dx}(x^n) = nx^{n-1}$$

for differentiating power functions. There is a related rule (which is actually a special case of the chain rule, in disguise) that you can use to differentiate functions of the form $[h(x)]^n$, that is, functions that are powers of other functions. According to this rule, you begin by computing $n[h(x)]^{n-1}$ and then multiply this expression by the derivative of $h(x)$.

The chain rule for powers

$$\frac{d}{dx}[h(x)]^n = n[h(x)]^{n-1}\frac{d}{dx}[h(x)]$$

Here are some examples.

EXAMPLE 4.2 ● Differentiate the function $f(x) = (2x^4 - x)^3$.

SOLUTION One way to do this problem is to rewrite the function as

$$f(x) = 8x^{12} - 12x^9 + 6x^6 - x^3$$

and then differentiate this polynomial term by term to get

$$f'(x) = 96x^{11} - 108x^8 + 36x^5 - 3x^2$$

But see how much easier it is to use the chain rule for powers. According to this rule,

$$f'(x) = 3(2x^4 - x)^2\frac{d}{dx}(2x^4 - x)$$

$$= 3(2x^4 - x)^2(8x^3 - 1)$$

And the answer even comes out in factored form! ●

EXAMPLE 4.3 ● Differentiate the function $f(x) = \dfrac{1}{(x + 3)^5}$.

SOLUTION We rewrite the function as

$$f(x) = (x + 3)^{-5}$$

and apply the chain rule to get

$$f'(x) = -5(x + 3)^{-6}\frac{d}{dx}(x + 3)$$

Since $d/dx\,(x + 3) = 1$, we conclude that

$$f'(x) = -\frac{5}{(x + 3)^6}$$

For practice, compute $f'(x)$ again, this time using the quotient rule. Which method do you prefer? ●

To see that the chain rule for powers is really nothing more than a special case of the chain rule, think of the function $y = [h(x)]^n$ as the function formed from $y = u^n$ and $u = h(x)$. Then,

$$\frac{dy}{dx} = \frac{d}{dx}[h(x)]^n \qquad \frac{dy}{du} = n[h(x)]^{n-1} \qquad \text{and} \qquad \frac{du}{dx} = \frac{d}{dx}[h(x)]$$

and the chain rule $dy/dx = (dy/du)(du/dx)$ can be rewritten as

$$\frac{d}{dx}[h(x)]^n = n[h(x)]^{n-1}\frac{d}{dx}[h(x)]$$

which is precisely the chain rule for powers.

Related rates In many problems, a quantity is given as a function of one variable which in turn can be written as a function of a second variable and the goal is to find the rate of change of the original quantity with respect to the second variable. Such problems are sometimes called **related-rates problems** and can be solved by means of the chain rule. Here is an example.

EXAMPLE 4.4 ● An environmental study of a certain suburban community suggests that the average daily level of carbon monoxide in the air will be $c(p) = \sqrt{0.5p^2 + 17}$ parts per million when the population is p thousand. It is estimated that t years from now the population of the community will be $p(t) = 3.1 + 0.1t^2$ thousand. At what rate will the carbon monoxide level be changing with respect to time 3 years from now?

SOLUTION The goal is to find dc/dt when $t = 3$. According to the chain rule,

$$\frac{dc}{dt} = \frac{dc}{dp}\frac{dp}{dt} = [\tfrac{1}{2}(0.5p^2 + 17)^{-1/2}(p)](0.2t)$$

Since $p = 3.1 + 0.1t^2$, it follows that $p = 4$ when $t = 3$. Hence, when $t = 3$,

$$\frac{dc}{dt} = \tfrac{1}{2}(25)^{-1/2}(4)(0.6) = 0.24 \text{ parts per million per year}$$ ●

Problems In Problems 1 through 10, use the chain rule to compute the derivative dy/dx.

1 $y = u^2 + 1$, $u = 3x - 2$
2 $y = 2u^2 - u + 5$, $u = 1 - x^2$
3 $y = \sqrt{u}$, $u = x^2 + 2x - 3$
4 $y = u^2$, $u = \sqrt{x}$
5 $y = 1/u^2$, $u = x^2 + 1$
6 $y = 1/u$, $u = 3x^2 + 5$
7 $y = 1/\sqrt{u}$, $u = x^2 - 9$
8 $y = u^2 + u - 2$, $u = 1/x$
9 $y = 1/(u - 1)$, $u = x^2$
10 $y = u^2$, $u = 1/(x - 1)$

In Problems 11 through 19, differentiate the given function.

11 $f(x) = (3x^4 - 7x^2 + 9)^5$
12 $h(t) = (t^5 - 4t^3 - 7)^8$
13 $y = \sqrt{5x^6 - 12}$
14 $f(t) = 3/(1 - t)^4$
15 $h(u) = 1/\sqrt{5u^3 + 2}$
16 $y = [(t + 1)/(t - 1)]^{2/3}$
17 $g(u) = (u^2 + 1)^3(2u - 1)^2$
18 $y = (1 - x)^5(x + 1)^8$
19 $h(x) = (x + 1)^5/(1 - x)^4$

In Problems 20 through 23, find an equation of the line that is tangent to the graph of f for the given value of x.

20 $f(x) = (3x^2 + 1)^2$; $x = -1$
21 $f(x) = (x^2 - 3)^5(2x - 1)^3$; $x = 2$
22 $f(x) = 1/(2x - 1)^6$; $x = 1$
23 $f(x) = [(x + 1)/(x - 1)]^3$; $x = 3$

24 Differentiate the function $f(x) = (3x^2 - 9x)^2$ by two different methods, first using the chain rule and then the product rule. Show that the two answers are really the same.

25 The gross annual earnings of a certain company were $f(t) = \sqrt{10t^2 + t + 236}$ thousand dollars t years after its formation in January, 1975. At what rate were the gross annual earnings growing in January, 1979?

26 At a certain factory, the total cost of manufacturing q units during the daily production run is $C(q) = 0.2q^2 + q + 900$ dollars. From experience it has been determined that approximately $q(t) = t^2 + 100t$ units are manufactured during the first t hours of a production run. Compute the rate at which the total manufacturing cost is changing with respect to time 1 hour after production commences.

27 It is estimated that t years from now the population of a certain suburban community will be $20 - 6/(t + 1)$ thousand. An environmental study indicates that the average daily level of carbon monoxide in the air will be $0.5\sqrt{p^2 + p + 58}$ parts per million when the population is p thousand. Find the rate at which the level of carbon monoxide will be changing with respect to time 2 years from now.

28 When electric blenders are sold for p dollars apiece, local consumers will buy $8{,}000/p$ blenders a month. It is estimated that t months from now, the price of the blenders will be $0.04t^{3/2} + 15$ dollars. Compute the rate at which the monthly demand for the blenders will be changing with respect to time 25 months from now. Will the demand be increasing or decreasing?

29 An importer of Brazilian coffee estimates that local consumers will buy approximately $4{,}374/p^2$ pounds of the coffee per week when the price is p dollars per pound. It is estimated that t weeks from now the price of Brazilian coffee will be $0.02t^2 + 0.1t + 6$ dollars per pound. At what rate will the weekly demand for the coffee be changing 10 weeks from now? Will the demand be increasing or decreasing?

5 • MAXIMA AND MINIMA

In Example 1.3, we used calculus to maximize a profit function like the one shown in Figure 5.1. We observed that the maximum value corresponded to the unique point on the graph at which the slope of the tangent was zero, set the derivative equal to zero, and solved for x.

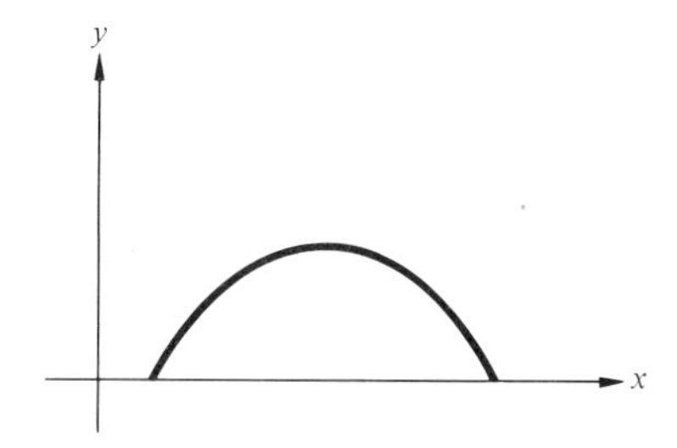

Figure 5.1 A profit function.

The simplicity of this example is misleading. In general, not every point at which the derivative of a function is zero is a peak. For example, the familiar function $f(x) = x^2$, shown in Figure 5.2*a* reaches its *lowest* point at $x = 0$ when the derivative, $f'(x) = 2x$, is zero. It is even possible for the derivative of a function to be zero at a point that is neither a maximum nor a minimum. For example, the derivative of the function $f(x) = x^3$ is $f'(x) = 3x^2$, which is zero

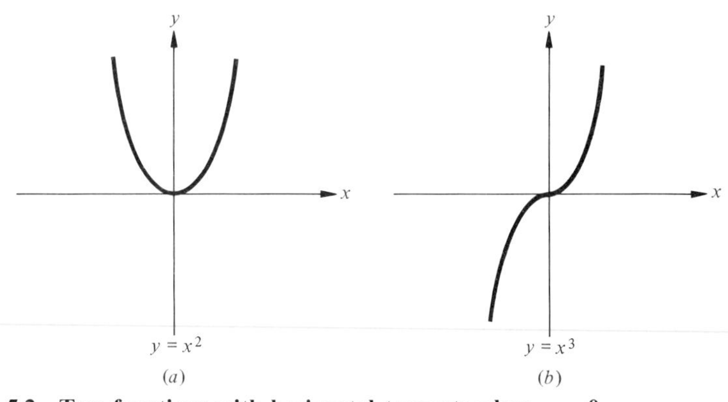

Figure 5.2 Two functions with horizontal tangents when $x = 0$.

when $x = 0$. The graph of f, shown in Figure 5.2b, "levels off" when $x = 0$, yet has neither a maximum nor a minimum at this point.

In this section we shall clarify the situation by developing a systematic procedure for locating and identifying maxima and minima. In the process, we shall discover how to use the derivative as a tool in curve sketching.

Relative maxima and minima

A **relative maximum** of a function is a peak, a point that is higher than any neighboring point on the graph. A **relative minimum** is the bottom of a valley, a point on the graph that is lower than any neighboring point. The function sketched in Figure 5.3 has a relative maximum at $x = b$ and relative minima at $x = a$ and $x = c$. Notice that a relative maximum need not be the highest point on a graph. It is maximal only *relative to* neighboring points. Similarly, a relative minimum need not be the lowest point on a graph.

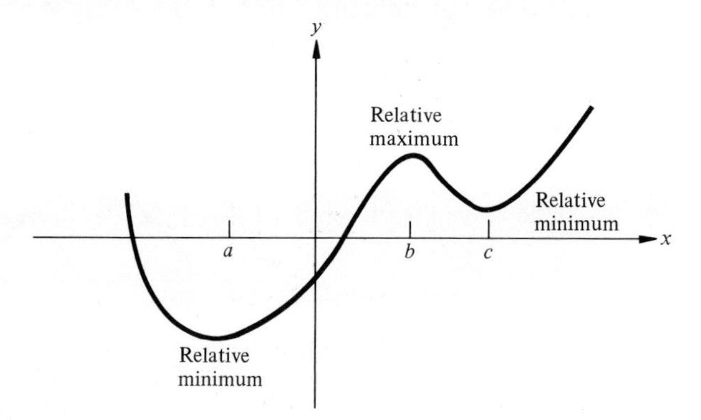

Figure 5.3 Relative maxima and minima.

A function is said to be **increasing** if its graph is rising as x increases, and **decreasing** if its graph is falling as x increases. The function in Figure 5.4 is increasing for $a < x < b$ and for $x > c$. It is decreasing for $x < a$ and for $b < x < c$.

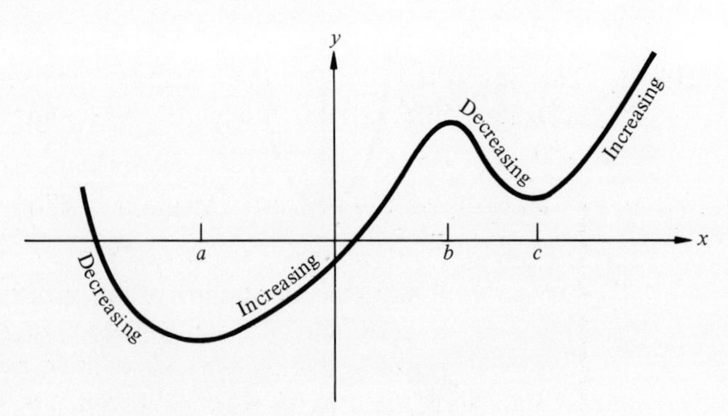

Figure 5.4 Increase and decrease of a function.

Knowledge of the intervals on which a function is increasing and decreasing leads to easy identification of its relative extrema, that is, its relative maxima and minima. A relative maximum occurs when a function stops increasing and starts decreasing. In Figure 5.4, this happens at $x = b$. A relative minimum occurs when a function stops decreasing and starts increasing. In Figure 5.4, this happens at $x = a$ and $x = c$.

The sign of the derivative

A function whose derivative is defined for all values of x is said to be **differentiable.** If you know the derivative of such a function, you can find out where the function is increasing or decreasing by checking the sign of the derivative. This is because the derivative is the slope of the tangent. When the derivative is positive, the slope of the tangent is positive and the function is increasing (Figure 5.5*a*). When the derivative is negative, the slope of the tangent is negative and the function is decreasing (Figure 5.5*b*).

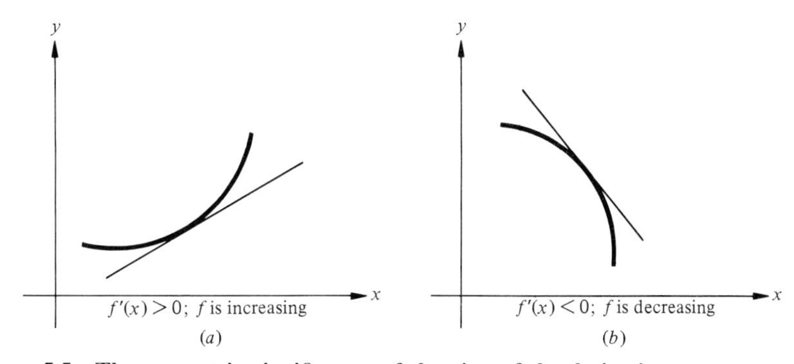

Figure 5.5 The geometric significance of the sign of the derivative.

A point at which the derivative of a function is zero is said to be a **critical point** of the function. Every relative extremum of a differentiable function is a critical point. However, as we have seen, not every critical point is necessarily a relative extremum.

Curve sketching

The preceding observations suggest the following general procedure for sketching differentiable functions and finding their relative extrema.

How to use the derivative to graph a differentiable function

Step 1. Compute the derivative $f'(x)$.
Step 2. Find the x coordinates of the critical points by setting $f'(x)$ equal to zero, and plot the corresponding points. (These are the only possible points at which relative extrema can occur.)
Step 3. To determine where the function is increasing or decreasing, check the sign of the derivative on the intervals whose endpoints

are the x coordinates of the critical points. Draw the graph so that it increases on the intervals on which the derivative is positive and decreases on the intervals on which the derivative is negative.

Here are two examples.

EXAMPLE 5.1 ● Determine where the polynomial $f(x) = 2x^3 + 3x^2 - 12x - 7$ is increasing and where it is decreasing, find its relative extrema and draw a graph.

SOLUTION We begin by computing and factoring the derivative:

$$f'(x) = 6x^2 + 6x - 12 = 6(x + 2)(x - 1)$$

From the factored form of the derivative we see that $f'(x) = 0$ when $x = -2$ and when $x = 1$. Since $f(-2) = 13$ and $f(1) = -14$, it follows that the critical points are $(-2, 13)$ and $(1, -14)$. We begin the sketch in Figure 5.6*a* by plotting these critical points.

To determine where the function is increasing and where it is decreasing, we check the sign of the derivative for $x < -2$, for $-2 < x < 1$, and for $x > 1$. When $x < -2$, both $x + 2$ and $x - 1$ are negative, so the derivative $f'(x) = 6(x + 2)(x - 1)$ is positive. Hence, f is increasing when $x < -2$.

When $-2 < x < 1$, $x + 2$ is positive while $x - 1$ is still negative. Hence the derivative is negative and f is decreasing when $-2 < x < 1$.

Finally, when $x > 1$, both $x + 2$ and $x - 1$ are positive. Hence the derivative is positive and f is increasing when $x > 1$.

We complete the sketch in Figure 5.6*b*. Notice that the function has a relative maximum at the critical point $(-2, 13)$ and a relative minimum at the critical point $(1, -14)$. ●

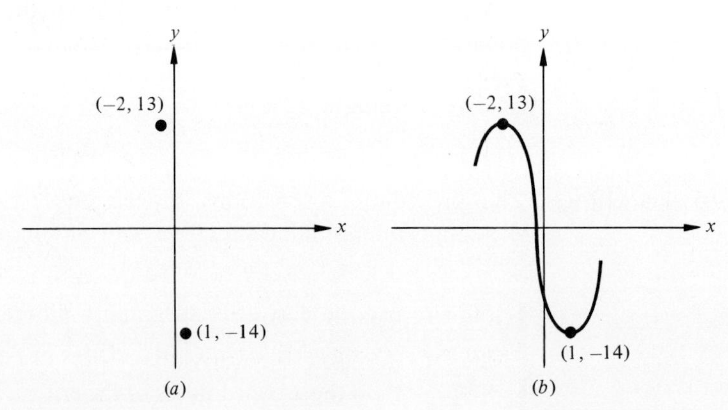

Figure 5.6 Steps leading to the graph of $y = 2x^3 + 3x^2 - 12x - 7$.

EXAMPLE 5.2 ● Determine where the rational function $f(x) = x^2/(x - 2)$ is increasing and where it is decreasing, find its relative extrema and draw a graph.

SOLUTION Using the quotient rule we find that

$$f'(x) = \frac{(x-2)(2x) - x^2(1)}{(x-2)^2} = \frac{x(x-4)}{(x-2)^2}$$

This derivative is zero when $x = 0$ and $x = 4$ and the corresponding critical points are $(0, 0)$ and $(4, 8)$.

To find where the function is increasing and where it is decreasing, we check the sign of the derivative on the intervals determined by the critical points. Since $f'(x)$ is undefined at $x = 2$, its sign for $0 < x < 2$ could be different from its sign for $2 < x < 4$, and so we check these two intervals separately. We find:

Interval	Sign of $f'(x)$	Increasing or decreasing
$x < 0$	positive	increasing
$0 < x < 2$	negative	decreasing
$2 < x < 4$	negative	decreasing
$x > 4$	positive	increasing

It follows that f must have a relative maximum at $(0, 0)$ and a relative minimum at $(4, 8)$. The graph is sketched in Figure 5.7. Notice that there is a discontinuity when $x = 2$. ●

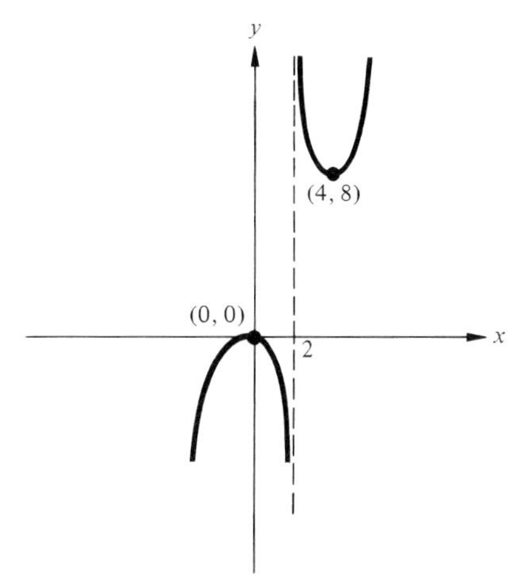

Figure 5.7 The graph of $y = x^2/(x - 2)$.

The procedure illustrated in the preceding examples is based on the fact that at the relative extrema of a differentiable function, the derivative is zero. If a function is not differentiable everywhere, the points at which its derivative is undefined may also be relative extrema. Here is an example.

EXAMPLE 5.3 ● Determine where the function $f(x) = x^{2/3}$ is increasing and where it is decreasing, find its relative extrema and draw a graph.

SOLUTION The derivative of f is $f'(x) = \frac{2}{3}x^{-1/3}$, which is positive for $x > 0$ and negative for $x < 0$. Hence f is increasing when $x > 0$ and decreasing when $x < 0$. The graph of f is shown in Figure 5.8. Notice that f has a relative minimum when $x = 0$, the value of x for which the derivative is undefined. The derivative is undefined because at (0, 0), the tangent to the graph of f is the y axis, whose slope is undefined. ●

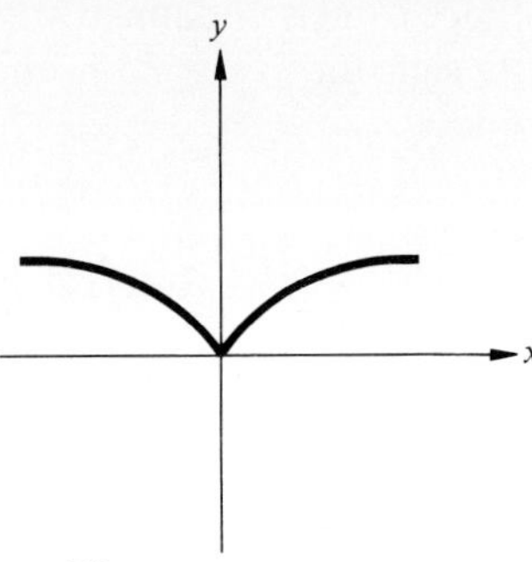

Figure 5.8 The graph of $y = x^{2/3}$.

Absolute maxima and minima In most practical optimization problems, the goal is to find the **absolute maximum** or **absolute minimum** of a particular function on some interval, rather than the relative maxima or minima. The absolute maximum of a function on an interval is the largest value of the function on the interval. The absolute minimum is the smallest value. These absolute extrema often coincide with relative extrema, but not always. For example, on the interval $a \leq x \leq b$, the function in Figure 5.9 attains its absolute maximum at the relative maximum. However, its absolute minimum on the interval occurs at $x = a$ which is not a relative minimum.

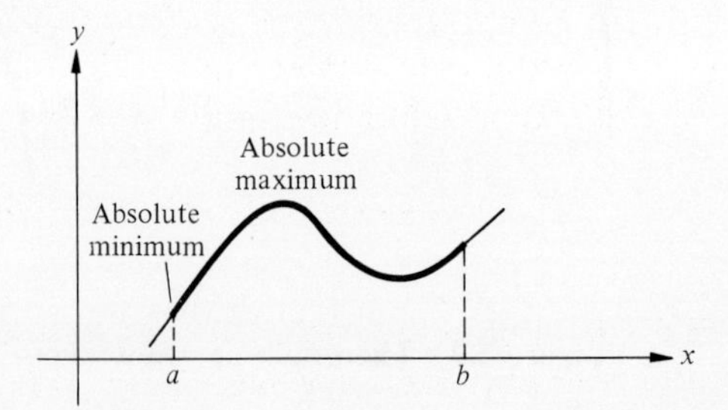

Figure 5.9 Absolute extrema.

An interval of the form $a \leq x \leq b$ is called a **closed interval.** In general, a function that is differentiable on a closed interval $a \leq x \leq b$ attains an absolute maximum and an absolute minimum on that interval. An absolute extremum can occur either at a relative extremum in the interval or at an endpoint $x = a$ or $x = b$. The possiblities are illustrated in Figure 5.10.

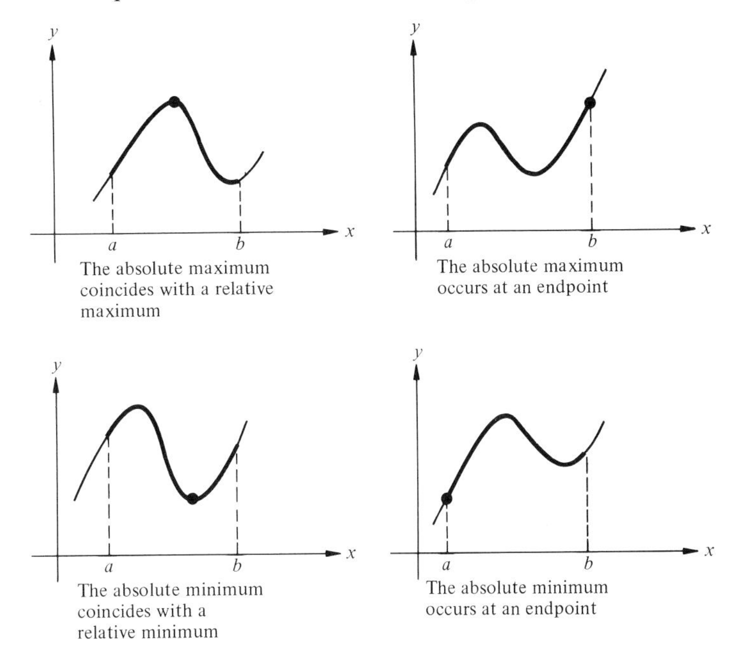

Figure 5.10 The location of absolute extrema on a closed interval.

These observations suggest the following simple procedure for locating and identifying absolute extrema.

How to find the absolute extrema of a differentiable function
To find the absolute extrema of a differentiable function f on the interval $a \leq x \leq b$, first locate all the critical points of f in the interval and compute the corresponding values of f. Then compute $f(a)$ and $f(b)$, the values of f at the endpoints. The largest and smallest of all these values are the absolute maximum and absolute minimum, respectively.

The procedure is illustrated in the following example.

EXAMPLE 5.4 ● For several weeks, the highway department has been recording the speed of freeway traffic flowing past a certain downtown exit. The data suggest that between the hours of 1:00 P.M. and 6:00 P.M. on a normal weekday, the speed of the traffic at the exit is approximately $S(t) = 2t^3 - 21t^2 + 60t + 40$ kilometers per hour, where t is the number of hours past noon. At what time between 1:00 P.M. and 6:00 P.M. is the traffic moving the fastest, and at what time is it moving the slowest?

SOLUTION The goal is to find the absolute maximum and absolute minimum of the function $S(t)$ on the interval $1 \le t \le 6$. From the derivative

$$S'(t) = 6t^2 - 42t + 60 = 6(t-2)(t-5)$$

you get the t coordinates of the critical points, $t = 2$ and $t = 5$, both of which lie in the interval $1 \le t \le 6$.

Now compute $S(t)$ for these values of t and at the endpoints $t = 1$ and $t = 6$ to get

$$S(1) = 81 \qquad S(2) = 92 \qquad S(5) = 65 \qquad S(6) = 76$$

Since the largest of these values is $S(2) = 92$ and the smallest is $S(5) = 65$, you can conclude that the traffic is moving fastest at 2:00 P.M. when its speed is 92 kilometers per hour and slowest at 5:00 P.M. when its speed is 65 kilometers per hour.

For reference, the graph of S is sketched in Figure 5.11 ●

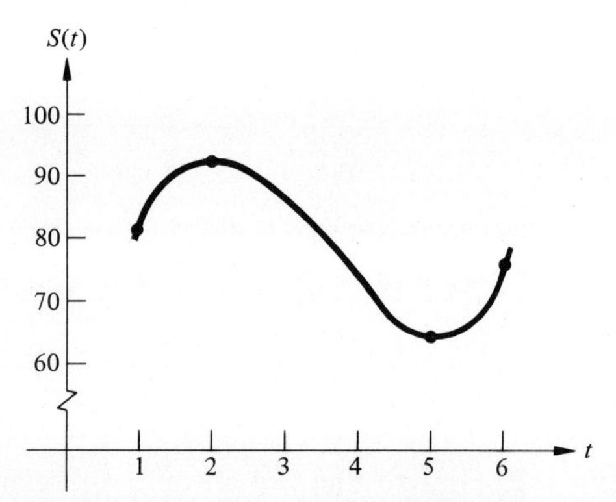

Figure 5.11 Traffic speed: $S(t) = 2t^3 - 21t^2 + 60t + 40$.

Problems

In Problems 1 through 3, specify where the function is increasing and where it is decreasing.

1 **2**

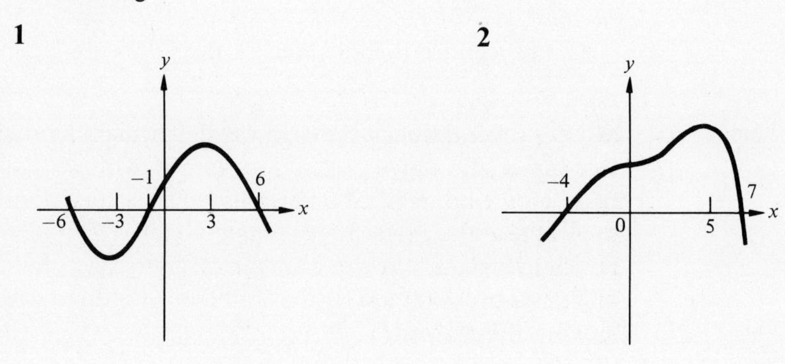

3

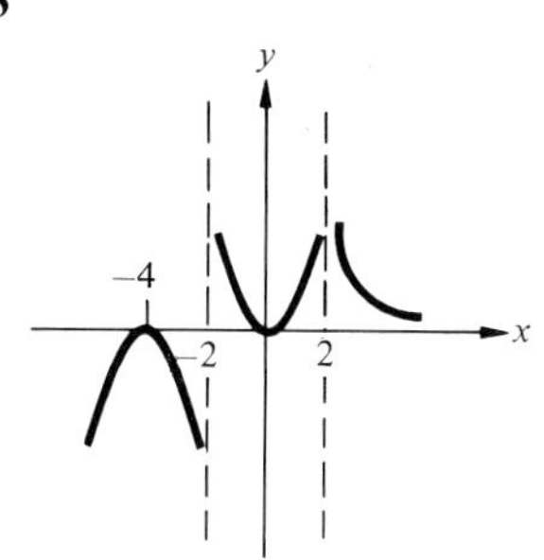

In Problems 4 through 6, sketch a graph of a function that has all of the given properties.

4 (a) $f'(x) > 0$ when $x < -5$ and when $x > 1$
(b) $f'(x) < 0$ when $-5 < x < 1$
(c) $f(-5) = 4$ and $f(1) = -1$

5 (a) $f'(x) < 0$ when $x < -1$
(b) $f'(x) > 0$ when $-1 < x < 3$ and when $x > 3$
(c) $f'(3) = 0$

6 (a) $f'(x) > 0$ when $x > 2$
(b) $f'(x) < 0$ when $x < 0$ and when $0 < x < 2$
(c) f has a discontinuity when $x = 0$

In Problems 7 through 22, determine where the given function is increasing and where it is decreasing, find its relative extrema and draw the graph.

7 $f(x) = x^3 + 3x^2 + 1$
8 $f(x) = \frac{1}{3}x^3 - 9x + 2$
9 $f(x) = x^3 - 3x - 4$
10 $f(x) = x^5 - 5x^4 + 100$
11 $f(x) = 3x^5 - 5x^3$
12 $f(x) = 324x - 72x^2 + 4x^3$
13 $f(x) = 2x^3 + 6x^2 + 6x + 5$
14 $f(x) = 10x^6 + 24x^5 + 15x^4 + 3$
15 $f(x) = (x^2 - 1)^5$
16 $f(x) = (x^2 - 1)^4$
17 $f(x) = (x^3 - 1)^4$
18 $f(x) = x^2/(x - 1)$
19 $f(x) = x + 1/x$
20 $f(x) = 2x + 18/x + 1$
21 $f(x) = (x^2 - 3x)/(x + 1)$
22 $f(x) = 6x^2 + 12{,}000/x$

In Problems 23 through 31, find the absolute maximum and absolute minimum of the given function on the specified interval.

23 $f(x) = x^3 + 3x^2 + 1;\ -3 \le x \le 2$
24 $f(x) = \frac{1}{3}x^3 - 9x + 2;\ 0 \le x \le 2$
25 $f(x) = x^5 - 5x^4 + 1;\ 0 \le x \le 5$
26 $f(x) = 3x^5 - 5x^3;\ -2 \le x \le 0$
27 $f(x) = 10x^6 + 24x^5 + 15x^4 + 3;\ -1 \le x \le 1$
28 $f(x) = (x^2 - 4)^5;\ -3 \le x \le 2$
29 $f(x) = x^2/(x - 1);\ -2 \le x \le -\frac{1}{2}$
30 $f(x) = x + 1/x;\ \frac{1}{2} \le x \le 3$
31 $f(x) = 1/(x^2 - 9);\ 0 \le x \le 2$

32 Find constants a, b, and c such that the graph of the function $f(x) = ax^2 + bx + c$ has a relative maximum at (5, 12) and crosses the y axis at (0, 3).

33 Find the equation of a function whose derivative is zero when $x = 2$, but that has neither a relative maximum nor a relative minimum when $x = 2$.

34 Use calculus to prove that the relative extremum of the quadratic function $y = ax^2 + bx + c$ occurs when $x = -b/2a$.

35 Use calculus to prove that the relative extremum of the quadratic function $y = (x - p)(x - q)$ occurs midway between its x intercepts.

36 Suppose that x years after its founding in 1960, a certain civil rights organization had a membership of $f(x) = 100(2x^3 - 45x^2 + 264x)$.

(a) At what time between 1960 and 1974 was the membership of the organization largest? What was the membership at that time?

(b) At what time between 1961 and 1974 was the membership the smallest? What was the membership at that time?

37 An all-news radio station has made a survey of the listening habits of local residents between the hours of 5:00 P.M. and midnight. The survey indicates that the percentage of the local adult population that is tuned in to the station x hours after 5:00 P.M. is $f(x) = \frac{1}{8}(-2x^3 + 27x^2 - 108x + 240)$.

(a) At what time between 5:00 P.M. and midnight are the most people listening to the station?

(b) At what time between 5:00 P.M. and midnight are the fewest people listening?

6 • PRACTICAL OPTIMIZATION PROBLEMS

The purpose of this section is to formulate a general procedure for solving practical optimization problems and to illustrate some of the techniques of successful problem solving.

The first step in solving such a problem is to decide precisely what you are to optimize. Once you have identified this quantity, choose a letter to represent it. Some people are most comfortable using the standard letter f for this purpose. Others find it helpful to choose a letter more closely related to the quantity, such as R for revenue or A for area.

Your goal is to represent the quantity to be optimized as a function of some other variable so that you can apply calculus. It is usually a good idea to express the desired function in words before trying to represent it mathematically.

Once the function has been expressed in words, the next step is to choose an appropriate variable. Sometimes the choice is obvious. In other problems you may be faced with a choice among several natural variables. When this happens, think ahead and try to choose the variable that leads to the simplest functional representation. In some problems the quantity to be optimized is expressed most naturally in terms of two variables. If so, you will have to find a way to write one of these variables in terms of the other.

The next step is to express the quantity to be optimized as a function of the variable you have chosen. In most problems, the function has a practical interpretation only when the variable lies in a certain interval. Once you have written the function and identified the appropriate interval, the difficult part is done and the rest is routine. To complete the problem, simply apply the techniques of calculus developed in Section 5 to optimize your function on the specified interval.

Here are two examples to illustrate this procedure.

EXAMPLE 6.1 ● A cable is to be run from a power plant on one side of a river 900 meters wide to a factory on the other side, 3,000 meters downstream. The cost of running the cable under the water is \$5 per meter while the cost over land is \$4 per meter. What is the most economical route over which to run the cable?

SOLUTION We begin by drawing a diagram (Figure 6.1) to help us visualize the situation.

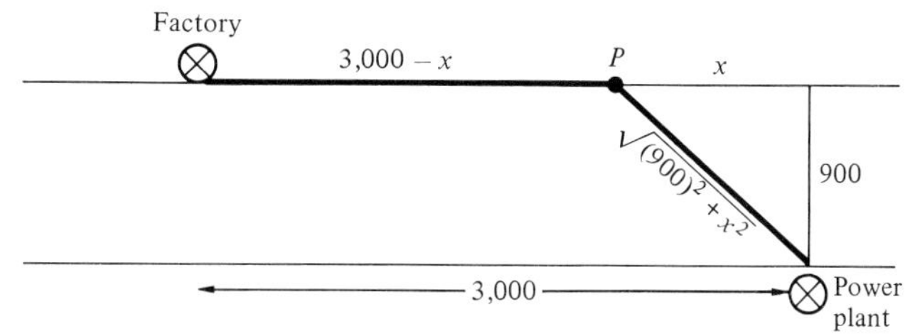

Figure 6.1 Relative positions of factory, river, and power plant.

Notice that in drawing the diagram in Figure 6.1 we have assumed that the cable should be run in a *straight line* from the power plant to some point P on the opposite bank. Do you see why this assumption is justified?

Our goal is to minimize the cost of installing the cable. If we let C denote this cost, we have

$$C = 5(\text{number of meters of cable under water}) + 4(\text{number of meters of cable over land})$$

We let x denote the distance from P to the point directly across the river from the power plant. Then, the distance over land from P to the factory is $3{,}000 - x$ and (by the pythagorean theorem) the distance under water from the power plant to the point P is $\sqrt{(900)^2 + x^2}$. Hence, the total cost function is

$$C(x) = 5\sqrt{(900)^2 + x^2} + 4(3{,}000 - x)$$

Since the distances x and $3{,}000 - x$ cannot be negative, $C(x)$ has a practical interpretation only on the interval $0 \leq x \leq 3{,}000$. Our goal, therefore, is to find the absolute minimum of the function C on this interval.

To find the critical points, we compute the derivative

$$C'(x) = \frac{5x}{\sqrt{(900)^2 + x^2}} - 4$$

and set it equal to zero to get

$$\frac{5x}{\sqrt{(900)^2 + x^2}} - 4 = 0$$

or

$$\sqrt{(900)^2 + x^2} = \tfrac{5}{4}x$$

Squaring both sides of this equation and solving for x we get

$$(900)^2 + x^2 = \tfrac{25}{16}x^2$$

or

$$x^2 = \tfrac{16}{9}(900)^2$$

from which we conclude that

$$x = \pm 1{,}200$$

Since $0 \leq x \leq 3{,}000$, we choose the positive value $x = 1{,}200$.

To find the absolute minimum of C on the interval $0 \leq x \leq 3{,}000$, we compare the value of C at the critical point with the values at the endpoints $x = 0$ and $x = 3{,}000$. Since

$$C(0) = 16{,}500 \qquad C(1{,}200) = 14{,}700 \qquad C(3{,}000) = 1{,}500\sqrt{109} > 15{,}000$$

we conclude that the installation cost is minimal if the cable reaches the opposite bank 1,200 meters downstream from the power plant. ●

In the next example, the quantity to be optimized is expressed most naturally in terms of *two* variables. Fortunately, additional information in the problem allows us to write one of these variables in terms of the other so that ultimately we have a function of just one variable.

EXAMPLE 6.2 ● The highway department is planning to build a picnic area for motorists along a major highway. It is to be rectangular with an area of 5,000 square meters and is to be fenced off on the three sides not adjacent to the highway. What is the least amount of fencing that will be needed to complete the job?

SOLUTION We label the sides of the picnic area as indicated in Figure 6.2 and let f denote the amount of fencing required.

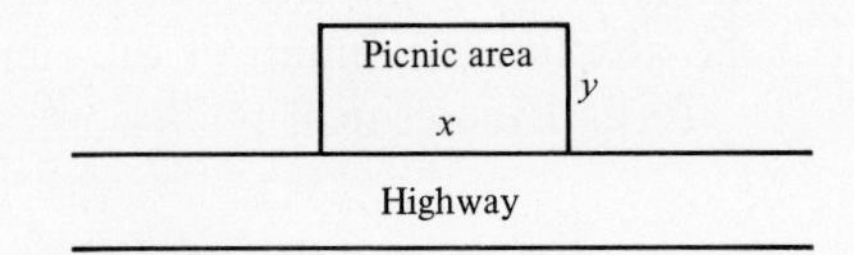

Figure 6.2 Rectangular picnic area.

Then,

$$f = x + 2y$$

Since we want to express f in terms of a single variable, we look for an equation relating x and y. The fact that the area is to be 5,000 square meters gives us this equation. In particular,

$$xy = 5{,}000 \qquad \text{or} \qquad y = \frac{5{,}000}{x}$$

Substituting $y = 5{,}000/x$ into our previous expression for f we get

$$f(x) = x + \frac{10{,}000}{x}$$

Since $f(x)$ has a practical interpretation for any positive value of x, our goal is to find the absolute minimum of f on the interval $x > 0$.

To find the critical points we set the derivative

$$f'(x) = 1 - \frac{10{,}000}{x^2}$$

equal to zero and solve for x getting

$$1 - \frac{10{,}000}{x^2} = 0$$

$$x^2 = 10{,}000$$

$$x = \pm 100$$

Only the positive value of x is of interest, so we take $x = 100$.

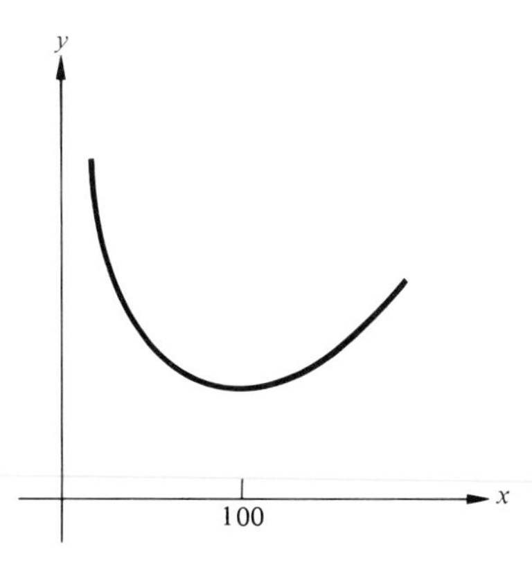

Figure 6.3 The graph of $y = x + \dfrac{10{,}000}{x}$ for $x > 0$.

Since the interval $x > 0$ is not a closed interval, we cannot find the absolute minimum of f on this interval by routinely comparing the values of f at critical points and endpoints. Instead, we observe that $f'(x) < 0$ if $0 < x < 100$, and $f'(x) > 0$ if $x > 100$. It follows that f is decreasing for $0 < x < 100$ and increasing for $x > 100$, as indicated in Figure 6.3 and that $x = 100$ does indeed correspond to the absolute minimum of $f(x)$ for $x > 0$.

Finally, since $f(100) = 200$, we conclude that at least 200 meters of fencing will be required to complete the job. ●

Problems

1 There are 320 meters of fencing available to enclose a rectangular field. How should this fencing be used so that the enclosed area is as large as possible?

2 Prove that of all rectangles with a given perimeter, the square has the largest area.

3 A city recreation department plans to build a rectangular playground having an area of 3,600 square meters and surround it by a fence. How can this be done using the least amount of fencing?

4 Prove that of all rectangles with a given area the square has the smallest perimeter.

5 A college bookstore can obtain the book *Social Groupings of the American Dragonfly* from the publisher at a cost of \$3 per book. The bookstore has been offering the book at a price of \$15 per copy and at this price has been selling 200 copies a month. The bookstore is planning to lower its price to stimulate sales and estimates that for each \$1 reduction in the price, 20 more books will be sold each month. At what price should the bookstore sell the book to generate the largest possible profit?

6 During the summer, members of a scout troup have been collecting used bottles that they plan to deliver to a glass company for recycling. So far, in 80 days the scouts have collected 24,000 kilograms of glass for which the glass company currently offers 1 cent per kilogram. However, because bottles are accumulating faster than they can be recycled, the company plans to reduce by 1 cent each day the price it will pay for 100 kilograms of used glass. Assume that the scouts can continue to collect bottles at the same rate and that transportation costs make more than one trip to the glass company unfeasible. What is the most profitable time for the scouts to conclude their summer project and deliver the bottles?

7 A bus company is willing to charter buses only to groups of 40 or more people. If a group contains exactly 40 people, each is charged \$60. In larger groups, everybody's fare is reduced by 50 cents for each person in excess of 40. What size group will generate the most revenue for the bus company?

8 A closed box with a square base is to have a volume of 250 cubic meters. The material for the top and bottom of the box costs \$2 per square meter and the material for the sides costs \$1 per square meter. Can the box be constructed for less than \$300?

9 A carpenter has been asked to build an open box with a square base. The sides of the box will cost \$3 per square meter and the base will cost \$4 per

square meter. What are the dimensions of the box of largest volume that can be constructed for \$48?

10 A truck is 975 kilometers due east of a car and is traveling west at a constant speed of 60 kilometers per hour. Meanwhile, the car is going north at a constant speed of 90 kilometers per hour. At what time will the car and truck be closest to each other? (*Hint:* You will simplify the calculation if you minimize the *square* of the distance between the car and truck rather than the distance itself. Can you explain why this simplification is justified?)

11 If the factory in Example 6.1 is only 1,000 meters downstream from the power plant, what is the most economical route over which to run the cable?

12 For the summer, the company that is installing the cable in Example 6.1 has hired a temporary employee with a Ph.D. in mathematics. The mathematician, recalling a problem from freshman calculus, asserts that no matter how far downstream the factory is located (beyond 1,200 meters), it would be most economical to have the cable reach the opposite bank 1,200 meters downstream from the power plant. The foreman, amused by the naïveté of his overeducated employee, replies: "Any fool can see that if the factory is further away, the cable should reach the opposite bank further downstream. It's just common sense." Who is right? And why?

13 It is noon, and the hero of a popular spy story (the same fellow who escaped from the diamond smugglers in Chapter 1, Section 3, Problem 13) is driving a Jeep through the sandy desert in the tiny principality of Alta Loma. He is 32 kilometers from the nearest point on a straight, paved road. Down the road 16 kilometers is a power plant in which a band of international saboteurs has placed a time bomb set to explode at 12:50 P.M. The Jeep can travel 48 kilometers per hour in the sand and 80 kilometers per hour on the paved road. If he arrives at the power station in the shortest possible time, how long will our hero have to defuse the bomb?

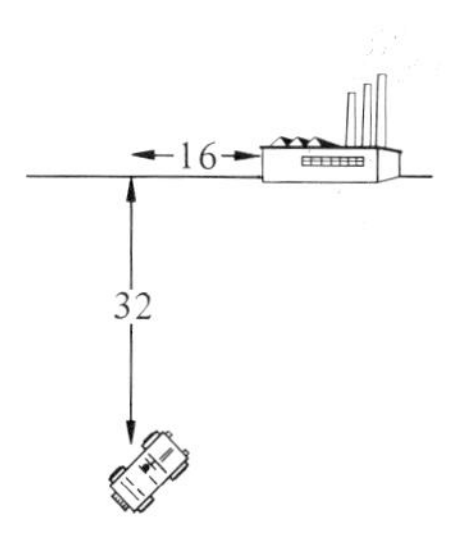

14 A printer receives an order to produce a rectangular poster containing 25 square centimeters of print surrounded by margins of 2 centimeters on each side and 4 centimeters on the top and bottom. What are the dimensions of the smallest piece of paper that can be used to make the poster? (*Hint:* An unwise choice of variables will make the calculations unnecessarily complicated.)

15 A plastics firm has received an order from the city recreation department to manufacture 8,000 special Styrofoam kickboards for its summer swimming program. The firm owns 10 machines, each of which can produce 30 kickboards an hour. The cost of setting up the machines to produce the kickboards is \$20 per machine. Once the machines have been set up, the operation is fully automated and can be supervised by a single foreman earning \$4.80 per hour.

(a) How many of the machines should be used to minimize the cost of production?

(b) How much will the foreman earn during the production run if the optimal number of machines is used?

(c) How much will it cost to set up the machines if the optimal number of machines is used?

16 A manufacturing firm receives an order for Q items. The firm owns several machines, each of which can produce n items an hour. The set-up cost is s dollars per machine and the operating cost is p dollars per hour.

(a) Derive a formula for the number of machines that should be used to keep production costs as small as possible.

(b) Prove that the production costs are minimal when the cost of setting up the machines is equal to the cost of running the machines.

17 An efficiency study of the morning shift at a certain factory indicates that an average worker who arrives on the job at 8:00 A.M. will have assembled $f(x) = -x^3 + 6x^2 + 15x$ transistor radios x hours later. The study indicates further that after a 15-minute coffee break, the average worker can assemble $g(x) = -\frac{1}{3}x^3 + x^2 + 23x$ radios in x hours. Determine the time between 8:00 A.M. and noon at which a 15-minute coffee break should be scheduled so that the average worker will assemble the maximum number of radios by lunchtime at 12:15 P.M.

18 A retailer has bought several cases of a certain imported wine. As the wine ages, its value initially increases, but eventually the wine will pass its prime and its value will decrease. Suppose that x years from now, the value of a case will be changing at a rate of $53 - 10x$ dollars per year. Suppose, in addition, that the storage rate will remain fixed at \$3 per case per year. When should the retailer sell the wine to obtain the largest possible profit?

19 Suppose the consumer demand for a certain commodity is $D(p) = mp + b$ units per month when the market price is p dollars per unit.

(a) Assume that $m < 0$ and $b > 0$ and sketch this demand function, labeling the points at which it intersects the coordinate axes. Explain in economic terms why the assumptions about the signs of m and b are reasonable.

(b) Express consumers' total monthly expenditure for the commodity as a function of p and sketch the graph of this function. Where does the graph cross the p axis?

(c) Use calculus to find an expression (in terms of m and b) for the market price at which consumer expenditure will be greatest. Show that this optimal price is the value of p that is midway between the origin and the p intercept of the demand curve.

20 An economic law states that profit is maximized when marginal revenue equals marginal cost. (Recall that marginal revenue and marginal cost are the derivatives of total revenue and total cost, respectively.)
   (a) Use the theory of relative extrema to explain why this law is true.
   (b) What assumptions about the shape of the profit curve are implicit in this law?

21 Suppose the total cost in dollars of manufacturing q units is given by the function $C(q) = 3q^2 + q + 48$.
   (a) Express the average manufacturing cost per unit as a function of q.
   (b) For what value of q is the average cost the smallest?
   (c) For what value of q is the average cost equal to the marginal cost? Compare this value with your answer in part (b).
   (d) On the same set of axes, graph the total-cost, marginal-cost, and average-cost functions.

22 An economic law states that average cost is smallest when average cost equals marginal cost.
   (a) Prove this law. (*Hint:* If $C(q)$ represents the total cost of manufacturing q units, the average cost is $A(q) = C(q)/q$. Average cost will be smallest when $A'(q) = 0$. Use the quotient rule to compute $A'(q)$, then set $A'(q)$ equal to zero and the economic law will emerge.)
   (b) What assumptions about the shape of the average-cost graph are implicit in this economic law?

7 • THE SECOND DERIVATIVE TEST

In this section, you will see how you can use the derivative of a function's derivative to classify the critical points of the function.

The second derivative

The second derivative
The *second derivative* of f is the derivative of its derivative f' and is denoted by the symbol f''.

The derivative f' is sometimes called the **first derivative** to distinguish it from the second derivative f''. If the function is denoted by y instead of f, the symbol d^2y/dx^2 is often used instead of f'' to denote the second derivative.

EXAMPLE 7.1 ● Compute the second derivative of the function $f(x) = 5x^4 - 3x^2 - 3x + 7$.

SOLUTION The first derivative is

$$f'(x) = 20x^3 - 6x - 3$$

To compute the second derivative, we simply differentiate again:

$$f''(x) = 60x^2 - 6$$ ●

EXAMPLE 7.2 ● Compute the second derivative of the function $y = (x^2 + 1)^5$.

SOLUTION Using the chain rule we find that

$$\frac{dy}{dx} = 5(x^2 + 1)^4(2x) = 10x(x^2 + 1)^4$$

Using the product rule we differentiate again to get

$$\begin{aligned}\frac{d^2y}{dx^2} &= 10x[4(x^2 + 1)^3(2x)] + 10(x^2 + 1)^4 \\ &= 80x^2(x^2 + 1)^3 + 10(x^2 + 1)^4 \\ &= 10(x^2 + 1)^3(9x^2 + 1)\end{aligned}$$ ●

Classification of critical points Here is a simple test involving the sign of the second derivative that you can use to classify critical points.

> **The second derivative test**
> Suppose $f'(a) = 0$.
>
> If $f''(a) > 0$, then f has a relative minimum at $x = a$.
> If $f''(a) < 0$, then f has a relative maximum at $x = a$.
>
> However, if $f''(a) = 0$, the test is inconclusive and f may have a relative maximum, a relative minimum, or no relative extremum at all at $x = a$.

The use of the second derivative test is illustrated in the following example.

EXAMPLE 7.3 ● Use the second derivative test to find the relative maxima and minima of the function $f(x) = 2x^3 + 3x^2 - 12x - 7$.

SOLUTION Since the derivative

$$f'(x) = 6x^2 + 6x - 12 = 6(x + 2)(x - 1)$$

is zero when $x = -2$ and $x = 1$, the corresponding points $(-2, 13)$ and $(1, -14)$ are the critical points of f. To test these points, we compute the second derivative

$$f''(x) = 12x + 6$$

and evaluate it at $x = -2$ and $x = 1$. Since

$$f''(-2) = -18 < 0$$

we conclude that $(-2, 13)$ is a relative maximum, and since

$$f''(1) = 18 > 0$$

we conclude that $(1, -14)$ is a relative minimum. ●

The function in the preceding example is the same one we analyzed in Example 5.1 using only the first derivative. Notice the relative ease with which we can now identify the extrema. Using the second derivative test, we compute $f''(x)$ only at the critical points themselves. Using the first derivative, we had to investigate the sign of $f'(x)$ over entire intervals.

There are, however, some disadvantages to the second derivative test. For example, for many functions (such as rational functions) the work involved in computing the second derivative is time consuming and may diminish the efficiency of the test. Moreover, if both $f'(a)$ and $f''(a)$ are zero, the second derivative test tells us nothing whatsoever about the nature of the critical point. This is illustrated in Figure 7.1 which shows the graphs of three functions whose first and second derivatives are both zero when $x = 0$.

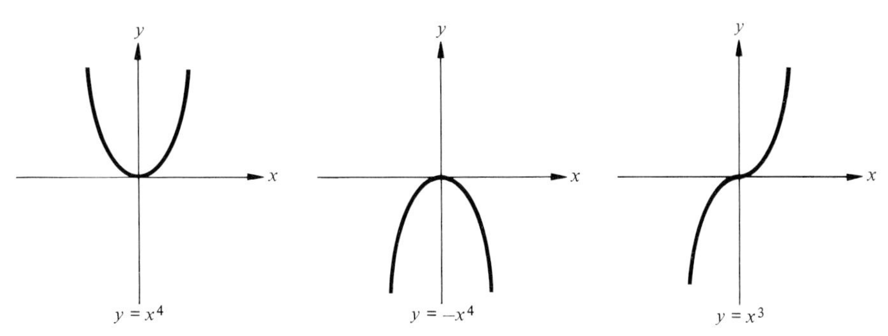

Figure 7.1 Three functions whose first and second derivatives are zero at $x = 0$.

Remember that the second derivative test is a test for *relative* extrema only and tells you nothing about the *absolute* extrema of a function. In most practical optimization problems, you will have to use something more than the second derivative test to verify that a particular critical point is actually the desired absolute extremum.

Concavity

To see the connection between relative extrema and the sign of the second derivative, suppose that the second derivative f'' is positive. This implies that the first derivative f' must be increasing. But f' is the slope of the tangent.

Hence the slope of the tangent is increasing, and so the graph of f is **concave upward** as shown in Figure 7.2*a*. In particular, it has a relative minimum at its critical point. On the other hand, if f'' is negative, then f' is decreasing. This implies that the slope of the tangent is decreasing, and so the graph of f is **concave downward** as shown in Figure 7.2*b*. In this case, the critical point is a relative maximum.

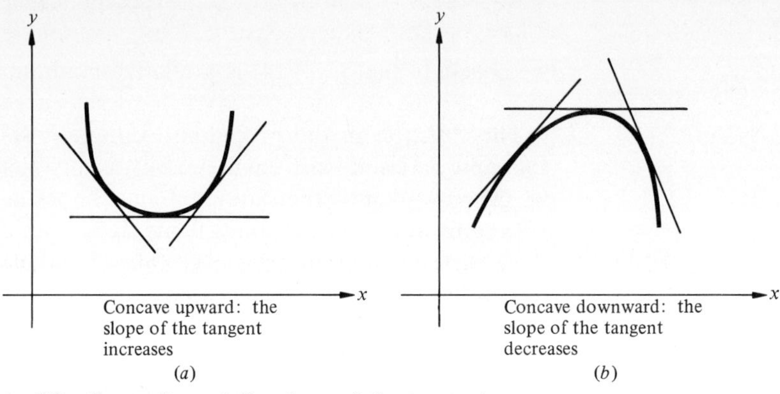

Figure 7.2 Concavity and the slope of the tangent.

Problems

In Problems 1 through 8, find the second derivative of the given function.

1 $f(x) = 5x^{10} - 6x^5 - 27x$
2 $y = \frac{2}{5}x^5 - 4x^3 + 9x - 6$
3 $y = x^2 - 1/x^2$
4 $f(x) = 5\sqrt{x} + 3/x^2 + 5x^{-1/3}$
5 $f(x) = (x^3 + 1)^5$
6 $y = \sqrt{1 - x^2}$
7 $y = x^2/(x - 3)$
8 $f(x) = (x + 1)/(x - 1)$

In Problems 9 through 15, use the second derivative test to find the relative maxima and minima of the given function.

9 $f(x) = x^4 - 2x^2 + 3$
10 $f(x) = x^2 - 10x + 6$
11 $f(x) = (x^2 - 9)^2$
12 $f(x) = x^3 + 3x^2 + 1$
13 $f(x) = x + 1/x$
14 $f(x) = 2x + 1 + 18/x$
15 $f(x) = x^2/(x - 2)$

16 At what point does the tangent to the curve $y = 2x^3 - 3x^2 + 6x$ have the smallest slope? What is the slope of the tangent at this point?

17 For what value of x in the interval $-1 \leq x \leq 4$ is the graph of the function $f(x) = 2x^2 - \frac{1}{3}x^3$ steepest? What is the slope of the tangent at this point?

18 An efficiency study of the morning shift at a certain factory indicates that an average employee who arrives on the job at 8:00 A.M. will have assembled $f(x) = -x^3 + 6x^2 + 15x$ transistor radios x hours later. At what time during the morning does the employee reach the **point of diminishing returns?** That is, at what time is the employee's rate of production greatest?

19 Use calculus to show that the graph of the quadratic function $y = ax^2 + bx + c$ is concave upward if a is positive, and concave downward if a is negative.

8 • PARTIAL DERIVATIVES

In many practical situations, the value of one quantity may depend on the values of several others. For example, the amount of water in a reservoir may depend on the amount of rainfall and on the amount of water consumed by local residents. The demand for butter may depend on the price of butter and on the price of margarine. The output of a factory may depend on the amount of capital invested in the plant and on the size of the labor force. Relationships of this sort can often be represented mathematically by functions having more than one independent variable. Here is a simple example.

EXAMPLE 8.1 ● A service station sells three grades of gasoline, regular, low-lead, and premium, at prices of 70, 73, and 75 cents a gallon, respectively. Express the station's total revenue from the sale of the gasoline as a function of the number of gallons of each grade sold.

SOLUTION We introduce three variables x_1, x_2, and x_3 to denote the number of gallons sold of regular, low-lead, and premium, respectively. Then,

$$\text{Total revenue} = 70x_1 + 73x_2 + 75x_3 \text{ cents}$$ ●

The functional notation that we have been using for functions of a single variable can be extended to describe functions of several variables. In Example 8.1, for instance, we could let f denote the total revenue and write $f(x_1, x_2, x_3) = 70x_1 + 73x_2 + 75x_3$. The use of this functional notation is illustrated in the next example.

EXAMPLE 8.2 ● Compute $f(2, 3)$ if $f(x_1, x_2) = x_1^2 - 5x_1x_2 + x_2$.

SOLUTION We substitute $x_1 = 2$ and $x_2 = 3$ into the formula for f to get

$$f(2, 3) = 2^2 - 5(2)(3) + 3 = -23$$ ●

In many problems involving functions of several variables, the goal is to find the rate of change of the function with respect to one of its variables when all the others are held constant. That is, the goal is to differentiate the function with respect to the particular variable in question while keeping all the other variables fixed. This process is known as **partial differentiation** and the resulting derivative is said to be a **partial derivative** of the function.

Partial derivatives
Suppose $f(x_1, x_2, \ldots, x_n)$ is a function of n variables. The *partial derivative* of f with respect to its jth variable x_j is denoted by f_{x_j} and is defined to be the function obtained by differentiating f with respect to x_j, treating all the other variables as constants.

No new rules are needed for the computation of partial derivatives. To compute f_{x_j}, simply differentiate f with respect to the single variable x_j, pretending that all the other variables are constants. Here is an example.

EXAMPLE 8.3 ● Find the partial derivatives f_x, f_y, and f_z if $f(x, y, z) = x^2 + 2xy^2 + yz^3$.

SOLUTION To compute f_x, we think of f as a function of x and differentiate the sum term by term, treating y and z as constants. We get

$$f_x(x, y, z) = 2x + 2y^2 + 0 = 2x + 2y^2$$

To compute f_y, we pretend that x and z are constants and differentiate with respect to y getting

$$f_y(x, y, z) = 0 + 4xy + z^3 = 4xy + z^3$$

To compute f_z, we treat x and y as the constants and differentiate with respect to z getting

$$f_z(x, y, z) = 0 + 0 + 3yz^2 = 3yz^2$$ ●

When a dependent variable such as z is used to denote a function $f(x_1, x_2, \ldots, x_n)$,the symbol $\partial z/\partial x_j$ is usually used instead of f_{x_j} to denote the partial derivative of the function with respect to x_j. This notation is used in the next example.

EXAMPLE 8.4 ● Find the partial derivatives $\partial z/\partial x$ and $\partial z/\partial y$ if $z = (x^2 + xy + y)^5$.

SOLUTION Holding y fixed and using the chain rule to differentiate z with respect to x we get

$$\frac{\partial z}{\partial x} = 5(x^2 + xy + y)^4(2x + y)$$

Holding x fixed and using the chain rule to differentiate z wth respect to y we get

$$\frac{\partial z}{\partial y} = 5(x^2 + xy + y)^4(x + 1)$$ ●

Marginal analysis In economics, the term marginal analysis refers to the practice of using a derivative to estimate the change in the value of a function resulting from a unit increase in one of its variables. In Section 3 you saw some examples of marginal analysis involving ordinary derivatives of functions of one variable. Here are two examples illustrating marginal analysis for functions of several variables.

EXAMPLE 8.5 ● Suppose the daily output Q of a factory depends on the amount K of capital (measured in thousand-dollar units) invested in the plant and equipment, and also on the size L of the labor force (measured in worker-hours). In economics, the partial derivatives $\partial Q/\partial K$ and $\partial Q/\partial L$ are known as the **marginal products** of capital and labor, respectively. Give economic interpretations of the two marginal products.

SOLUTION The marginal product of labor $\partial Q/\partial L$ is the rate at which output Q changes with respect to labor L for a fixed level K of capital investment. Hence $\partial Q/\partial L$ is approximately the change in output that will result if capital investment is held fixed and labor is increased by 1 worker-hour.

Similarly, the marginal product of capital $\partial Q/\partial K$ is approximately the change in output that will result if the size of the labor force is held fixed and capital investment is increased by \$1,000. ●

EXAMPLE 8.6 ● It is estimated that the weekly output at a certain plant is given by the function $f(x, y) = 1{,}200x + 500y + x^2y - x^3 - y^2$ units, where x is the number of skilled workers and y the number of unskilled workers employed at the plant. Currently, the work force consists of 30 skilled workers and 60 unskilled workers. Use marginal analysis to estimate the change in the weekly output that will result from the addition of one more skilled worker if the number of unskilled workers is not changed.

SOLUTION The partial derivative

$$f_x(x, y) = 1{,}200 + 2xy - 3x^2$$

is the rate of change of output with respect to the number of skilled workers. For any values of x and y, this is an approximation to the number of additional units that will be produced each week if the number of skilled workers is increased from x to $x + 1$ while the number of unskilled workers is kept fixed at y. In particular, if the work force is increased from 30 skilled and 60 unskilled workers to 31 skilled and 60 unskilled workers, the resulting change in output is approximately

$$f_x(30, 60) = 1{,}200 + 2(30)(60) - 3(30)^2 = 2{,}100 \text{ units}$$

For practice, compute the change in output exactly by subtracting appropriate values of f. ●

Problems In Problems 1 through 4, compute the indicated values of the given function.

1 $f(x_1, x_2, x_3) = x_1(x_2 - 1)^2 + 3x_1x_3$; $f(1, 0, -2)$, $f(0, 8, 4)$
2 $g(x, y) = (x + y)/(x - y)$; $g(2, -2)$, $g(-1, 3)$
3 $h(r, s) = \sqrt{s^2 - r^2}$; $h(0, 1)$, $h(-1, 3)$
4 $f(x, y, z) = x + y^2 + \sqrt{1 - z}$; $f(1, 2, 0)$, $f(-1, 3, -3)$

5 A manufacturer can produce electric typewriters at a cost of $80 apiece, and manual typewriters at a cost of $20 apiece.
(a) Express the manufacturer's total monthly production cost as a function of the number of electric typewriters and the number of manual typewriters produced each month.
(b) Compute the total monthly cost if 500 electric and 800 manual typewriters are produced each month.
(c) The manufacturer wants to increase the output of electric typewriters by 100 a month from the level in part (b). What corresponding change should be made in the monthly output of the manual typewriters so that the total monthly cost will not change?

6 Suppose the manufacturer in Problem 5 wishes to keep production costs fixed at $56,000 a month.
(a) Write an equation relating the monthly output of electric typewriters and the monthly output of manual typewriters that must be satisfied if this goal is to be achieved.
(b) Draw the graph of the equation in part (a). (This graph is said to be a **level curve** of the total cost function.)
(c) Determine the maximum monthly output of electric typewriters compatible with the manufacturer's goal of holding monthly costs fixed at $56,000. Explain how this information can be obtained from the level curve in part (b).

7 Using x skilled and y unskilled workers, a manufacturer can produce $Q(x, y) = 10x^2y$ units a day. Currently there are 20 skilled workers and 40 unskilled workers on the job.
(a) How many units are currently being produced each day?
(b) By how much will the daily production level change if one more skilled worker is added to the current work force?
(c) By how much will the daily production level change if one more unskilled worker is added to the current work force?
(d) By how much will the daily production level change if one more skilled worker *and* one more unskilled worker are added to the current work force?

In Problems 8 through 14, compute all the partial derivatives of the given function.

8 $f(x, y) = 2xy^5 + 3x^2y + x^2 + 5$
9 $z = x\sqrt{y} + 3x^{-2/3}y^5 + 6y$
10 $f(x, y, z) = x^2yz + 3xy^2 + xz^5$
11 $f(x, y) = (3x + 2y)^5$
12 $w = (3x + 2y + z^2)^9$
13 $z = (u + v)/(u - v)$
14 $f(x_1, x_2) = 10x_1/(2 + x_1) + 20x_2/(5 + x_2) - x_1 - x_2$
15 The monthly demand for a certain brand of toasters is given by a function $f(x, y)$, where x is the amount of money (measured in units of $1,000) spent

on advertising and y is the selling price (in dollars) of the toasters. Give economic interpretations of the partial derivatives f_x and f_y. Under normal economic conditions, what will be the sign of each of these partial derivatives?

16 At a certain factory, the daily output is $Q = 2K^{1/2}L^{1/3}$ units, where K denotes the capital investment measured in units of \$1,000 and L the size of the labor force measured in worker-hours. Suppose the current capital investment is \$900,000 and that 1,000 worker-hours of labor are used each day. Use marginal analysis to estimate the effect of an additional capital investment of \$1,000 on the daily output if the size of the labor force is not changed.

17 A bicycle dealer has found that if the price of 10-speed bicycles is x dollars and the price of gasoline is $10y$ cents per gallon, approximately $f(x, y) = 200 - 10\sqrt{x} + 4(y + 2)^{3/2}$ bicycles will be sold each month. Currently the bicycles sell for \$121 apiece and the price of gasoline is 70 cents a gallon. Use marginal analysis to estimate the effect on the monthly sale of bicycles if the price of gasoline is increased to 80 cents a gallon while the price of the bicycles is held fixed.

18 A publishing house has found that in a certain city *each* of its salespeople will sell approximately $r^2/(2{,}000p) + s^2/100 - s$ sets of encyclopedias a month, where s denotes the total number of salespeople employed, p the price of a set of the encyclopedias, and r the amount of money spent each month on local advertising. Currently the publisher employs 10 salespeople, spends \$6,000 a month on local advertising, and sells the encyclopedias for \$800 a set. The cost of producing the encyclopedias is \$80 a set and each salesperson earns \$600 a month. Use marginal analysis to estimate the change in the publisher's total monthly profit that will result if one more salesperson is hired.

19 Two competing brands of power lawnmowers are sold in the same town. The price of the first brand is x dollars per mower, the price of the second brand is y dollars per mower, and the average per capita income of the community is z dollars a year. The local demand for the first brand of mowers is given by a function $f(x, y, z)$.

(a) How would you expect the demand for the first brand of mowers to be affected by an increase in x? By an increase in y? By an increase in z?

(b) Translate your answers to part (a) into conditions on the signs of the partial derivatives of f.

(c) If $f(x, y, z) = a + bx + cy + dz$, what can you say about the signs of the coefficients b, c, and d if your conclusions in part (a) are to hold?

20 In economics, two commodities are said to be **substitute commodities** if the demand Q_1 for the first increases as the price p_2 of the second increases, and if the demand Q_2 for the second increases as the price p_1 of the first increases.

(a) Give an example of a pair of substitute commodities.

(b) If two commodities are substitutes, what must be true of the partial derivatives $\partial Q_1/\partial p_2$ and $\partial Q_2/\partial p_1$?

(c) Suppose the demand functions for two commodities are $Q_1 = 3{,}000 + 400/(p_1 + 3) + 50p_2$ and $Q_2 = 2{,}000 - 100p_1 + 500/(p_1 + 4)$. Are the commodities substitutes?

21 Two commodities are said to be **complementary commodities** if the demand Q_1 for the first decreases as the price p_2 of the second increases, and if the demand Q_2 for the second decreases as the price p_1 of the first increases.

(a) Give an example of a pair of complementary commodities.

(b) If two commodities are complementary, what must be true of the partial derivatives $\partial Q_1/\partial p_2$ and $\partial Q_2/\partial p_1$?

(c) Suppose the demand functions for two commodities are $Q_1 = 2{,}000 + 400/(p_1 + 3) - 50p_2$ and $Q_2 = 2{,}000 - 100p_2 + 500/(p_1 + 4)$. Are the commodities complementary?

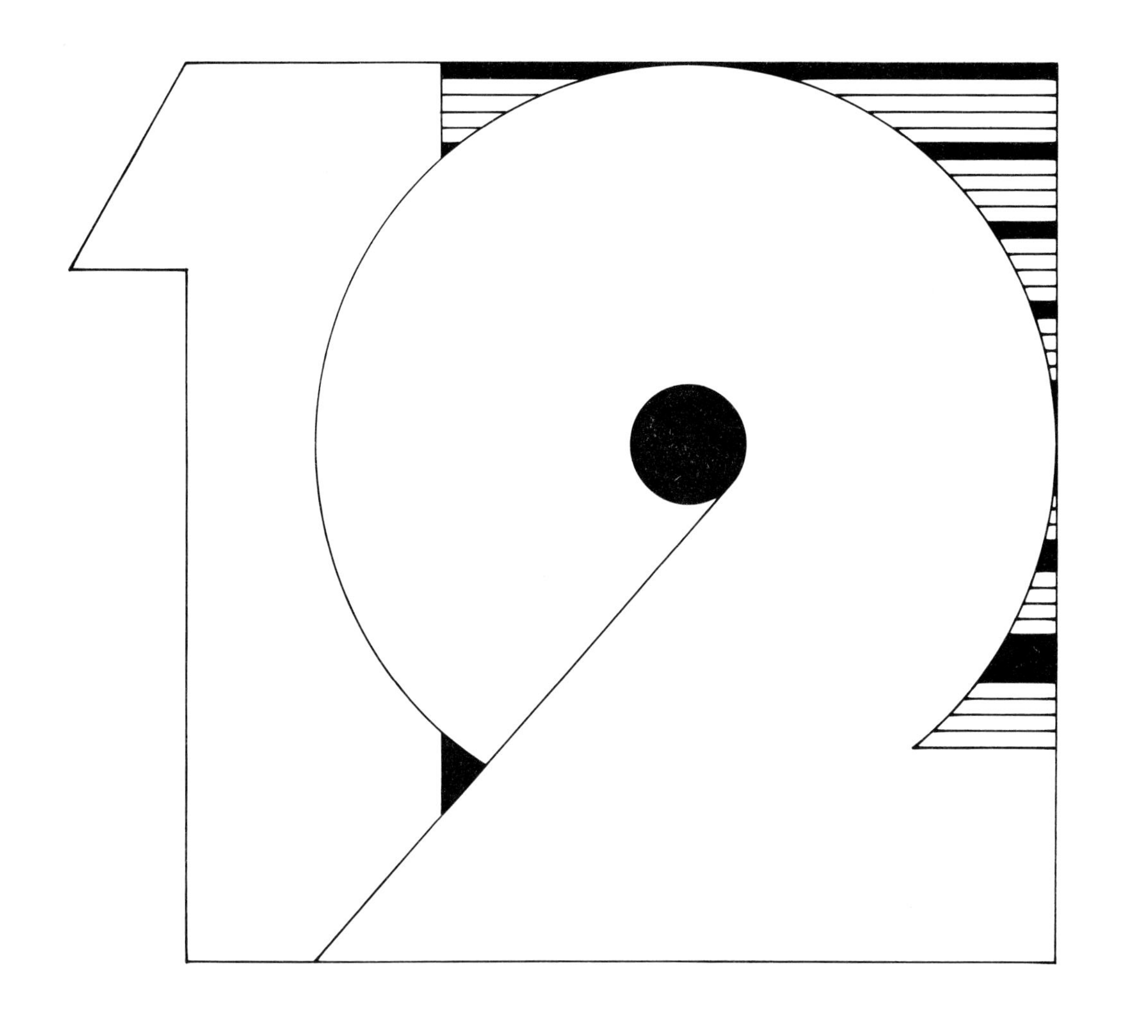

EXPONENTIAL AND LOGARITHMIC FUNCTIONS

1 • EXPONENTIAL FUNCTIONS

First impressions can be deceiving. Take the algebraic expression $(1 + 1/n)^n$ for instance. At first glance, this expression may look no more interesting than many other algebraic expressions. Yet, the number it approaches as n increases without bound turns out to be one of the most important and useful numbers in mathematics.

The following table (accurate to two decimal places) may give you some feel for what happens to the expression $(1 + 1/n)^n$ as n increases.

n	1	2	3	5	10	50	100
$(1 + 1/n)^n$	2	2.25	2.37	2.49	2.59	2.69	2.70

By means of techniques beyond the scope of this book, it can be shown that as n increases without bound, $(1 + 1/n)^n$ approaches an irrational number, traditionally denoted by the letter e, whose value is approximately 2.718.

The number e
As n increases without bound, $(1 + 1/n)^n$ approaches the number e, which is approximately 2.718.

Functions involving powers of e play a central role in applied mathematics. They are used in demography to forecast population size, in finance to calculate the value of investments, in archaeology to date ancient artifacts, in psychology to study learning phenomena, in public health to analyze the spread of epidemics, and in industry to estimate the reliability of products. You will see some of these applications in Section 2. To illustrate how the number e might arise in practice, here is a brief discussion of compound interest. (A more thorough discussion of this topic was presented in Chapter 9.)

Compound interest

Suppose a sum of money is invested and the interest is compounded only once. If P is the initial investment (the principal) and r is the interest rate (expressed as a decimal), the balance A after the interest is added will be

$$A = P + Pr = P(1 + r) \text{ dollars}$$

That is, the balance at the end of an interest period is obtained by multiplying the balance at the beginning of the period by the expression $1 + r$, where r is the interest rate per period.

At most banks, interest is compounded more than once a year. The interest that is added to the account during one period will itself earn interest during the subsequent periods. If the annual interest rate is r and interest is compounded k times, the year is divided into k equal interest periods and the interest rate during each is r/k. To compute the balance at the end of any period, we simply multiply the balance at the beginning of that period by the expression $1 + r/k$.

Hence the balance at the end of the first period is $P(1 + r/k)$. The balance at the end of the second period is $P(1 + r/k)(1 + r/k) = P(1 + r/k)^2$. The balance at the end of the third period is $P(1 + r/k)^2(1 + r/k) = P(1 + r/k)^3$. At the end of 1 year, the interest has been compounded k times and the balance is $P(1 + r/k)^k$ and at the end of t years, the interest has been compounded kt times and the balance is given by the function

$$A(t) = P\left(1 + \frac{r}{k}\right)^{kt} \text{ dollars}$$

As the frequency with which the interest is compounded increases, the corresponding balance $A(t)$ also increases. Hence a bank that compounds interest frequently may attract more customers than one that offers the same interest rate but that compounds interest less often. The question arises: What happens to the balance at the end of t years as the frequency with which the interest is compounded increases without bound? That is, what will the balance be at the end of t years if interest is compounded not quarterly, not monthly, not daily, but *continuously?* In mathematical terms: What happens to the expression $P(1 + r/k)^{kt}$ as k increases without bound? The answer turns out to involve the number e. Here is the argument.

To simplify the calculation, let $n = k/r$. Then, $k = nr$, and so

$$P\left(1 + \frac{r}{k}\right)^{kt} = P\left(1 + \frac{1}{n}\right)^{nrt} = P\left[\left(1 + \frac{1}{n}\right)^{n}\right]^{rt}$$

Since n increases without bound as k does, and since $(1 + 1/n)^n$ approaches e as n increases without bound, it follows that $P(1 + r/k)^{kt}$ approaches Pe^{rt} as k increases without bound.

Here is a summary of the situation.

Compound interest

If P dollars are invested at an annual interest rate r, the balance $A(t)$ after t years will be

$$A(t) = P\left(1 + \frac{r}{k}\right)^{kt} \text{ dollars}$$

if the interest is compounded k times a year, and

$$A(t) = Pe^{rt} \text{ dollars}$$

if the interest is compounded continuously.

The following numerical example illustrates what happens to a bank balance as the interest is compounded with increasing frequency. The calculations were done on a pocket calculator.

EXAMPLE 1.1 ● Suppose \$1,000 is invested at an annual interest rate of 6 percent. Compute the balance after 10 years if the interest is compounded

(a) quarterly (b) monthly (c) continuously

SOLUTION (a) To compute the balance after 10 years if the interest is compounded quarterly, we use the formula $A(t) = P(1 + r/k)^{kt}$, with $t = 10$, $P = 1{,}000$, $r = 0.06$, and $k = 4$. We find that

$$A(10) = 1{,}000\left(1 + \frac{0.06}{4}\right)^{40} = \$1{,}814.02$$

(b) This time we take $t = 10$, $P = 1{,}000$, $r = 0.06$, and $k = 12$ and find that

$$A(10) = 1{,}000\left(1 + \frac{0.06}{12}\right)^{120} = \$1{,}819.40$$

(c) Using the formula $A(t) = Pe^{rt}$, with $t = 10$, $P = 1{,}000$, and $r = 0.06$, we find that

$$A(10) = 1{,}000e^{0.6} = \$1{,}822.12$$

●

Exponential functions The function $A(t) = Pe^{rt}$ is closely related to a class of functions called **exponential functions.** An exponential function is a function of the form $f(x) = a^x$, where a is a positive constant. In an exponential function, the independent variable x is the **exponent** of a positive constant a known as the **base** of the function. Thus, an exponential function is different from a **power function** $f(x) = x^n$ in which the base is the variable and the exponent is the constant.

Recall that for rational values of x, a^x is defined by the following conditions. (These conditions, as well as other algebraic properties of exponents are reviewed in the Appendix.)

Definition of a^x for rational values of x

Integer Powers: If n is a positive integer,

$$a^n = a \cdot a \cdots a$$

where the product $a \cdot a \cdots a$ contains n factors.

Fractional Powers: If n and m are positive integers,

$$a^{n/m} = (\sqrt[m]{a})^n$$

where $\sqrt[m]{}$ denotes the positive mth root.

Negative Powers:

$$a^{-x} = \frac{1}{a^x}$$

Zero Power:

$$a^0 = 1$$

(It is possible to extend the definition of a^x to the set of all real numbers. This extension is discussed in more advanced texts.)

Graphing techniques for exponential functions are illustrated in the next example.

EXAMPLE 1.2 ● Sketch the function $f(x) = a^x$ if $0 < a < 1$ and if $a > 1$.

SOLUTION In both cases, the y intercept is $(0, 1)$ since $f(0) = a^0 = 1$. There are no x intercepts since (for positive a) a^x is always positive. To determine the behavior of the graph as x increases or decreases without bound, we consider the two cases $0 < a < 1$ and $a > 1$ separately.

If $0 < a < 1$, the value of the product $a^n = a \cdot a \cdots a$ approches zero as the number n of factors increases. This suggests that a^x approaches zero as x increases without bound. On the other hand, the value of the product $a^{-n} = (1/a)(1/a) \cdots (1/a)$ increases without bound (since $1/a > 1$) as the number n of factors increases. This suggests that a^x increases without bound as x decreases without bound. A graph with these features is sketched in Figure 1.1a.

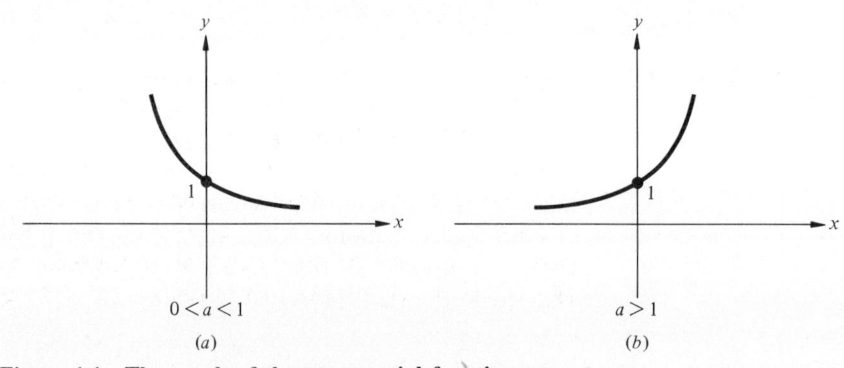

Figure 1.1 The graph of the exponential function $y = a^x$.

If $a > 1$, a^x increases without bound as x increases without bound and a^x approaches zero as x decreases without bound. The corresponding graph is sketched in Figure 1.1b. ●

Since $e > 1$, the graph of the special exponential function $y = e^x$ resembles the sketch in Figure 1.1b.

The laws of exponents Exponential functions obey the following useful laws.

Laws of exponents
The product law: $a^r a^s = a^{r+s}$
The quotient law: $a^r/a^s = a^{r-s}$
The power law: $(a^r)^s = a^{rs}$

Some elementary calculations based on these laws are worked out in the algebra review in the Appendix. The example presented here illustrates a useful computational trick based on the power law for exponents.

EXAMPLE 1.3 ● Find $f(6)$ if $f(x) = e^{kx}$ and $f(2) = 5$.

SOLUTION You do not have to know the value of k or of e to solve this problem! The fact that $f(2) = 5$ tells you that

$$e^{2k} = 5$$

and using the power law for exponents you can rewrite the expression for $f(6)$ in terms of this quantity. You get

$$f(6) = e^{6k} = (e^{2k})^3 = 5^3 = 125$$ ●

Problems

1 Program a computer or a programmable calculator to evaluate $(1 + 1/n)^n$ for $n = 1{,}000, 2{,}000, \ldots, 50{,}000$.

2 Program a computer or a programmable calculator to evaluate $(1 + 1/n)^n$ for $n = -1{,}000, -2{,}000, \ldots, -50{,}000$. On the basis of these calculations, what can you conjecture about the behavior of $(1 + 1/n)^n$ as n *decreases* without bound?

3 Learn how to use your calculator to find powers of e. In particular, find e^2, e^{-2}, $e^{0.05}$, $e^{-0.05}$, e^0, e, $\sqrt{e}$, and $1/\sqrt{e}$.

4 Suppose \$1,000 is invested at an annual interest rate of 7 percent. Compute the balance after 10 years if the interest is compounded.
(a) annually (b) quarterly (c) monthly (d) continuously

5 A sum of money is invested at a certain fixed interest rate, and the interest is compounded continuously. After 10 years the money has doubled. How will the balance at the end of 20 years compare with the initial investment?

6 (a) Solve the equation $A = Pe^{rt}$ for P. (*Hint:* Multiply both sides by e^{-rt}.)
(b) How much money should be invested today at an annual interest rate of 6 percent compounded continuously so that 10 years from now it will be worth \$10,000? [*Hint:* Use the result from part (a).]

7 Sketch the curves $y = 2^x$ and $y = 3^x$ on the same set of axes.

8 Sketch the curves $y = 2^{-x}$ and $y = 3^{-x}$ on the same set of axes.

In Problems 9 through 14, sketch the given function.

9 $f(x) = e^x$

10 $f(x) = e^{-x}$

11 $f(x) = 2 - 3e^x$ **12** $f(x) = 3 - 2e^x$
13 $f(x) = 5 - 3e^{-x}$ **14** $f(x) = 3 - 5e^{-x}$
15 Find $f(2)$ if $f(x) = e^{kx}$ and $f(1) = 20$.
16 Find $f(9)$ if $f(x) = e^{kx}$ and $f(3) = 2$.
17 Find $f(8)$ if $f(x) = Ae^{kx}$, $f(0) = 20$, and $f(2) = 40$.
18 Find $f(4)$ if $f(x) = 50 - Ae^{-kx}$, $f(0) = 20$, and $f(2) = 30$.
19 Find $f(2)$ if $f(x) = 50 - Ae^{kx}$, $f(0) = 30$, and $f(4) = 5$.

2 • EXPONENTIAL MODELS

Functions involving powers of e play a central role in applied mathematics. Here is a sampling of practical situations from the social, managerial, and natural sciences described mathematically in terms of such functions.

Exponential growth

A quantity $Q(t)$ that increases according to a law of the form $Q(t) = Q_0e^{kt}$, where Q_0 and k are positive constants, is said to experience **exponential growth.** For example, if interest is compounded continuously, the resulting bank balance $A(t) = Pe^{rt}$ grows exponentially. Also, in the absence of environmental constraints, population increases exponentially. As you will see later in the book, quantities that grow exponentially are characterized by the fact that their rate of growth is proportional to their size.

To sketch the function $Q(t) = Q_0e^{kt}$, we observe that $Q(t)$ is always positive, that $Q(0) = Q_0$, and that $Q(t)$ increases without bound as t increases without bound and approaches zero as t decreases without bound. A sketch is drawn in Figure 2.1.

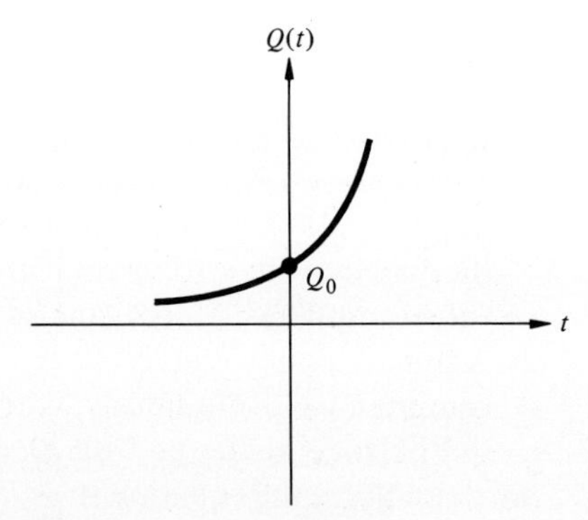

Figure 2.1 Exponential growth: $Q(t) = Q_0e^{kt}$.

EXAMPLE 2.1 ● Biologists have determined that under ideal conditions, the number of bacteria in a culture grows exponentially. Suppose that 2,000 bacteria are initially present in a certain culture and that 6,000 are present 20 minutes later. How many bacteria will be present at the end of 1 hour?

SOLUTION We let $Q(t)$ denote the number of bacteria present after t minutes. Since the number of bacteria grows exponentially, and since 2,000 bacteria were initially present, we know that Q is a function of the form

$$Q(t) = 2{,}000e^{kt}$$

Since 6,000 bacteria are present after 20 minutes, it follows that

$$6{,}000 = 2{,}000e^{20k} \qquad \text{or} \qquad e^{20k} = 3$$

To find the number of bacteria present at the end of 1 hour, we compute $Q(60)$ using the power law for exponents and get

$$Q(60) = 2{,}000e^{60k} = 2{,}000(e^{20k})^3 = 2{,}000(3)^3 = 54{,}000$$ ●

Exponential decay

A quantity $Q(t)$ that decreases according to a law of the form $Q(t) = Q_0e^{-kt}$ is said to experience **exponential decay.** Radioactive substances decay exponentially. In general, any quantity that decreases at a rate proportional to its size decays exponentially. A sketch of the function $Q(t) = Q_0e^{-kt}$ is shown in Figure 2.2.

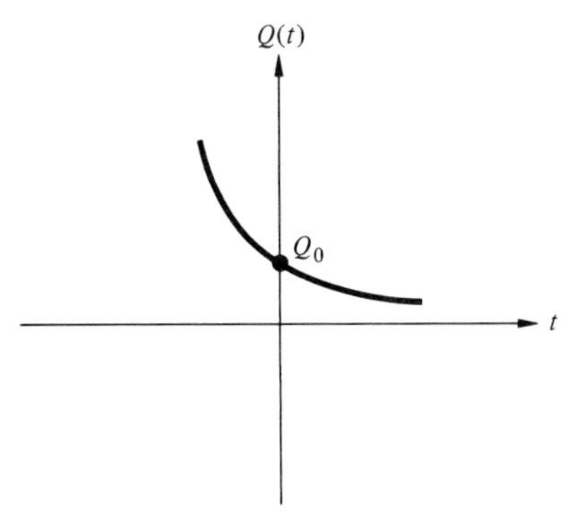

Figure 2.2 Exponential decay: $Q(t) = Q_0e^{-kt}$.

EXAMPLE 2.2 ● A certain industrial machine depreciates so that its value after t years is given by a function of the form $Q(t) = Q_0e^{-0.04t}$. After 20 years, the machine is worth \$8,986.58. What was its original value?

SOLUTION Our goal is to find Q_0. Since $Q(20) = 8{,}986.58$, we have

$$Q_0e^{-0.8} = 8{,}986.58$$

Multiplying both sides of this equation by $e^{0.8}$ we get

$$Q_0 = 8{,}986.58e^{0.8} = \$20{,}000$$ ●

Learning curves The graph of a function of the form $Q(t) = C - Ae^{-kt}$, where C, A, and k are positive constants, is sometimes called a **learning curve.** The name arose when psychologists discovered that functions of this form often describe the relationship between the efficiency with which an individual performs a task and the amount of training or experience the individual has had.

To sketch the function $Q(t) = C - Ae^{-kt}$, we observe that $Q(0) = C - A$, that $Q(t)$ approaches C as t increases without bound (since Ae^{-kt} approaches zero), and that $Q(t)$ decreases without bound as t does. A sketch is drawn in Figure 2.3. The behavior of the graph as t increases without bound reflects the fact that eventually an individual will approach peak efficiency and additional training will have little effect on performance.

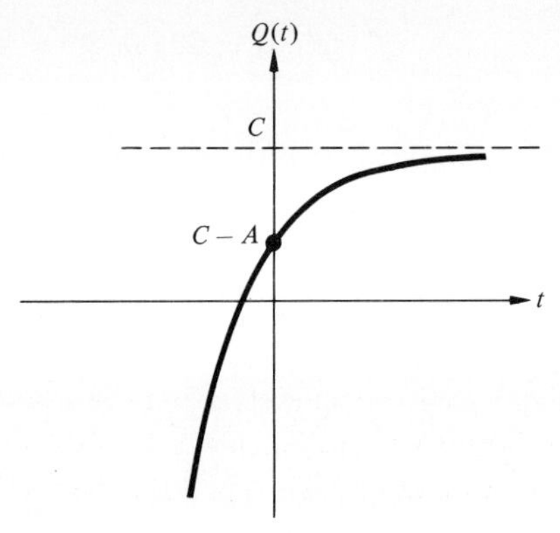

Figure 2.3 A learning curve: $Q(t) = C - Ae^{-kt}$.

EXAMPLE 2.3 ● The rate at which a postal clerk can sort mail is a function of the clerk's experience. The postmaster of a large city estimates that after t months on the job, the average clerk can sort $Q(t) = 700 - 400e^{-0.5t}$ letters per hour.

(a) How many letters can a new employee sort per hour?
(b) How many letters can a clerk with 6 months experience sort per hour?
(c) Approximately how many letters will the average clerk ultimately be able to sort per hour?

SOLUTION (a) The number of letters a new employee can sort per hour is

$$Q(0) = 700 - 400 = 300$$

(b) After 6 months, the average clerk can sort

$$Q(6) = 700 - 400e^{-0.5(6)} = 700 - 400e^{-3} = 680$$

letters per hour.

(c) As t increases without bound, $Q(t)$ approaches 700. Hence, the average clerk will ultimately be able to sort approximately 700 letters per hour. ●

Logistic curves The graph of a function of the form $Q(t) = A/(B + Ce^{-kt})$, where A, B, C, and k are positive constants, is an S-shaped or **sigmoidal curve.** The term **logistic curve** is also used to refer to such a curve. A sketch of the function $Q(t) = A/(B + Ce^{-kt})$ is shown in Figure 2.4. Notice that the curve crosses the vertical axis at a height of $A/(B + C)$ and that $Q(t)$ approaches A/B as t increases without bound.

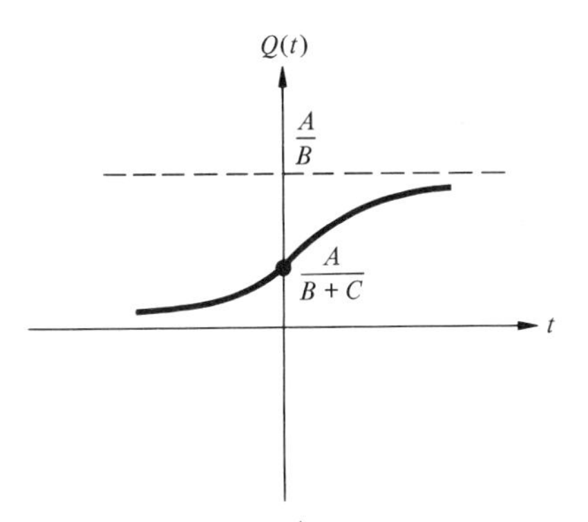

Figure 2.4 A logistic curve: $Q(t) = \dfrac{A}{B + Ce^{-kt}}$.

Logistic curves are rather accurate models of population growth when environmental factors impose an upper bound on the possible size of the population. They also describe the spread of epidemics and rumors in a community. Here is a typical example.

EXAMPLE 2.4 ● Public health records indicate that t weeks after the outbreak of a rare form of influenza, approximately $Q(t) = 80/(4 + 76e^{-1.2t})$ thousand people had caught the disease.

(a) How many people had the disease when it first broke out?
(b) How many had caught the disease by the end of the second week?
(c) If the trend continues, approximately how many people in all will contract the disease?

SOLUTION (a) Since

$$Q(0) = \frac{80}{4 + 76} = 1$$

we conclude that initially 1,000 people had the disease.

(b) Since

$$Q(2) = \frac{80}{4 + 76e^{-1.2(2)}} = 7.343$$

we conclude that by the end of the second week, 7,343 people had caught the disease.

(c) Since $Q(t)$ approaches $80/4 = 20$ as t increases without bound, we conclude that approximately 20,000 people will eventually contract the disease. ●

Problems

1 It is projected that t years from now, the population of a certain country will be $P(t) = 50e^{0.02t}$ million.
(a) What is the current population of the country?
(b) What will the population be 30 years from now?

2 How much money should be invested today at an annual rate of 7 percent compounded continuously so that 20 years from now it will be worth $20,000? (Recall that the balance after t years is $A(t) = Pe^{rt}$, where r is the interest rate expressed as a decimal and P is the initial deposit.)

3 It is estimated that the population of a certain country grows exponentially. If the population was 60 million in 1974 and 90 million in 1979, what will the population be in 1989?

4 The following data were compiled by a medical student during the first 10 minutes of an experiment designed to study the growth of bacteria.

Number of minutes	0	10
Number of bacteria	5,000	8,000

Assuming that the number of bacteria grows exponentially, how many bacteria will be present after 30 minutes?

5 The gross national product (GNP) of a certain country was 100 billion dollars in 1965 and 180 billion dollars in 1975. Assuming that the GNP is growing exponentially, what will the GNP be in 1985?

6 The population density x miles from the center of a certain city is $Q(x) = 12e^{-0.07x}$ thousand people per square mile.
(a) What is the population density at the center of the city?
(b) What is the population density 10 miles from the center of the city?

7 The amount of a certain radioactive substance remaining after t years is given by a function of the form $Q(t) = Q_0e^{-0.0001t}$. At the end of 5,000 years, 2,000 grams of the substance remain. How many grams were present initially?

8 A radioactive substance decays exponentially. If 500 grams of the substance were present initially and 400 grams are present 50 years later, how many grams will be present after 200 years?

9 A statistical study indicates that the fraction of the electric toasters manufactured by a certain company that are still in working condition after t years of use is approximately $f(t) = e^{-0.2t}$.
(a) What fraction of the toasters can be expected to work for at least 3 years?
(b) What fraction of the toasters can be expected to fail during their 3d year of use?
(c) What fraction of the toasters can be expected to fail before 1 year of use?

10 A manufacturer of toys has found that the fraction of its plastic battery-operated toy oil tankers that sink in less than t days is approximately $f(t) = 1 - e^{-0.03t}$.

(a) Sketch this reliability function. What happens to the graph as t increases without bound?

(b) What fraction of the tankers can be expected to float for at least 10 days?

(c) What fraction of the tankers can be expected to sink between the 15th and 20th days?

11 Psychologists believe that when a person is asked to recall a set of facts, the number of facts recalled after t minutes is given by a function of the form $Q(t) = A(1 - e^{-kt})$, where k is a positive constant and A is the total number of relevant facts in the person's memory.

(a) Sketch the function Q.

(b) What happens to the graph of Q as t increases without bound? Explain this behavior in practical terms.

12 When professors select texts for their courses, they usually choose from among the books already on their shelves. For this reason, most publishers send complimentary copies of new texts to professors teaching related courses. The mathematics editor at a major publishing house estimates that if x thousand complimentary copies are distributed, the first-year sales of a certain new mathematics text will be approximately $f(x) = 20 - 15e^{-0.2x}$ thousand copies.

(a) Sketch this sales function.

(b) How many copies can the editor expect to sell in the first year if no complimentary copies are sent out?

(c) How many copies can the editor expect to sell in the first year if 10,000 complimentary copies are sent out?

(d) If the editor's estimate is correct, what is the most optimistic projection for the first-year sales of the text?

13 A cool drink is removed from a refrigerator on a hot summer day and placed in a room whose temperature is 30°C. According to a law of physics, the temperature of the drink t minutes later is given by a function of the form $f(t) = 30 - Ae^{-kt}$. If the temperature of the drink was 10°C when it left the refrigerator and 15°C after 20 minutes, what will the temperature of the drink be after 40 minutes?

14 When a certain industrial machine is t years old, its resale value will be $V(t) = 4{,}800e^{-t/5} + 400$ dollars.

(a) Sketch the function V. What happens to the value of the machine as t increases without bound?

(b) How much was the machine worth when it was new?

(c) How much will the machine be worth after 10 years?

15 Public health records indicate that t weeks after the outbreak of a rare form of influenza, approximately $f(t) = 6/(3 + 9e^{-0.8t})$ thousand people had caught the disease.

(a) How many people had the disease initially?

(b) How many had caught the disease by the end of 3 weeks?
(c) If the trend continues, approximately how many people in all will contract the disease?

16 It is estimated that t years from now, the population of a certain country will be $P(t) = 80/(8 + 12e^{-0.06t})$ million.
(a) What is the current population?
(b) What will the population be 50 years from now?
(c) What will happen to the population in the long run?

3 • THE NATURAL LOGARITHM

In many practical problems, a number x is known and the goal is to find the corresponding number y such that $x = e^y$. This number y is called the **natural logarithm** of x and is denoted by the symbol $\ln x$. The letter l in the symbol ln stands for "logarithm" and the n for "natural." The word "ln" is virtually unpronounceable. Most people read it as "log" or "Lynn."

The natural logarithm
Corresponding to each positive number x there is a unique power y such that $x = e^y$. This power y is called the *natural logarithm* of x and is denoted by $\ln x$. Thus, $y = \ln x$ if and only if $x = e^y$.

EXAMPLE 3.1 ● Find:

(a) $\ln e$ (b) $\ln 1$

SOLUTION (a) Ln e is the unique number y such that $e = e^y$. Clearly, this number is $y = 1$. Hence $\ln e = 1$.
(b) Ln 1 is the unique number y such that $1 = e^y$. Since $e^0 = 1$, it follows that $\ln 1 = 0$. ●

The relationship between e^x and $\ln x$

The next example establishes two important identities that show that logarithmic and exponential functions have a certain "neutralizing" effect on each other.

EXAMPLE 3.2 ● Simplify the following expressions.

(a) $e^{\ln x}$ (b) $\ln e^x$

SOLUTION (a) Ln x is the unique number y for which $x = e^y$. Hence, $e^{\ln x} = e^y = x$.
(b) Ln e^x is the unique number y for which $e^x = e^y$. Clearly, this number is x itself. Hence, $\ln e^x = x$. ●

The two identities derived in Example 3.2 show that the functions $\ln e^x$ and $e^{\ln x}$ leave the variable x unchanged. In general, two functions f and g for which

$f[g(x)] = x$ and $g[f(x)] = x$ are said to be **inverses** of one another. Thus, the exponential function $y = e^x$ and the logarithmic function $y = \ln x$ are inverses of one another.

The inverse relationship between e^x and $\ln x$
For any value of x,

$$\ln e^x = x \qquad \text{and} \qquad e^{\ln x} = x$$

The next example illustrates how you can use the inverse relationship between e^x and $\ln x$ to solve equations.

EXAMPLE 3.3 ● Solve each of the following equations for x.

(a) $3 = e^{20x}$ (b) $2 \ln x = 1$

SOLUTION (a) We take the natural logarithm of each side of the equation to get

$$\ln 3 = \ln e^{20x} = 20x$$

Solving for x and using a calculator (or the natural logarithm table in the Appendix) to find $\ln 3$ we get

$$x = \frac{\ln 3}{20} = \frac{1.0986}{20} = 0.0549$$

(b) To isolate $\ln x$ on the left-hand side of the equation, we divide both sides by 2 getting

$$\ln x = \tfrac{1}{2}$$

We then apply the exponential function to both sides of the equation and conclude that

$$e^{\ln x} = e^{1/2} \qquad \text{or} \qquad x = \sqrt{e} = 1.649$$ ●

It can be shown in general that if $f(x)$ and $g(x)$ are inverses of one another, the graph of g can be obtained by reflecting the graph of f across the line $y = x$. It follows that the graph of the logarithmic function $y = \ln x$ can be obtained in this manner from the graph of the exponential function $y = e^x$. The graphs of the functions are shown in Figure 3.1. Notice that $\ln x$ is defined only for positive values of x, that $\ln 1 = 0$, that $\ln x$ increases without bound as x increases without bound, and that $\ln x$ decreases without bound as x approaches zero from the right.

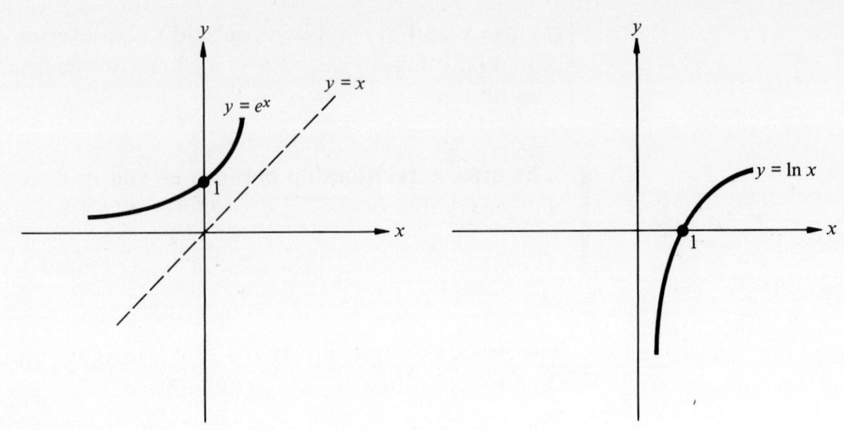

Figure 3.1 The graphs of $y = e^x$ and $y = \ln x$.

Properties of the natural logarithm

The laws of exponents can be used to derive the following important properties of logarithms.

> **Properties of logarithms**
> The logarithm of a product: $\ln uv = \ln u + \ln v$.
> The logarithm of a quotient: $\ln u/v = \ln u - \ln v$.
> The logarithm of a power: $\ln u^v = v \ln u$.

The use of these properties is illustrated in the next example.

EXAMPLE 3.4 ●

(a) Find $\ln \sqrt{ab}$ if $\ln a = 3$ and $\ln b = 7$.
(b) Show that $\ln 1/x = -\ln x$.
(c) Find x if $2^x = e^3$.

SOLUTION (a) $$\ln \sqrt{ab} = \ln (ab)^{1/2} = \tfrac{1}{2}\ln ab = \tfrac{1}{2}(\ln a + \ln a) = \tfrac{1}{2}(3 + 7) = 5$$

(b) $$\ln \frac{1}{x} = \ln 1 - \ln x = 0 - \ln x = -\ln x$$

(c) We take the natural logarithm of each side of the equation $2^x = e^3$ to get

$$x \ln 2 = 3 \qquad \text{or} \qquad x = \frac{3}{\ln 2} = 4.328$$ ●

Logarithms with other bases

You may be wondering how natural logarithms are related to the logarithms you studied in high school. The following discussion will show you the connection.

The graph of the exponential function $y = a^x$ (shown in Figure 3.2 for $a > 1$) suggests that to each positive number y there corresponds a unique number x such that $y = a^x$. This power x is called the **logarithm of y to the base a** and is denoted by $\log_a y$. Thus $x = \log_a y$ if and only if $y = a^x$.

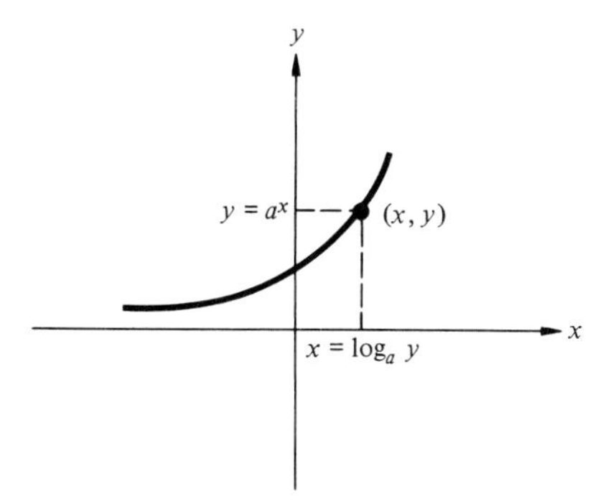

Figure 3.2 The graph of the function $y = a^x$ for $a > 1$.

The logarithms you used in high school algebra to simplify numerical calculations were logarithms to the base 10. For numerical work, 10 is a particularly convenient base for logarithms because the standard decimal representation of numbers is based on powers of 10. Natural logarithms are logarithms to the base e. Because of the importance of the special exponential function e^x, natural logarithms arise frequently in applied work.

Investment growth

In the next example, you will see how to use logarithms to determine how much time it takes a quantity that grows exponentially to reach a specified size.

EXAMPLE 3.5 ● A sum of money is deposited in a bank offering interest at an annual rate of 6 percent compounded continuously. How long will it take for the value of this investment to double?

SOLUTION The balance after t years will be $A(t) = Pe^{0.06t}$ dollars, where P is the initial deposit. Our goal is to find the value of t for which $A(t) = 2P$. That is, we wish to solve the equation

$$2P = Pe^{0.06t}$$

for t. Dividing by P and taking the logarithm of each side we get

$$\ln 2 = 0.06t \qquad \text{or} \qquad t = \frac{\ln 2}{0.06} = 11.55$$

We conclude that it takes approximately $11\frac{1}{2}$ years for the money to double.

Notice that the initial deposit P was eliminated from the equation at an early stage, and that the final answer is independent of this quantity. ●

Exponential curve fitting

In the next example you will see how to use logarithms to fit an exponential curve to a set of data.

EXAMPLE 3.6 ● Urban analysts believe that the population density x miles from the center of a city is given by a function of the form $Q(x) = Ae^{-kx}$. Find this function if the population density at the center of the city is 15 thousand people per square mile and the density 10 miles from the center is 9 thousand per square mile.

SOLUTION For convenience we express the density in units of 1,000 people per square mile. The fact that $Q(0) = 15$ tells us that $A = 15$. The fact that $Q(10) = 9$ tells us that

$$9 = 15e^{-10k} \qquad \text{or} \qquad \tfrac{3}{5} = e^{-10k}$$

Taking the logarithm of each side we get

$$\ln \tfrac{3}{5} = -10k \qquad \text{or} \qquad k = -\frac{\ln \frac{3}{5}}{10} = 0.051$$

We conclude that $Q(x) = 15e^{-0.051x}$. ●

Carbon dating Carbon 14 (^{14}C) is a radioactive isotope of carbon that is widely used to date ancient fossils and artifacts. Here is an outline of the technique.

The carbon dioxide in the air contains ^{14}C as well as carbon 12 (^{12}C), a nonradioactive isotope. Scientists have found that the ratio of ^{14}C to ^{12}C in the air has remained approximately constant throughout history. Living plants absorb carbon dioxide from the air and so the ratio of ^{14}C to ^{12}C in a living plant is the same as that in the air itself. When a plant dies, the absorption of carbon dioxide ceases. The ^{12}C already in the plant remains while the ^{14}C decays, and the ratio of ^{14}C to ^{12}C decreases exponentially. The ratio of ^{14}C to ^{12}C in a fossil t years after it was alive is approximately $R(t) = R_0e^{-kt}$, where $k = (\ln 2)/5{,}730$ and R_0 is the ratio of ^{14}C to ^{12}C in the atmosphere. By comparing $R(t)$ with R_0, scientists can estimate the age of the fossil.

EXAMPLE 3.7 ● An archaeologist has found a fossil in which the ratio of ^{14}C to ^{12}C is $\frac{1}{5}$ the ratio found in the atmosphere. Approximately how old is the fossil?

SOLUTION The age of the fossil is the value of t for which $R(t) = \frac{1}{5}R_0$, that is, for which

$$\tfrac{1}{5}R_0 = R_0e^{-kt}$$

Dividing by R_0 and taking logarithms we get

$$\ln \tfrac{1}{5} = -kt \qquad \text{or} \qquad t = -\frac{\ln \frac{1}{5}}{k} = \frac{\ln 5}{k}$$

Since $k = (\ln 2)/5{,}730$, we find that

$$t = \frac{5{,}730 \ln 5}{\ln 2} = 13{,}305$$

and we conclude that the fossil is approximately 13,305 years old.

Problems

1 Learn how to use your calculator to find natural logarithms. In particular, find $\ln 1$, $\ln 2$, $\ln e$, $\ln 5$, $\ln \frac{1}{5}$, and $\ln e^2$. What happens if you try to find $\ln 0$ or $\ln -2$? Why?

In Problems 2 through 7, evaluate the given expression (without using tables or a calculator).

2 $\ln e^3$

3 $\ln \sqrt{e}$

4 $e^{\ln 5}$

5 $e^{2 \ln 3}$

6 $e^{3 \ln 2 - 2 \ln 5}$

7 $\ln (e^3 \sqrt{e}/e^{1/3})$

In Problems 8 through 19, solve the given equation for x.

8 $2 = e^{0.06x}$

9 $Q_0/2 = Q_0 e^{-1.2x}$

10 $3 = 2 + 5e^{-4x}$

11 $-2 \ln x = b$

12 $-\ln x = t/50 + C$

13 $5 = 3 \ln x - \frac{1}{2} \ln x$

14 $\ln x = \frac{1}{3}(\ln 16 + 2 \ln 2)$

15 $\ln x = 2(\ln 3 - \ln 5)$

16 $3^x = e^2$

17 $a^k = e^{kx}$

18 $a^{x+1} = b$

19 $x^{\ln x} = e$

20 Find $\ln (1/\sqrt{ab^3})$ if $\ln a = 2$ and $\ln b = 3$.

21 Find $(1/a) \ln (\sqrt{b}/c)^a$ if $\ln b = 6$ and $\ln c = -2$.

22 How quickly will money triple if it is invested at an annual interest rate of 6 percent compounded continuously?

23 Money deposited in a certain bank doubles every 13 years. The bank compounds interest continuously. What annual interest rate does the bank offer?

24 How quickly will money double if it is invested at an annual interest rate of 7 percent and interest is compounded
(a) continuously (b) annually (c) quarterly

25 A certain bank offers an interest rate of 6 percent a year compounded annually. A competing bank compounds its interest continuously. What (nominal) interest rate should the competing bank offer so that the *effective* interest rates of the two banks will be equal? (See Chapter 9, Section 1 for the definition of the effective interest rate.)

The **half-life** of a radioactive substance is the time it takes for 50 percent of a sample of the substance to decay. Problems 26 through 28 deal with this concept.

26 The amount of a certain radioactive substance remaining after t years is given by a function of the form $Q(t) = Q_0e^{-0.003t}$. Find the half-life of the substance.

27 Radium decays exponentially. Its half-life is 1,690 years. How long will it take for a 50-gram sample of radium to be reduced to 5 grams?

28 A radioactive substance decays exponentially. Show that the amount $Q(t)$ of the substance remaining after t years is $Q(t) = Q_0(\frac{1}{2})^{t/\lambda}$, where Q_0 is the amount present initially and λ is the half-life of the substance.

29 The mathematics editor at a major publishing house estimates that if x thousand complimentary copies are distributed to instructors, the first-year sales of a new mathematics text will be approximately $f(x) = 20 - 15e^{-0.2x}$ thousand copies. According to this estimate, approximately how many complimentary copies should the editor send out to generate first-year sales of 12,000 copies?

30 A medical student studying the growth of bacteria in a certain culture has compiled the following data.

Number of minutes	0	20
Number of bacteria	6,000	9,000

Use this data to find an exponential function of the form $Q(t) = Q_0e^{kt}$ expressing the number of bacteria in the culture as a function of time.

31 An economist has compiled the following data on the gross national product (GNP) of a certain country.

Year	1965	1975
GNP (in billions)	100	180

Use this data to predict the GNP in 1986 if the GNP is growing
(a) linearly (b) exponentially

32 An efficiency expert hired by a manufacturing firm has compiled the following data relating workers' output to their experience.

Experience (months)	0	6
Output (units per hour)	300	410

The expert believes that the output Q is related to experience t by a function of the form $Q(t) = 500 - Ae^{-kt}$. Find the function of this form that fits the data.

33 An archaeologist has found a fossil in which the ratio of ^{14}C to ^{12}C is $\frac{1}{3}$ the ratio found in the atmosphere. Approximately how old is the fossil?

4 • DIFFERENTIATION OF LOGARITHMIC AND EXPONENTIAL FUNCTIONS

Both the logarithmic function $y = \ln x$ and the exponential function $y = e^x$ turn out to have rather simple derivatives.

Differentiation of logarithmic functions

The derivative of $\ln x$ is simply $1/x$.

The derivative of ln x

$$\frac{d}{dx}(\ln x) = \frac{1}{x}$$

One proof of this formula is based on the fact that $(1 + 1/n)^n$ approaches e as n increases without bound, and can be found in most calculus texts. Here are two examples that illustrate the use of the formula.

EXAMPLE 4.1 ● Differentiate the function $f(x) = x \ln x$.

SOLUTION We combine the product rule with the formula for the derivative of $\ln x$ to get

$$f'(x) = x\left(\frac{1}{x}\right) + \ln x = 1 + \ln x$$ ●

EXAMPLE 4.2 ● Differentiate the function $y = \ln(2x^3 + 1)$.

SOLUTION Letting $y = \ln u$ and $u = 2x^3 + 1$ and applying the chain rule we get

$$\frac{dy}{dx} = \frac{dy}{du}\frac{du}{dx} = \frac{1}{u}(6x^2) = \frac{6x^2}{2x^3 + 1}$$ ●

The preceding example illustrates the following general rule for differentiating the logarithm of a differentiable function.

The chain rule for logarithmic functions

$$\frac{d}{dx}[\ln h(x)] = \frac{h'(x)}{h(x)}$$

That is, to differentiate $\ln h$, simply divide the derivative of h by h itself.

Here is another example.

EXAMPLE 4.3 ● Differentiate the function $f(x) = \ln(x^2 + 1)^3$.

SOLUTION As a preliminary step, we simplify $f(x)$ using one of the properties of logarithms to get

$$f(x) = 3 \ln (x^2 + 1)$$

Now we apply the chain rule for logarithms and find that

$$f'(x) = 3 \frac{2x}{x^2 + 1} = \frac{6x}{x^2 + 1}$$

Convince yourself that the final answer would have been the same if we had not made the initial simplification of $f(x)$. ●

Differentiation of exponential functions

The derivative of e^x is *not* xe^{x-1}. Before you proceed, be sure you see the fallacy in applying the power rule $d/dx\,(x^n) = nx^{n-1}$ to the function e^x.

To get the correct formula for the derivative of e^x, we differentiate both sides of the equation

$$\ln e^x = x$$

with respect to x. Using the chain rule for logarithms to differentiate $\ln e^x$ we get

$$\frac{\frac{d}{dx}(e^x)}{e^x} = 1 \qquad \text{or} \qquad \frac{d}{dx}(e^x) = e^x$$

That is, e^x is its own derivative!

The derivative of e^x

$$\frac{d}{dx}(e^x) = e^x$$

You should have no trouble combining the formula for the derivative of e^x with the chain rule to get the following formula for the derivative of $e^{h(x)}$ where h is a differentiable function of x.

Chain rule for exponential functions

$$\frac{d}{dx}[e^{h(x)}] = h'(x)e^{h(x)}$$

That is, to compute the derivative of $e^{h(x)}$, simply multiply $e^{h(x)}$ by the derivative of the exponent $h(x)$.

Here are some examples that illustrate the use of these rules.

EXAMPLE 4.4 ● Differentiate the function $f(x) = xe^x$.

SOLUTION By the product rule,

$$f'(x) = xe^x + e^x = (x + 1)e^x$$

●

EXAMPLE 4.5 ● Differentiate the function $f(x) = e^{x^2+1}$.

SOLUTION By the chain rule,

$$f'(x) = 2xe^{x^2+1}$$

●

Exponential growth In Section 2, you saw that a quantity $Q(t)$ that increases according to a law of the form $Q(t) = Q_0e^{kt}$, where Q_0 and k are positive constants, is said to experience exponential growth. In the next example, we use the chain rule for exponential functions to calculate the rate of change of a quantity that grows exponentially.

EXAMPLE 4.6 ● It is projected that t years from now, the population of a certain city will be $P(t) = 200e^{0.04t}$ thousand. At what rate will the population of the city be changing 5 years from now?

SOLUTION The rate of change of the population is the derivative

$$P'(t) = 0.04(200)e^{0.04t} = 8e^{0.04t}$$

Since

$$P'(5) = 8e^{0.2} = 9.771$$

we conclude that 5 years from now, the population of the city will be increasing at a rate of 9,771 people per year. ●

The next example shows that if a quantity grows exponentially, its rate of growth is proportional to its size. The converse of this fact will be established in Chapter 13.

EXAMPLE 4.7 ● If $Q(t) = Q_0e^{kt}$, find an expression for the rate of change of Q with respect to t.

SOLUTION The rate of change of Q is the derivative

$$Q'(t) = kQ_0e^{kt} = kQ(t)$$

This says that the rate of change $Q'(t)$ is proportional to $Q(t)$ itself and that the constant k that appeared in the exponent of $Q(t)$ is the constant of proportionality. ●

Optimization Now that you know how to differentiate exponential functions, you can use calculus to find maximum and minimum values of functions involving powers of e. Here is an example.

EXAMPLE 4.8 ● The consumer demand for a certain commodity is $D(p) = 5{,}000e^{-0.02p}$ units per month when the market price is p dollars per unit. Determine the market price that will result in the greatest possible consumer expenditure.

SOLUTION The consumer expenditure for the commodity is the price per unit times the number of units sold. That is,

$$E(p) = pD(p) = 5{,}000pe^{-0.02p}$$

Since only nonnegative values of p are meaningful in this context, the goal is to maximize $E(p)$ for $p \geq 0$.

The derivative of E is

$$\begin{aligned} E'(p) &= 5{,}000[-0.02pe^{-0.02p} + e^{-0.02p}] \\ &= 5{,}000e^{-0.02p}(1 - 0.02p) \end{aligned}$$

Since $e^{-0.02p}$ is never zero, $E'(p) = 0$ if and only if

$$1 - 0.02p = 0 \quad \text{or} \quad p = \frac{1}{0.02} = 50$$

To verify that $p = 50$ is really the optimal price you are seeking, check the sign of $E'(p)$ for $0 < p < 50$ and for $p > 50$. Since $5{,}000e^{-0.02p}$ is always

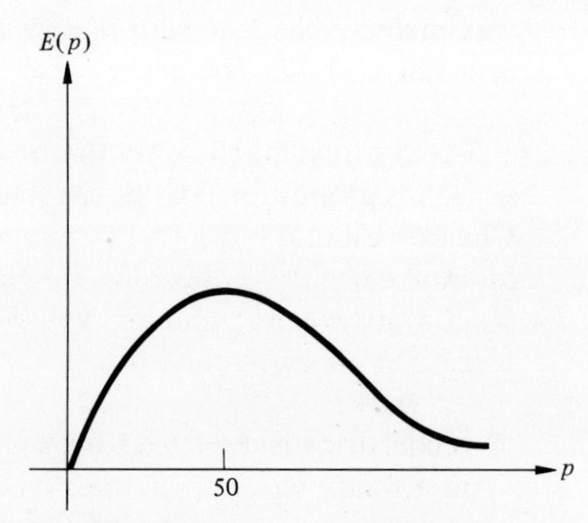

Figure 4.1 Consumer expenditure: $E(p) = 5{,}000\,pe^{-0.02p}$.

positive, it follows that $E'(p)$ is positive if $0 < p < 50$ and negative if $p > 50$. Hence E is increasing for $0 < p < 50$ and decreasing for $p > 50$ as shown in Figure 4.1, and you can conclude that consumer expenditure will be greatest when the market price is \$50 per unit. ●

Problems

In Problems 1 through 20, differentiate the given function.

1 $f(x) = e^{5x}$

2 $f(x) = 3e^{4x+1}$

3 $f(x) = e^{x^2+2x-1}$

4 $f(x) = e^{1/x}$

5 $f(x) = 30 + 10e^{-0.05x}$

6 $f(x) = x^2e^x$

7 $f(x) = xe^{-x^2}$

8 $f(x) = x/e^x$

9 $f(x) = (1 + e^x)^2$

10 $f(x) = \ln x^3$

11 $f(x) = \ln (2x^2 + x - 3)$

12 $f(x) = \ln \sqrt{x^2 + 1}$

13 $f(x) = x^2 \ln x$

14 $f(x) = (\ln x)/x$

15 $f(x) = x/\ln x$

16 $f(x) = \ln [(x + 1)/(x - 1)]$

17 $f(x) = e^x \ln x$

18 $f(x) = \ln e^{2x}$

19 $f(x) = e^{x+\ln x}$

20 $f(x) = e^{\ln x+2 \ln 3x}$

21 It is projected that t years from now, the population of a certain country will be $P(t) = 50e^{0.02t}$ million.
(a) At what rate will the population be changing 10 years from now?
(b) By how much will the population actually change during the 11th year?
(c) Explain why the answers to parts (a) and (b) are not identical.

22 A certain industrial machine depreciates so that its value after t years is $V(t) = 20{,}000e^{-0.4t}$ dollars.
(a) At what rate is the machine depreciating after 5 years?
(b) By how much will the value of the machine actually decrease during the 6th year?
(c) Explain why the answers to parts (a) and (b) are not identical.

23 The mathematics editor at a major publishing house estimates that if x thousand complimentary copies are distributed to professors, the first-year sales of a certain new text will be $f(x) = 20 - 15e^{-0.2x}$ thousand copies. Currently, the editor is planning to distribute 10,000 complimentary copies.
(a) Use marginal analysis to estimate the increase in first-year sales that will result if 1,000 additional complimentary copies are distributed.
(b) Calculate the actual increase in first-year sales that will result from the distribution of the additional 1,000 complimentary copies. Is the estimate in part (a) a good one?

24 Public health records indicate that t weeks after the outbreak of a rare form of influenza, approximately $Q(t) = 80/(4 + 76e^{-1.2t})$ thousand people had caught the disease. At what rate was the disease spreading at the end of the 2d week?

25 A cool drink is removed from a refrigerator on a hot summer day and placed in a room whose temperature is 30°C. According to a law of physics, the temperature of the drink t minutes later is given by a function of the form

$f(t) = 30 - Ae^{-kt}$. Show that the rate of change of the temperature of the drink with respect to time is proportional to the difference between the temperature of the room and that of the drink.

26 The consumer demand for a certain commodity is $D(p) = 3{,}000e^{-0.01p}$ units per month when the market price is p dollars per unit. Express consumers' total monthly expenditure for the commodity as a function of p, and determine the market price that will result in the largest possible consumer expenditure.

27 A manufacturer can produce radios at a cost of \$5 apiece and estimates that if they are sold for x dollars apiece, consumers will buy approximately $1{,}000e^{-0.1x}$ radios per week. At what price should the manufacturer sell the radios to maximize profit?

In Problems 28 through 30, rewrite the given function as an appropriate power of e and then use the chain rule to find dy/dx.

28 $y = 2^x$ **29** $y = x^x$ **30** $y = x^{\ln x}$

31 Prove the power rule $d/dx(x^n) = nx^{n-1}$. (*Hint:* Write x^n as $e^{n \ln x}$.)

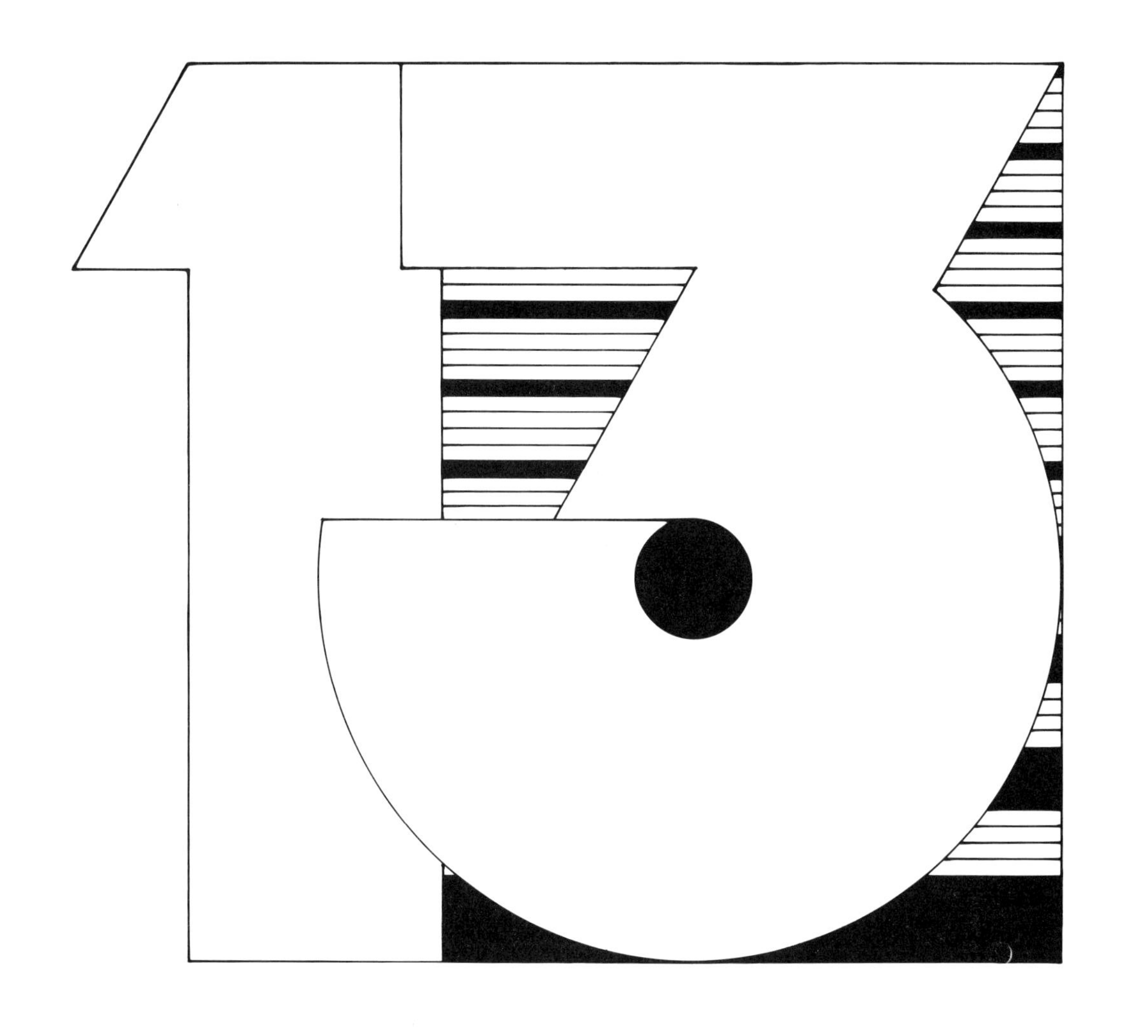

INTEGRAL CALCULUS

1 • ANTIDERIVATIVES

In many problems, the derivative of a function is known and the goal is to find the function itself. For example, a sociologist who knows the rate at which the population is growing may wish to use this information to predict future population levels. Or, an economist who knows the rate of inflation may wish to estimate future prices.

The process of obtaining a function from its derivative is called **antidifferentiation** or **integration.**

Antiderivative
A function F whose derivative equals f is said to be an **antiderivative** (or **indefinite integral**) of f.

EXAMPLE 1.1 ● Verify that $F(x) = \frac{1}{3}x^3 + 5x + 2$ is an antiderivative of $f(x) = x^2 + 5$.

SOLUTION We differentiate F and find that

$$F'(x) = x^2 + 5 = f(x)$$

as required. ●

A function has more than one antiderivative. For example, $\frac{1}{3}x^3$ is an antiderivative of x^2. But so is $\frac{1}{3}x^3 + 12$, since the derivative of the constant 12 is zero. In general, if F is one antiderivative of f, any function obtained by adding a constant to F is also an antiderivative of f. In fact, it turns out that all the antiderivatives of f can be obtained by adding constants to a given antiderivative.

The antiderivatives of a function
If F and G are antiderivatives of f, then there is a constant C such that $G(x) = F(x) + C$.

Notation

It is customary to write

$$\int f(x)\,dx = F(x) + C$$

to express the fact that every antiderivative of $f(x)$ is of the form $F(x) + C$. For example, we express the fact that every antiderivative of x^2 is of the form $\frac{1}{3}x^3 + C$ by writing

$$\int x^2\,dx = \tfrac{1}{3}x^3 + C$$

The symbol $\int$ is called an **integral sign** and tells us that the function following it is to be integrated. At first, the symbol dx may seem mysterious. Its role is to indicate that x is the variable with respect to which the integration is to be performed. Analogous notation is used if the function is expressed in terms of a variable other than x. For example, $\int t^2\,dt = \frac{1}{3}t^3 + C$. In the expression $\int px^2\,dx$, the dx tells us that x rather than p is the variable. Thus, $\int px^2\,dx = \frac{1}{3}px^3 + C$. (Do you see how to evaluate $\int px^2\,dp$?)

Techniques of integration According to the power rule $d/dx\,(x^n) = nx^{n-1}$, you differentiate a power function by reducing the power by 1 and multiplying by the original power. Here is the rule stated in reverse as a rule for integrating power functions.

Integration of power functions
For $n \neq -1$,

$$\int x^n\,dx = \frac{1}{n+1}x^{n+1} + C$$

That is, to integrate x^n (for $n \neq -1$), increase the power of x by 1 and divide by the new power.

To convince yourself that this rule is valid, simply differentiate $x^{n+1}/(n+1)$ and observe that you get x^n. The rule holds for all values of n except, of course, for $n = -1$, in which case $1/(n+1)$ is undefined. This special case will be discussed shortly. In the meantime, see if you can think of an antiderivative of x^{-1} on your own.

The rule for integrating power functions is illustrated in the next example.

EXAMPLE 1.2 ● Find the following integrals.

(a) $\int x^{3/5}\,dx$ (b) $\int 1\,dx$ (c) $\int 1/\sqrt{x}\,dx$

SOLUTION (a) Increasing the power of x by 1 and then dividing by the new power we get

$$\int x^{3/5}\,dx = \frac{5}{8}x^{8/5} + C$$

(b) Since $1 = x^0$, we conclude that

$$\int 1\,dx = x + C$$

(c) We begin by rewriting $1/\sqrt{x}$ as $x^{-1/2}$. Then,

$$\int 1/\sqrt{x}\,dx = \int x^{-1/2}\,dx = 2x^{1/2} + C = 2\sqrt{x} + C$$ ●

Were you able to discover an antiderivative of x^{-1}? That is, were you able to think of a function whose derivative is $1/x$? The natural logarithm $\ln x$ is such a function and so it appears that $\int 1/x\, dx = \ln x + C$. Actually, this is true only when x is positive, since $\ln x$ is not defined for negative values of x. When x is negative, it turns out that the function $\ln |x|$ is an antiderivative of $1/x$. To see this, observe that if x is negative, $|x| = -x$ and so

$$\frac{d}{dx}(\ln |x|) = \frac{d}{dx}[\ln(-x)] = \left(\frac{1}{-x}\right)(-1) = \frac{1}{x}$$

Since $|x| = x$ when x is positive, we can summarize the situation using a single formula as follows.

The integral of $1/x$

$$\int \frac{1}{x}\, dx = \ln |x| + C$$

Integration of the exponential function e^x is trivial since e^x is its own derivative.

The integral of e^x

$$\int e^x\, dx = e^x + C$$

It is easy to rewrite the constant-multiple rule and the sum rule for differentiation as rules for integration.

The constant-multiple rule for integrals
For any constant c,

$$\int cf(x)\, dx = c \int f(x)\, dx$$

That is, the integral of a constant times a function is equal to the constant times the integral of the function.

The sum rule for integrals

$$\int [f(x) + g(x)]\, dx = \int f(x)\, dx + \int g(x)\, dx$$

That is, the integral of a sum is the sum of the individual integrals.

EXAMPLE 1.3 ● Find $\int (3e^x + 2/x - \frac{1}{2}x^2)\,dx$.

SOLUTION

$$\int \left(3e^x + \frac{2}{x} - \frac{1}{2}x^2\right) dx = 3\int e^x\,dx + 2\int \frac{1}{x}\,dx - \frac{1}{2}\int x^2\,dx$$

$$= 3e^x + 2\ln|x| - \tfrac{1}{6}x^3 + C$$ ●

Notice that instead of writing a separate constant for each of the three antiderivatives computed in Example 1.3, we simply added a single constant C at the end.

You may have noticed that we have not stated general rules for the integration of products and quotients. This is because there are no such general rules. There is one rule, however, that you can use to integrate products of a very special type. This rule, which is the integral version of the chain rule, will be discussed in Section 2.

Practical applications Here are two problems in which the rate of change of a quantity is known and the goal is to find an expression for the quantity itself. Since the rate of change is the derivative of the quantity, we find the expression for the quantity itself by antidifferentiation.

EXAMPLE 1.4 ● It is estimated that x months from now the population of a certain town will be changing at a rate of $2 + 6\sqrt{x}$ people per month. The current population is 5,000. What will the population be 9 months from now?

SOLUTION Let $P(x)$ denote the population of the town x months from now. Then the derivative of P is the rate of change of the population with respect to time. That is,

$$\frac{dP}{dx} = 2 + 6\sqrt{x}$$

It follows that the population function P is an antiderivative of $2 + 6\sqrt{x}$. That is,

$$P(x) = \int (2 + 6\sqrt{x})\,dx = 2x + 4x^{3/2} + C$$

for some constant C. To determine C, we use the information that at present (when $x = 0$), the population is 5,000. That is,

$$5{,}000 = 2(0) + 4(0)^{3/2} + C \quad \text{or} \quad C = 5{,}000$$

Hence

$$P(x) = 2x + 4x^{3/2} + 5{,}000$$

and we conclude that in 9 months, the population will be

$$P(9) = 2(9) + 4(27) + 5{,}000 = 5{,}126$$ ●

EXAMPLE 1.5 ● A manufacturer has found that marginal cost is $q + 3$ dollars per unit when q units have been produced. The total cost of producing the first 4 units is \$520. What is the total cost of producing the first 8 units?

SOLUTION Recall that the marginal cost is the derivative of the total cost function C. Hence C must be an antiderivative of $q + 3$. That is,

$$C(q) = \int (q + 3)\, dq = \tfrac{1}{2}q^2 + 3q + K$$

for some constant K. (The letter K is used for the constant to avoid confusion with the cost function C.)

The value of K is determined by the fact that $C(4) = 520$. In particular,

$$520 = \tfrac{1}{2}(4)^2 + 3(4) + K \qquad \text{or} \qquad K = 500$$

Hence,

$$C(q) = \tfrac{1}{2}q^2 + 3q + 500$$

and the cost of producing the first 8 units is $C(8) = \$556$ ●

Problems

In Problems 1 through 14, find the indicated integral.

1 $\int x^{3/4}\, dx$

2 $\int 1/t^2\, dt$

3 $\int \sqrt{u}\, du$

4 $\int 5\, dx$

5 $\int (3t^2 - 5t + 2)\, dt$

6 $\int (2e^u + 6/u - 7u^2)\, du$

7 $\int (x^{1/2} - 3x^{2/3} + 6)\, dx$

8 $\int (2u^{-1/2} + 3/u^2 - 5\sqrt{u})\, du$

9 $\int (ax^2 + bx + c)\, dx$

10 $\int u(2u + 1)\, du$

11 $\int x^2(x^2 + 3x + 2)\, dx$

12 $\int (2t^2 + 3t + 1)/t\, dt$

13 $\int x^3(2x + 1/x)\, dx$

14 $\int (u + 1)/u^3\, du$

15 It is estimated that x months from now the population of a certain town will be changing at a rate of $4 + 5x^{2/3}$ people per month. The current population is 10,000. What will the population be 8 months from now?

16 In a certain section of the country, the price of large Grade A eggs is currently \$1.60 a dozen. Studies indicate that x weeks from now, the price wil be changing at a rate of $0.2 + 0.003x^2$ cents per week. How much will eggs cost 10 weeks from now?

17 The resale value of a certain industrial machine decreases at a rate that changes with time. When the machine is t years old, the rate at which its value is changing is $220(t - 10)$ dollars per year. If the machine was bought new for \$12,000, how much will it be worth 10 years later?

18 An object is moving so that its speed after t minutes is $3 + 2t + 6t^2$ meters per minute. How far does the object travel during the second minute?

19 A manufacturer has found that marginal cost is $6q + 1$ dollars per unit when q units have been produced. The total cost (including overhead) of producing the first unit is \$130. What is the total cost of producing the first 10 units?

20 A manufacturer estimates marginal revenue to be $100q^{-1/2}$ dollars per unit when output is q units. The corresponding marginal cost has been found to be $0.4q$ dollars per unit. Suppose the manufacturer's profit is \$520 when the level of production is 16 units. What is the manufacturer's profit when the level of production is 25 units?

21 The marginal profit (marginal revenue minus marginal cost) of a certain company is $100 - 2q$ dollars per unit when q units are produced. If the company's profit is \$700 when 10 units are produced, what is the company's maximum possible profit?

22 Find the function whose tangent has slope $\frac{1}{3}x^2 + 2x + 5$ for each value of x and whose graph passes through the point $(1, \frac{1}{9})$.

23 Find the function whose tangent has slope e^x for each value of x and whose graph passes through the point $(0, 2)$.

24 Find a function whose graph has a relative minimum when $x = 1$ and a relative maximum when $x = 4$.

2 • INTEGRATION BY SUBSTITUTION

The integral version of the chain rule is known as **integration by substitution.** Before we develop this technique, let us examine a typical application of the chain rule.

Using the chain rule to differentiate the function $(x^2 + 3x + 5)^9$ we get

$$\frac{d}{dx}[(x^2 + 3x + 5)^9] = 9(x^2 + 3x + 5)^8(2x + 3)$$

Notice that this derivative is a product and that one of its factors, $2x + 3$, is the derivative of an expression, $x^2 + 3x + 5$, that occurs in the other. More precisely, the product is of the form

$$g(u)\frac{du}{dx}$$

where, in this case, $g(u) = 9u^8$ and $u = x^2 + 3x + 5$.

You can integrate many products of the form $g(u)\,du/dx$ by applying the chain rule in reverse. Specifically, if G is an antiderivative of g, then

$$\int g(u)\frac{du}{dx}\,dx = G(u) + C$$

since, by the chain rule,

$$\frac{d}{dx}[G(u)] = G'(u)\frac{du}{dx} = g(u)\frac{du}{dx}$$

Integration by substitution

$$\int g(u)\frac{du}{dx}\,dx = G(u) + C$$

where G is an antiderivative of g.

That is, to integrate a product of the form $g(u)\;du/dx$, find $\int g(u)\,du$ and then replace u in the answer by the corresponding expression involving the variable x.

Here is a typical example.

EXAMPLE 2.1 ● Find $\int 9(x^2 + 3x + 5)^8(2x + 3)\,dx$.

SOLUTION We notice that $2x + 3$ is the derivative of $x^2 + 3x + 5$, and this suggests that we let $u = x^2 + 3x + 5$ and $g(u) = 9u^8$. Then,

$$9(x^2 + 3x + 5)^8(2x + 3) = g(u)\frac{du}{dx}$$

and we can use the method of substitution to integrate this product.

First we integrate $g(u) = 9u^8$ to get

$$\int 9u^8\,du = u^9 + C$$

and then we replace u by $x^2 + 3x + 5$ in the answer to conclude that

$$\int 9(x^2 + 3x + 5)^8(2x + 3)\,dx = (x^2 + 3x + 5)^9 + C$$

●

The product to be integrated in the next example is not exactly of the form $g(u)\, du/dx$. However, it is a constant multiple of such a function and can be integrated by combining the method of substitution with the constant-multiple rule.

EXAMPLE 2.2 • Find $\int x^3 e^{x^4+2}\, dx$.

SOLUTION The derivative of $x^4 + 2$ is not x^3, but rather $4x^3$, and so we cannot apply the method of substitution to the product $x^3 e^{x^4+2}$ directly. We can, however, apply substitution to the related product $4x^3 e^{x^4+2}$ to get

$$\int 4x^3 e^{x^4+2}\, dx = e^{x^4+2} + C$$

Then, using the constant-multiple rule, we find that

$$\int x^3 e^{x^4+2}\, dx = \tfrac{1}{4} \int 4x^3 e^{x^4+2}\, dx = \tfrac{1}{4} e^{x^4+2} + C$$ •

Change of variables Integration by substitution may be thought of as a technique for simplifying an integral by changing the variable of integration. In particular, we start with $\int g(u)\,(du/dx)\, dx$, an integral in which the variable of integration is x, and transform it into $\int g(u)\, du$, a simpler integral in which the variable of integration is u. In this transformation, the expression $(du/dx)\, dx$ in the original integral is replaced in the simplified integral by the symbol du. You can remember this relationship between $(du/dx)\, dx$ and du by pretending that du/dx is a quotient and writing

$$\frac{du}{dx}\, dx = du$$

These observations suggest an alternative version of the method of substitution in which the variable u is formally substituted for an appropriate expression in x and the original intergral is transformed into one in which the variable of integration is u. The expression for du is found by computing the derivative du/dx and then multiplying by dx. To illustrate this procedure, let us reconsider the integral in Example 2.1.

EXAMPLE 2.3 • Find $\int 9(x^2 + 3x + 5)^8(2x + 3)\, dx$.

SOLUTION We let $u = x^2 + 3x + 5$. Then, $du/dx = 2x + 3$ and so $du = (2x + 3)\, dx$. Substituting $u = x^2 + 3x + 5$ and $du = (2x + 3)\, dx$ we get

$$\int 9(x^2 + 3x + 5)^8(2x + 3)\, dx = \int 9u^8\, du = u^9 + C$$
$$= (x^2 + 3x + 5)^9 + C$$ •

Many people find this formal method of substitution more appealing than the original method. They like the fact that it involves straightforward mechanical manipulation of symbols, and they appreciate the convenience of the notation. This formal method is also somewhat more versatile, as we shall see presently. However, for most of the integrals you will encounter, both methods work well and you should feel free to use the one with which you are most comfortable.

Here is another example involving the constant-multiple rule.

EXAMPLE 2.4 ● Find $\int (3x + 5)^6\, dx$.

SOLUTION Let $u = 3x + 5$. Then $du/dx = 3$ and so $du = 3\, dx$. Hence,

$$\int (3x + 5)^6\, dx = \tfrac{1}{3}\int u^6\, du = \tfrac{1}{3}(\tfrac{1}{7}u^7) + C$$

$$= \tfrac{1}{21}u^7 + C = \tfrac{1}{21}(3x + 5)^7 + C$$ ●

The next example is designed to show the versatility of the formal method of substitution. It deals with an integral that does not seem to be of the form $\int g(u)\,(du/dx)\, dx$, but that can be simplified dramatically nevertheless by a clever change of variables.

EXAMPLE 2.5 ● Find $\int \frac{x}{x + 1}\, dx$

SOLUTION There seems to be no easy way to integrate this quotient as it stands. But watch what happens if we make the substitution $u = x + 1$. Then $du = dx$ and $x = u - 1$. Hence,

$$\int \frac{x}{x + 1}\, dx = \int \frac{u - 1}{u}\, du = \int 1\, du - \int \frac{1}{u}\, du$$

$$= u - \ln|u| + C = x + 1 - \ln|x + 1| + C$$ ●

Problems In Problems 1 through 19, find the indicated integral.

1 $\int (2x + 6)^5\, dx$

2 $\int \sqrt{4x - 1}\, dx$

3 $\int \frac{1}{3x + 5}\, dx$

4 $\int e^{1-x}\, dx$

5 $\int [(x - 1)^5 + 3(x - 1)^2 + 6(x - 1) + 5]\, dx$

6 $\int 2xe^{x^2-1}\, dx$

7 $\int x(x^2 + 1)^5\, dx$

8 $\int 3x\sqrt{x^2+8}\,dx$

9 $\int x^2(x^3+1)^{3/4}\,dx$

10 $\int xe^{x^2}\,dx$

11 $\int x^5e^{1-x^6}\,dx$

12 $\int \frac{2x^4}{x^5+1}\,dx$

13 $\int \frac{x^2}{(x^3+5)^2}\,dx$

14 $\int (x+1)(x^2+2x+5)^{12}\,dx$

15 $\int (3x^2-1)e^{x^3-x}\,dx$

16 $\int \frac{3x^4+12x^3+6}{x^5+5x^4+10x+12}\,dx$

17 $\int \frac{10x^3-5x}{\sqrt{x^4-x^2+6}}\,dx$

18 $\int \frac{\ln 5x}{x}\,dx$

19 $\int \frac{1}{x\ln x}\,dx$

In Problems 20 through 24, use an appropriate change of variables to find the indicated integral.

20 $\int \frac{x}{x-1}\,dx$

21 $\int x\sqrt{x+1}\,dx$

22 $\int \frac{x}{(x-5)^6}\,dx$

23 $\int (x+1)(x-2)^9\,dx$

24 $\int \frac{x+3}{(x-4)^2}\,dx$

25 Find a function whose tangent has slope $x\sqrt{x^2+5}$ for each value of x and whose graph passes through the point (2, 10).

26 Find a function whose tangent has slope $2x/(1-3x^2)$ for each value of x and whose graph passes through the point (0, 5).

27 A tree has been transplanted, and after x years is growing at a rate of $1+1/(x+1)^2$ meters per year. After 2 years it has reached a height of 5 meters. How tall was it when it was transplanted?

28 It is estimated that x years from now the value of an acre of farm land will be increasing at a rate of $0.4x^3/\sqrt{0.2x^4+8{,}000}$ dollars per year. If the land is currently worth \$500 per acre, how much will it be worth in 10 years?

3 • DIFFERENTIAL EQUATIONS

Any equation that contains a derivative is called a **differential equation.** For example, the equations $dy/dx = 3x^2 + 6$, $dP/dt = kP$, and $(dy/dx)^2 + 3(dy/dx) + 2y = e^x$ are all differential equations.

Many practical situations, especially those involving rates, can be described mathematically by differential equations. For example, the assumption that population grows at a rate proportional to its size can be expressed by the differential equation $dP/dt = kP$, where P denotes the population size, t stands for time, and k is the constant of proportionality. In economics, statements about marginal cost and marginal revenue can be formulated as differential equations.

Any function that satisfies a differential equation is said to be a **solution** of that equation. Here is an example to illustrate this concept.

EXAMPLE 3.1 ● Verify that the function $y = e^x - x$ is a solution of the differential equation $dy/dx - y = x - 1$.

SOLUTION We must subtract the function y from its dervative dy/dx and show that the result is equal to $x - 1$. Since

$$\frac{dy}{dx} = e^x - 1$$

we get

$$\frac{dy}{dx} - y = (e^x - 1) - (e^x - x) = x - 1$$

as required. ●

You can easily check that the function $y = 3e^x - x$ is also a solution of the differential equation in Example 3.1. In fact, every function of the form $y = Ce^x - x$, where C is a constant, is a solution of this equation. Moreover, it turns out that every solution of the equation is of this form. For this reason, the function $y = Ce^x - x$ is said to be the **general solution** of the differential equation in Example 3.1. The solution obtained by replacing C by a specific number is sometimes called a **particular solution.**

EXAMPLE 3.2 ● Find the particular solution of the differential equation $dy/dx - y = x - 1$ that satisfies the condition that $y = 4$ when $x = 0$.

SOLUTION We use the given condition to determine the numerical value of the constant C in the general solution $y = Ce^x - x$. Substituting $y = 4$ and $x = 0$ into the general solution we get $C = 4$ and conclude that the desired particular solution is $y = 4e^x - x$. ●

Differential equations of the form $dy/dx = f(x)$

Every time you find an integral, you are actually solving a special type of differential equation. The differential equation in this case is of the form $dy/dx = f(x)$, and its general solution is $y = F(x) + C$, where F is an antiderivative of f.

Differential equations of the form $dy/dx = f(x)$ are particularly easy to solve because the derivative of the quantity in question is given explicitly as a function of the independent variable. (Notice that the differential equation $dP/dt = kP$ describing population growth is not of this form, because the derivative of P is expressed in terms of P itself rather than t.) Here is a practical problem involving a differential equation of the form $dy/dx = f(x)$.

EXAMPLE 3.3 ● The resale value of a certain industrial machine decreases over a 10-year period at a rate that depends on the age of the machine. When the machine is x years old, the rate at which its value is changing is $220(x - 10)$ dollars per year.

(a) Express the value of the machine as a function of its age and initial value.
(b) If the machine was originally worth \$12,000, how much will it be worth when it is 10 years old?

SOLUTION (a) We let $V(x)$ denote the value of the machine when it is x years old. The derivative dV/dx is equal to the rate $220(x - 10)$ at which the value is changing. Hence, we begin with the differential equation

$$\frac{dV}{dx} = 220(x - 10) = 220x - 2{,}200$$

To find V we solve this differential equation by integration:

$$V(x) = \int (220x - 2{,}200)\,dx = 110x^2 - 2{,}200x + C$$

Notice that C is equal to $V(0)$, the initial value of the machine. A more descriptive symbol for this constant is V_0. Using this notation we conclude that

$$V(x) = 110x^2 - 2{,}200x + V_0$$

(b) If $V_0 = 12{,}000$, then $V(x) = 110x^2 - 2{,}200x + 12{,}000$ and the value after 10 years is

$$V(10) = 11{,}000 - 22{,}000 + 12{,}000 = \$1{,}000$$ ●

Separable differential equations

Many useful differential equations can be formally rewritten so that all the terms containing the independent variable appear on one side of the equation, and all the terms containing the dependent variable appear on the other. Differential equations with this special property are said to be **separable** and can be solved by the following procedure involving two integrations.

Separable differential equations
A differential equation that can be written in the form

$$g(y)\,dy = f(x)\,dx$$

is said to be *separable*. Its general solution is obtained by integrating both sides of this equation. That is,

$$\int g(y)\,dy = \int f(x)\,dx$$

A proof that this procedure works can be found in most calculus texts. Here is an example to illustrate how the procedure is used.

EXAMPLE 3.4 ● The population of a certain country is growing at a rate that is proportional to its size. Express the population as a function of time.

SOLUTION We let $P(t)$ denote the population at time t and let k be the constant of proportionality. Since the rate of change of P with respect to time is the derivative dP/dt, we begin with the differential equation

$$\frac{dP}{dt} = kP$$

To separate the variables, we pretend that the derivative dP/dt is actually a quotient and write

$$\frac{1}{P}\,dP = k\,dt$$

Integrating both sides of this equation we get

$$\int \frac{1}{P}\,dP = \int k\,dt$$

or

$$\ln|P| = kt + C$$

Since the population size P is positive, we may dispense with the absolute value sign and write

$$\ln P = kt + C$$

Solving this equation for P we get

$$P(t) = e^{kt+C} = e^{C}e^{kt}$$

Finally, since $e^C = P(0)$, we introduce P_0 to stand for the constant e^C and write

$$P(t) = P_0 e^{kt}$$

●

Exponential growth Notice that the equation $P(t) = P_0 e^{kt}$ in the preceding example is precisely the equation describing exponential growth. The argument in Example 3.4 can be used to show that any quantity that grows at a rate that is proportional to its size grows exponentially.

Problems

1 Verify that the function $y = Ce^{kx}$ is a solution of the differential equation $dy/dx = ky$.

2 Verify that the function $y = \sqrt{2x \ln x - 2x + C}$ is a solution of the differential equation $dy/dx = (\ln x)/y$.

In Problems 3 through 8, write a differential equation describing the given situation. (Do not solve the equation.)

3 The number of bacteria in a culture grows at a rate that is proportional to the number present.

4 The population of a certain town increases at a constant rate.

5 A sample of radium decays at a rate that is proportional to its size.

6 The rate at which the temperature of an object changes is proportional to the difference between its own temperature and the temperature of the surrounding medium.

7 An investment grows at a rate equal to 7 percent of its size.

8 The rate at which a flu epidemic spreads through a community is proportional to the product of the number of residents who have already caught the disease and the number who are susceptible to the disease but who have not yet caught it.

In Problems 9 through 12, find the general solution of the given differential equation.

9 $\dfrac{dy}{dx} = 3x^2 + 5x - 6$

10 $\dfrac{dP}{dt} = \sqrt{t} + e^{-t}$

11 $\dfrac{dV}{dx} = \dfrac{2}{x + 1}$

12 $\dfrac{dA}{dt} = 2te^{t^2+5}$

In Problems 13 through 15, find the particular solution of the given differential equation that satisfies the given condition.

13 $\dfrac{dy}{dx} = e^{5x}$; $y = 1$ when $x = 0$

14 $\dfrac{dy}{dx} = 5x^4 - 3x^2 - 2$; $y = 4$ when $x = 1$

15 $\frac{dV}{dt} = 16t(t^2 + 1)^3$; $V = 1$ when $t = 0$

16 The resale value of a certain industrial machine decreases at a rate that depends on the age of the machine. When the machine is t years old, the rate at which its value is changing is $-960e^{-t/5}$ dollars per year.

(a) Express the value of the machine in terms of its age and initial value.

(b) If the machine was originally worth \$5,200, how much will it be worth when it is 10 years old?

17 At a certain factory, the marginal cost is $3(q - 4)^2$ dollars per unit when the level of output is q units.

(a) Express the total production cost in terms of the overhead (the cost of producing no units) and the number of units produced.

(b) What is the cost of producing 14 units if the overhead is \$436?

18 Population statistics indicate that x years after 1970 a certain county was growing at a rate of approximately $1{,}500x^{-1/2}$ people per year. In 1979 the population of the county was 39,000.

(a) What was the population in 1970?

(b) If this pattern of population growth continues in the future, how many people will be living in the county in 1995?

19 In a certain section of the country, the price of chicken is currently \$3.00 per kilogram. It is expected that x weeks from now the price will be increasing at a rate of $3\sqrt{x + 1}$ cents per week. How much will chicken cost 8 weeks from now?

20 In a certain Los Angeles suburb, a reading of air pollution levels taken at 7:00 A.M. shows the ozone level to be 0.25 parts per million. A 12-hour forecast of air conditions indicates that t hours later the ozone level will be changing at a rate of $(0.24 - 0.03t)/\sqrt{36 + 16t - t^2}$ parts per million per hour.

(a) Express the ozone level as a function of t.

(b) At what time will the peak ozone level occur? What will the ozone level be at this time?

In Problems 21 through 26, find the general solution of the given separable differential equation.

21 $\frac{dy}{dx} = y^2$

22 $\frac{dy}{dx} = e^y$

23 $\frac{dy}{dx} = e^{x+y}$

24 $\frac{dy}{dx} = \frac{x}{y}$

25 $\frac{dy}{dx} = \frac{y}{x}$

26 $\frac{dy}{dx} = 80 - y$ (for $y < 80$)

In Problems 27 through 29, find the particular solution of the given differential equation that satisfies the given condition.

27 $\dfrac{dy}{dx} = 0.05y$; $y = 500$ when $x = 0$

28 $\dfrac{dy}{dx} = \dfrac{x}{y^2}$; $y = 3$ when $x = 2$

29 $\dfrac{dy}{dx} = 5(8 - y)$ (for $y < 8$); $y = 6$ when $x = 0$

30 The balance of a certain savings account grows at a rate equal to 7 percent of its size. Express the balance as a function of time if the initial balance was $5,000.

31 The rate of growth of bacteria in a certain culture is proportional to the number of bacteria present. If 5,000 bacteria are present initially and 8,000 are present 10 minutes later, how many bacteria will be present after 1 hour?

32 By solving an appropriate separable differential equation, show that a quantity that decays at a rate proportional to its size decays exponentially.

33 The residents of a certain community have voted to discontinue the fluoridation of their water supply. The local reservoir currently holds 200 million gallons of fluoridated water that contains 1,600 pounds of fluoride. The fluoridated water is flowing out of the reservoir at a rate of 4 million gallons per day and is being replaced at the same rate by unfluoridated water. At all times, the remaining fluoride is evenly distributed in the reservoir.

(a) Express the amount of fluoride in the reservoir as a function of time.

(b) When will the fluoride content of the water be reduced to 400 pounds?

Hint: Begin with a differential equation expressing the following relationship:

$$\text{Rate of change of fluoride with respect to time} = -\begin{pmatrix}\text{number of}\\ \text{gallons}\\ \text{leaving the}\\ \text{reservoir}\\ \text{per day}\end{pmatrix}\begin{pmatrix}\text{number of}\\ \text{pounds}\\ \text{of fluoride}\\ \text{per gallon}\end{pmatrix}$$

4 • THE DEFINITE INTEGRAL

If F is an antiderivative of f, and a and b are numbers, the difference $F(b) - F(a)$ is said to be a **definite integral** of f. The solutions of many applied problems are definite integrals. Here is a simple example.

EXAMPLE 4.1 ● At a certain factory, the marginal cost is $6q + 1$ dollars per unit when q units are produced. What is the cost of producing the 10th unit?

SOLUTION Let $C(q)$ be the total cost of producing q units. Then

$$\frac{dC}{dq} = 6q + 1$$

and $C(q)$ is the antiderivative

$$C(q) = \int (6q + 1)\,dq = 3q^2 + q + C_0$$

where the constant C_0 is the overhead; that is, the cost when $q = 0$.

The cost of producing the 10th unit is the difference between the cost of producing 10 units and the cost of producing 9 units. That is, the cost of the 10th unit is the definite integral

$$C(10) - C(9) = (300 + 10 + C_0) - (243 + 9 + C_0) = \$58$$

Notice that the answer is independent of the overhead C_0 which appeared in the expressions for both $C(10)$ and $C(9)$ and cancelled out. ●

Because computations involving definite integrals arise frequently, the following convenient notation has been developed.

The definite integral

$$\int_a^b f(x)\,dx = F(b) - F(a)$$

where F is an antiderivative of f.

In terms of this notation, the cost in Example 4.1 of producing the 10th unit is $\int_9^{10} (6q + 1)\,dq$. The symbol $\int_a^b f(x)\,dx$ is read "the (definite) integral of $f(x)$ from a to b." The numbers a and b are called the **limits of integration.** In computations involving definite integrals, it is often convenient to use the symbol $F(x)\,|_a^b$ to stand for $F(b) - F(a)$. The use of this notation is illustrated in the next example.

EXAMPLE 4.2 ● A study indicates that x months from now the population of a certain town will be increasing at the rate of $2 + 6\sqrt{x}$ people per month. By how much will the population of the town increase during the next 4 months?

SOLUTION Let $P(x)$ denote the population of the town x months from now. The amount by which the population will increase during the next 4 months is the definite integral

$$\begin{aligned} P(4) - P(0) &= \int_0^4 (2 + 6\sqrt{x})\,dx \\ &= (2x + 4x^{3/2} + C)\Big|_0^4 = (40 + C) - (0 + C) \\ &= 40 \end{aligned}$$

●

Notice that in the evaluation of a definite integral, the constant C is eventually eliminated by the subtraction. You may, therefore, omit the constant altogether when evaluating definite integrals.

The evaluation of definite integrals by substitution

In the next example, we illustrate the use of the method of substitution to evaluate a definite integral.

EXAMPLE 4.3 ● Evaluate $\int_0^1 8x(x^2 + 1)^3\,dx$.

SOLUTION We let $u = x^2 + 1$. Then $du = 2x\,dx$ and so

$$\int 8x(x^2 + 1)^3\,dx = \int 4u^3\,du = u^4$$

The limits of integration, 0 and 1, refer to the variable x and not to u. We may, therefore, proceed in one of two ways. Either we can rewrite the antiderivative in terms of x, or we can find the values of u that correspond to $x = 0$ and $x = 1$.

If we choose the first alternative, we find that

$$\int 8x(x^2 + 1)^3\,dx = (x^2 + 1)^4$$

and so

$$\int_0^1 8x(x^2 + 1)^3\,dx = (x^2 + 1)^4\Big|_0^1 = 16 - 1 = 15$$

If we choose the second alternative, we use the fact that $u = x^2 + 1$ to conclude that $u = 1$ when $x = 0$ and $u = 2$ when $x = 1$. Hence,

$$\int_0^1 8x(x^2 + 1)^3\,dx = u^4\Big|_1^2 = 16 - 1 = 15$$

Probably the most efficient approach is to adopt the second alternative and write the solution compactly as follows.

$$\int_0^1 8x(x^2 + 1)^3\,dx = \int_1^2 4u^3\,du = u^4\Big|_1^2 = 16 - 1 = 15$$ ●

Here is a summary of the method of substitution for definite integrals.

Integration by substitution

$$\int_a^b g[u(x)]\frac{du}{dx}\,dx = \int_{u(a)}^{u(b)} g(u)\,du$$

Here is one more example.

EXAMPLE 4.4 ● Evaluate $\int_1^e (\ln x)/x\, dx$.

SOLUTION We let $u = \ln x$. Then $du = (1/x)\,dx$, $u(1) = 0$, and $u(e) = 1$. Hence

$$\int_1^e \frac{\ln x}{x}\,dx = \int_0^1 u\,du = \frac{1}{2}u^2\Big|_0^1 = \frac{1}{2}$$ ●

Area and integration There is a surprising connection between definite integrals and the geometric concept of area. If $f(x)$ is continuous and nonnegative on the interval $a \leq x \leq b$, and R is the region under the graph of f between $x = a$ and $x = b$ shown in Figure 4.1, then the area of R is simply the definite integral $\int_a^b f(x)\,dx$.

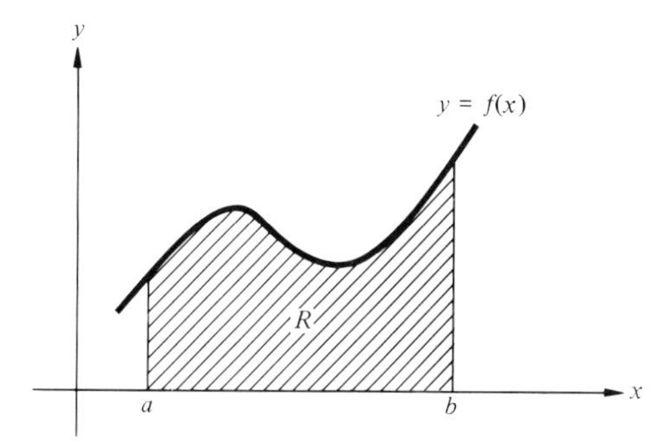

Figure 4.1 The area under the curve $y = f(x)$.

> **The area under a curve**
> If $f(x)$ is continuous and nonnegative on the interval $a \leq x \leq b$, and R is the region bounded by the graph of f, the vertical lines $x = a$ and $x = b$, and the x axis, then
>
> $$\text{Area of } R = \int_a^b f(x)\,dx$$

The use of this formula is illustrated in the next example for a region whose area we already know.

EXAMPLE 4.5 ● Find the area of the region bounded by the lines $y = 2x$, $x = 2$, and the x axis.

SOLUTION The region in question is the triangle in Figure 4.2 and its area is clearly 4. To compute the area using calculus, we apply the integral formula with $f(x) = 2x$. Since the region is bounded on the right by the line $x = 2$, we take $b = 2$. On the left, the boundary consists of the single point (0, 0) which is part of the

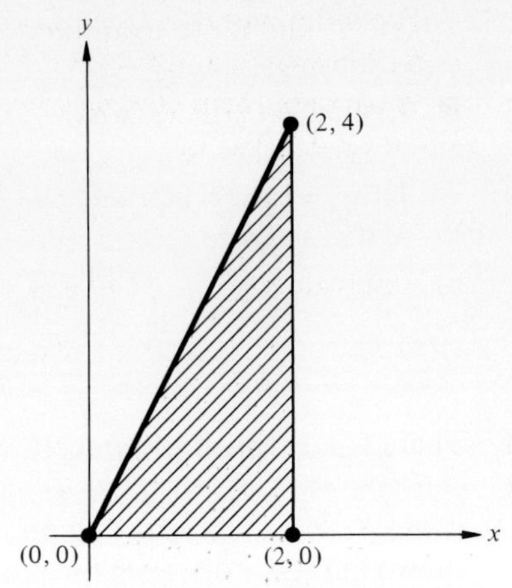

Figure 4.2 The area under $y = 2x$ from $x = 0$ to $x = 2$.

vertical line $x = 0$. Hence, we take $a = 0$ and find, as expected, that

$$\text{Area} = \int_0^2 2x\,dx = x^2\Big|_0^2 = 4$$

Here is a geometric argument that will show you the connection between definite integrals and areas.

Suppose $f(x)$ is continuous and nonnegative on the interval $a \leq x \leq b$. For any value of x in this interval, let $A(x)$ denote the area of the region under the graph of f between a and x, as shown in Figure 4.3.

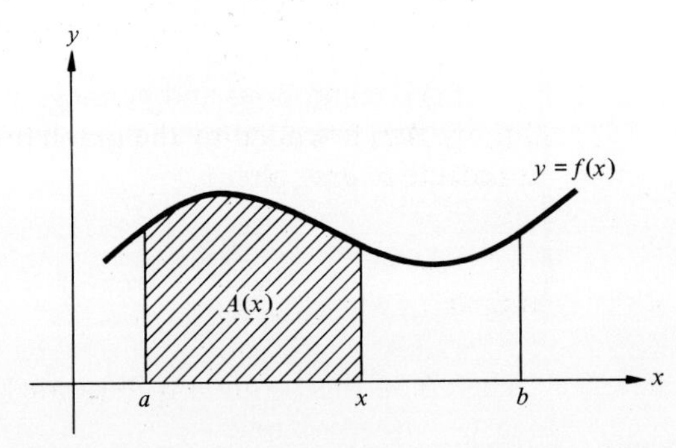

Figure 4.3 The area $A(x)$.

Our goal is to show that $A(b) = \int_a^b f(x)\,dx$. The key step is to establish that the derivative $A'(x)$ of the area function is equal to $f(x)$. To do this, we consider the quotient

$$\frac{A(x + \Delta x) - A(x)}{\Delta x}$$

The difference $A(x + \Delta x) - A(x)$ in the numerator is just the area under the curve between x and $x + \Delta x$. If Δx is small, this area is approximately the same as the area of the rectangle whose height is $f(x)$ and width is Δx, as indicated in Figure 4.4. That is,

$$A(x + \Delta x) - A(x) \approx f(x)\,\Delta x$$

or, equivalently,

$$\frac{A(x + \Delta x) - A(x)}{\Delta x} \approx f(x)$$

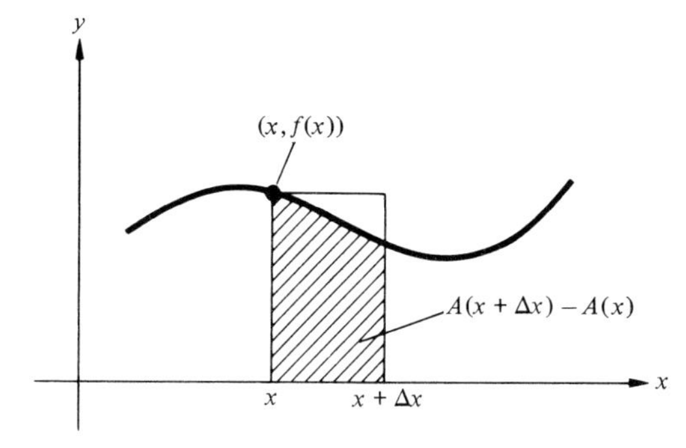

Figure 4.4 The area $A(x + \Delta x) - A(x)$ and an approximating rectangle.

As Δx approaches zero, the error resulting from this approximation approaches zero and it follows that

$$\frac{A(x + \Delta x) - A(x)}{\Delta x} \text{ approaches } f(x) \text{ as } \Delta x \text{ approaches zero}$$

But, by the definition of the derivative, this quotient approaches $A'(x)$ as Δx approaches zero. Hence,

$$A'(x) = f(x)$$

and so

$$\int_a^b f(x)\,dx = A(b) - A(a)$$

But, $A(a)$ is the area under the curve between $x = a$ and $x = a$, which is clearly zero. Hence,

$$\int_a^b f(x)\,dx = A(b)$$

and the formula is verified.

Here is one more example illustrating the use of integrals to compute areas.

EXAMPLE 4.6 ● Find the area of the region bounded by the curve $y = -x^2 + 2x + 3$ and the x axis.

SOLUTION From the factorization

$$y = -(x - 3)(x + 1)$$

we obtain the graph in Figure 4.5 and see that the region in question is bounded on the left by $x = -1$ and on the right by $x = 3$. Hence,

$$\text{Area} = \int_{-1}^{3} (-x^2 + 2x + 3)\, dx = (-\tfrac{1}{3}x^3 + x^2 + 3x)\Big|_{-1}^{3} = \tfrac{32}{3}$$ ●

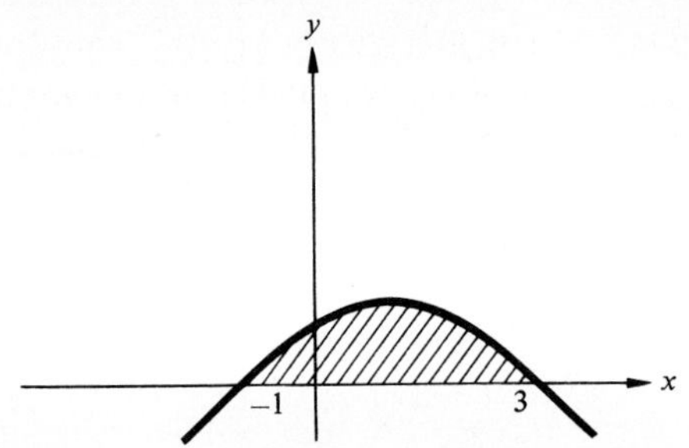

Figure 4.5 The area under the curve $y = -x^2 + 2x + 3$ from $x = -1$ to $x = 3$.

Problems In Problems 1 through 14, evaluate the given definite integral.

1 $\int_{0}^{1} (x^4 - 3x^3 + 1)\, dx$

2 $\int_{-1}^{0} (3x^5 - 3x^2 + 2x - 1)\, dx$

3 $\int_{2}^{5} (2 + 2u + 3u^2)\, du$

4 $\int_{1}^{9} (\sqrt{t} - 1/\sqrt{t})\, dt$

5 $\int_{1}^{3} (1 + 1/x + 1/x^2)\, dx$

6 $\int_{\ln 1/2}^{\ln 2} (e^u - e^{-u})\, du$

7 $\int_{-3}^{0} (2x + 6)^4\, dx$

8 $\int_{1}^{2} (2x - 4)^5\, dx$

9 $\int_{0}^{4} \frac{1}{\sqrt{6u + 1}}\, du$

10 $\int_{-1}^{1} 2te^{t^2-1}\, dt$

11 $\int_{1}^{2} \frac{x^2}{(x^3 + 5)^2}\, dx$

12 $\int_{-1}^{1} (u^3 + u)\sqrt{u^4 + 2u^2 + 6}\, du$

13 $\int_{1}^{2} (u + 1)(u - 2)^9\, du$

14 $\int_{e}^{e^2} \frac{1}{t \ln t}\, dt$

15 A study indicates that x months from now the population of a certain town will be increasing at a rate of $5 + 3x^{2/3}$ people per month. By how much will the population of the town increase over the next 8 months?

16 An object is moving so that its speed after t minutes is $5 + 2t + 3t^2$ meters per minute. How far does the object travel during the 2d minute?

17 The resale value of a certain industrial machine decreases over a 10-year period at a rate that changes with time. When the machine is x years old, the rate at which its value is changing is $220(x - 10)$ dollars per year. By how much does the machine depreciate during the 2d year?

18 The promoters of a county fair estimate that t hours after the gates open at 9:00 A.M. visitors will be entering the fair at a rate of $-4(t + 2)^3 + 54(t + 2)^2$ people per hour. How many people will enter the fair between 10:00 A.M. and noon?

19 A certain oil well that yields 400 barrels of crude oil a month will run dry in 2 years. The price of crude oil is currently \$18 per barrel and is expected to rise at a steady rate of 3 cents per barrel per month. If the oil is sold as soon as it is extracted from the ground, what is the total revenue the owner can expect from the well during the 2-year period?

20 Use calculus to find the area of the triangle bounded by the line $y = 4 - 3x$ and the coordinate axes.

21 Use calculus to find the area of the triangle with vertices $(-4, 0)$, $(2, 0)$, and $(2, 6)$.

22 Use calculus to find the area of the rectangle with vertices $(1, 0)$, $(-2, 0)$, $(-2, 5)$, and $(1, 5)$.

23 Use calculus to find the area of the trapezoid bounded by the lines $y = x + 6$ and $x = 2$ and the coordinate axes.

24 Find the area of the region bounded by the curve $y = \sqrt{x}$, the lines $x = 4$ and $x = 9$, and the x axis.

25 Find the area of the region bounded by the curve $y = 4x^3$, the x axis, and the line $x = 2$.

26 Find the area of the region bounded by the curve $y = 1 - x^2$ and the x axis.

27 Find the area of the region bounded by the curve $y = -x^2 - 6x - 5$ and the x axis.

28 Find the area of the region bounded by the curve $y = e^x$, the lines $x = 0$ and $x = \ln \frac{1}{2}$, and the x axis.

29 Find the area of the region bounded by the curve $y = x^2 - 2x$ and the x axis. (*Hint:* Reflect the region across the x axis and integrate the corresponding function.)

30 Find the area of the region bounded by the curves $y = x^3$ and $y = x^2$. (*Hint:* Subtract two areas.)

5 • THE FUNDAMENTAL THEOREM OF CALCULUS

The purpose of this section is to examine an important relationship between integrals and sums known as the **fundamental theorem of calculus.** The following geometric argument based on the interpretation of definite integrals as areas will show you why this relationship holds.

Suppose that $f(x)$ is nonnegative and continuous on the interval $a \leq x \leq b$. We can approximate the area under the graph of f between $x = a$ and $x = b$ as follows: First, we divide the interval $a \le x \le b$ into n equal subintervals of width Δx and let x_j denote the beginning of the jth subinterval. Next, we draw n rectangles such that the base of the jth rectangle is the jth subinterval and the height of the jth rectangle is $f(x_j)$. The situation is illustrated in Figure 5.1.

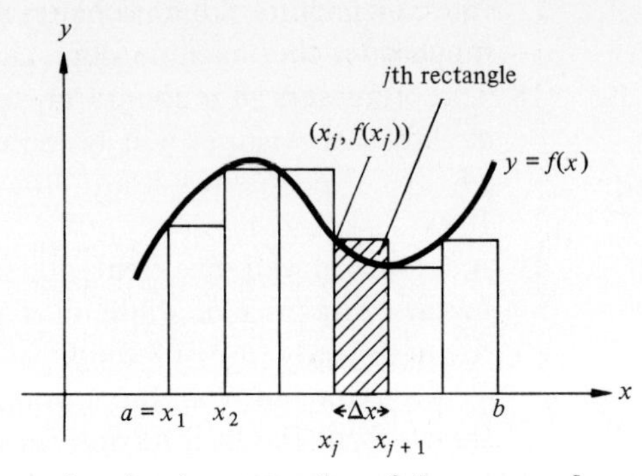

Figure 5.1 An approximation by rectangles of the area under a curve.

The area of the jth rectangle is $f(x_j)\,\Delta x$ and is an approximation to the area under the curve from $x = x_j$ to $x = x_{j+1}$. The sum of the areas of all n rectangles is

$$f(x_1)\,\Delta x + f(x_2)\,\Delta x + \cdots + f(x_n)\,\Delta x = \sum_{j=1}^{n} f(x_j)\,\Delta x$$

(If the use of the symbol Σ in this context is unfamiliar, see the discussion of summation notation in the algebra review in the Appendix.) This sum is an approximation to the total area under the curve from $x = a$ to $x = b$ and hence an approximation to the corresponding definite integral $\int_a^b f(x)\,dx$. That is,

$$\sum_{j=1}^{n} f(x_j)\,\Delta x \approx \int_a^b f(x)\,dx$$

As Figure 5.2 suggests, the sum of the areas of the rectangles approaches the actual area under the curve as the number of rectangles increases. That is,

$$\sum_{j=1}^{n} f(x_j)\,\Delta x \text{ approaches } \int_a^b f(x)\,dx$$

as n increases without bound.

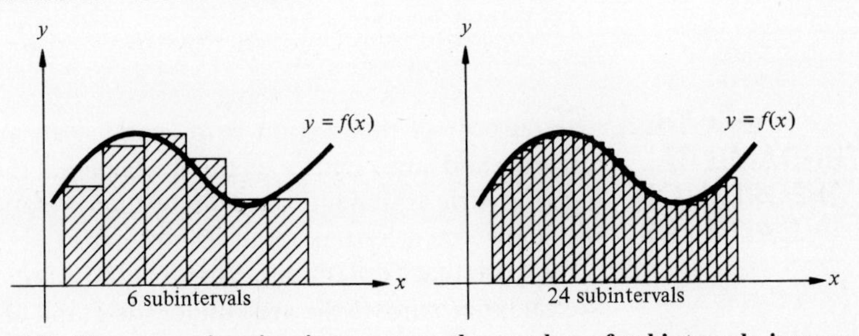

Figure 5.2 The approximation improves as the number of subintervals increases.

This is the relationship between sums and integrals that we were seeking. Although we have established it only for nonnegative functions, it actually holds for any function that is continuous on the interval $a \le x \le b$.

The fundamental theorem of calculus
Suppose f is continuous on the interval $a \le x \le b$ which is divided into n equal subintervals with x_j denoting the beginning of the jth subinterval. Then

$$\sum_{j=1}^{n} f(x_j)\,\Delta x \text{ approaches } \int_a^b f(x)\,dx$$

as n increases without bound.

Actually this is a somewhat restricted version of a more general characterization of definite integrals. The relationship between definite integrals and sums still holds if $f(x_j)$ in the jth term of the sum is replaced by $f(x_j')$, where x_j' is any point whatsoever in the jth subinterval. Moreover, the n subintervals need not have equal width, as long as the width of the largest eventually approaches zero as n increases. For most applications, however, the restricted characterization is quite sufficient, and we will have no use for the more general result.

Here is an example illustrating the use of the fundamental theorem of calculus.

EXAMPLE 5.1 ● A certain oil well that yields 300 barrels of crude oil a month will run dry in 3 years. It is estimated that t months from now the price of crude oil will be $P(t) = 18 + 0.3\sqrt{t}$ dollars per barrel. If the oil is sold as soon as it is extracted from the ground, what will be the total future revenue from the well?

SOLUTION To approximate the revenue during the 36-month period, we divide the interval $0 \le t \le 36$ into n equal subintervals of length Δt and let t_j denote the beginning of the jth subinterval. Then, $300\,\Delta t$ barrels of crude oil are produced during each subinterval. Moreover, if Δt is small, the price of crude oil throughout the jth subinterval will be approximately $P(t_j)$ dollars per barrel, the price that was in effect at the beginning of the interval. Hence,

$$\text{Revenue from } j\text{th subinterval} \approx 300P(t_j)\,\Delta t \qquad \text{dollars}$$

and

$$\text{Total revenue} \approx \sum_{j=1}^{n} 300P(t_j)\,\Delta t \qquad \text{dollars}$$

As n increases, the length Δt of the subintervals decreases and the approximation improves. In fact, $\Sigma_{j=1}^{n} 300P(t_j)\,\Delta t$ approaches the total revenue as n increases without bound. But, according to the fundamental theorem of calculus, $\sum_{j=1}^{n} 300P(t_j)\,\Delta t$ approaches $\int_a^b 300P(t)\,dt$ as n increases without bound, where $a = 0$ and $b = 36$. Hence

$$\text{Total revenue} = \int_0^{36} 300P(t)\,dt = 300\int_0^{36} (18 + 0.3\sqrt{t})\,dt$$

$$= 300(18t + 0.2t^{3/2})\Big|_0^{36}$$

$$= 300[18(36) + 0.2(6)^3] - 0 = \$207{,}360$$

●

The problem in Example 5.1 could have been formulated as a differential equation and solved without the use of the fundamental theorem of calculus. (For practice, try it.) The problem in the next example, however, cannot be solved easily without this theorem. In this example, a function f gives the fraction of individuals in a group that can be expected to remain in the group for any specified period of time. The rate at which new members arrive is known and the problem is to predict the size of the group at some future time. Problems of this type arise in many fields, including sociology, demography, and ecology.

EXAMPLE 5.2 ● A new county mental health clinic has just opened. Statistics compiled at similar facilities suggest that the fraction of patients who will still be receiving treatment at the clinic t months after their initial visit is given by the function $f(t) = e^{-t/20}$. The clinic initially accepts 300 people for treatment, and plans to accept new patients at a rate of 10 per month. Approximately how many people will be receiving treatment at the clinic 15 months from now?

SOLUTION Since $f(15)$ is the fraction of patients whose treatment continues at least 15 months, it follows that of the current 300 patients, only $300f(15)$ will still be receiving treatment 15 months from now.

To approximate the number of *new* patients who will be receiving treatment 15 months from now, we divide the 15-month time interval into n equal subintervals of length Δt and let t_j denote the beginning of the jth subinterval. Since new patients are accepted at a rate of 10 per month, the number of new patients accepted during the jth subinterval is $10\,\Delta t$. Fifteen months from now, approximately $15 - t_j$ months will have elapsed since these $10\,\Delta t$ new patients had their initial visits, and so approximately $f(15 - t_j)(10\,\Delta t)$ of them will still be receiving treatment at that time. It follows that the total number of new patients still receiving treatment 15 months from now can be approximated by the sum

$$\sum_{j=1}^{n} 10f(15 - t_j)\,\Delta t$$

Adding this to the number of current patients who will still be receiving treatment in 15 months, we get

$$N \approx 300f(15) + \sum_{j=1}^{n} 10f(15 - t_j)\,\Delta t$$

where N is the total number of patients who will be receiving treatment 15 months from now.

As n increases, the approximation improves. In fact,

$$300f(15) + \sum_{j=1}^{n} 10f(15 - t_j)\,\Delta t \text{ approaches } N$$

as n increases without bound. But, according to the fundamental theorem of calculus,

$$\sum_{j=1}^{n} 10f(15 - t_j)\,\Delta t \text{ approaches } \int_0^{15} 10f(15 - t)\,dt$$

as n increases without bound. It follows that

$$N = 300f(15) + \int_0^{15} 10f(15 - t)\,dt$$

Since $f(t) = e^{-t/20}$, we have $f(15) = e^{-3/4}$ and $f(15 - t) = e^{-(15-t)/20} = e^{-3/4}e^{t/20}$. Hence

$$N = 300e^{-3/4} + 10e^{-3/4}\int_0^{15} e^{t/20}\,dt = 300e^{-3/4} + 10e^{-3/4}\left(20e^{t/20}\Big|_0^{15}\right)$$

$$= 200 + 100e^{-3/4} = 247.24$$

That is, 15 months from now, the clinic will be treating approximately 247 patients. ●

Problems

Solve Problems 1 through 9 twice, first using the fundamental theorem of calculus and then by solving an appropriate differential equation.

1 An object is moving so that its speed after t minutes is $f(t) = 1 + 4t + 3t^2$ meters per minute. How far does the object travel during the third minute?

2 A tree has been transplanted and after x years is growing at a rate of $f(x) = 0.5 + 1/(x + 1)^2$ meters per year. By how much does the tree grow during the second year?

3 It is estimated that t years from now the value of a certain parcel of land will be increasing at a rate of $r(t)$ dollars per year. Find an expression for the amount by which the value of the land will increase during the next 5 years.

4 The promoters of a county fair estimate that t hours after the gates open at 9:00 A.M., visitors will be entering the fair at a rate of $r(t)$ people per hour. Find an expression for the number of people who will enter the fair between 11:00 A.M. and 1:00 P.M.

5 A bicycle manufacturer expects that x months from now consumers will be buying 5,000 bicycles a month at a price of $P(x) = 80 + 3\sqrt{x}$ dollars per bicycle. What is the total revenue the manufacturer can expect from the sale of the bicycles over the next 16 months?

6 A bicycle manufacturer expects that x months from now consumers will be buying $f(x) = 5{,}000 + 50\sqrt{x}$ bicycles per month at a price of $P(x) = 80 + 3\sqrt{x}$ dollars per bicycle. What is the total revenue the manufacturer can expect from the sale of the bicycles over the next 16 months?

7 A manufacturer expects that x months from now consumers will be buying $n(x)$ lamps per month at a price of $p(x)$ dollars per lamp. Find an expression for the total revenue the manufacturer can expect from the sale of the lamps over the next 12 months.

8 Suppose that t months from now an oil well will be producing crude oil at a rate of $r(t)$ barrels per month and that the price of crude oil will be $p(t)$ dollars per barrel. Assuming that the oil is sold as soon as it is extracted from the ground, find an expression for the total revenue from the oil well over the next 2 years.

9 The hero of a popular spy story has decided to retire in the oil-rich sheikdom of Monrovia. There he buys an oil well that yields 400 barrels of crude oil a month and that will run dry in 5 years. The price of Monrovian crude oil is currently \$18 per barrel, and is expected to rise at a constant rate of 3 cents per barrel per month over the next 5 years. If the spy sells the oil as soon as it is extracted from the ground, what is the total future revenue he will get from his well?

10 The operators of a new computer dating service estimate that the fraction of people who will retain their memberships in the service for at least t months is given by the function $f(t) = e^{-t/10}$. There are 8,000 charter members and the operators expect to attract 200 new members per month. How many members will the service have 10 months from now?

11 Let $f(t)$ denote the fraction of the membership of a certain group that will remain in the group for at least t years. Suppose that the group has just been formed with an initial membership of P and that t years from now new members will be added to the group at the rate of $r(t)$ per year. Find an expression for the size of the group 10 years from now.

12 A national consumers' association has compiled statistics suggesting that the fraction of its members who are still active t months after joining the group is given by the function $f(t) = 0.1 + 0.9e^{-t/8}$ (for $t \leq 60$). A new local chapter has 150 charter members and expects to attract new members at the rate of 10 per month.

(a) How many active members can this chapter expect to have at the end of 8 months?

(b) Why would it be unreasonable to suppose that the fraction of members active for more than t months might be represented by the given function $f(t)$ for *all* values of t?

APPENDIX

ALGEBRA REVIEW

The real numbers

An **integer** is a "whole number," either positive or negative. For example, 1, 2, 875, -15, -83, and 0 are integers, while $\frac{2}{3}$, 8.71, and $\sqrt{2}$ are not. A **rational number** is a number that can be expressed as the quotient of two integers. That is, a rational number is a number of the form m/n, where m and n are integers. For example, $\frac{2}{3}$, $\frac{3}{8}$, $-6\frac{1}{2}$, and 0.25 are rational numbers. Every integer is a rational number, since it can be expressed as itself divided by 1.

There are some numbers that cannot be expressed as the quotient of two integers. For example $\sqrt{2}$ and $-\sqrt{5}$ are not rational numbers. Neither is the number π, which you have seen in geometry. A number that is not rational is said to be an **irrational number.**

The integers, rational numbers, and irrational numbers together form the **real numbers** and can be visualized geometrically as points on a number line (Figure 1).

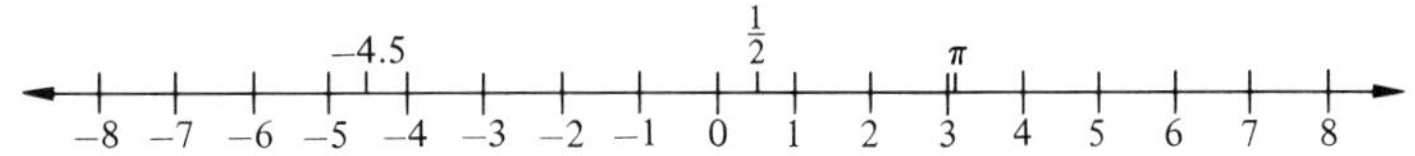

Figure 1 The number line.

Inequalities and intervals

Using the number line to represent the real numbers, we say that x is **greater than** y $(x > y)$ if x is to the right of y on the number line, and x is **less than** y $(x < y)$ if x is to the left of y on the number line (Figure 2). For example,

$$5 > 2 \qquad -16 < 0 \qquad -8.16 < -1.24 \qquad \text{and} \qquad \sqrt{9} > \sqrt{7}$$

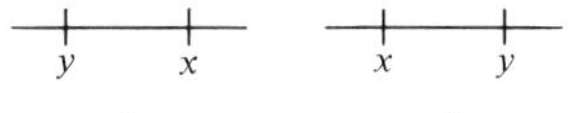

$y \quad x$ $\qquad$ $x \quad y$

$x > y$ $\qquad$ $x < y$

Figure 2 Inequalities.

The symbol $\geq$ denotes **greater than or equal to,** and so the expression $x \geq y$ means x is greater than or possibly equal to y. Similarly, the symbol $\leq$ denotes **less than or equal to.** For example,

$$8 \geq 7.88 \qquad 14 \geq 6.1 \qquad \text{and} \qquad -5 \leq -5$$

An **interval** is a set of real numbers that can be represented on the number line by a line segment. Inequalities can be used to describe intervals. For example, the interval $a \leq x \leq b$ consists of all numbers x that are between a and b, including a and b themselves. Three intervals are illustrated in Figure 3.

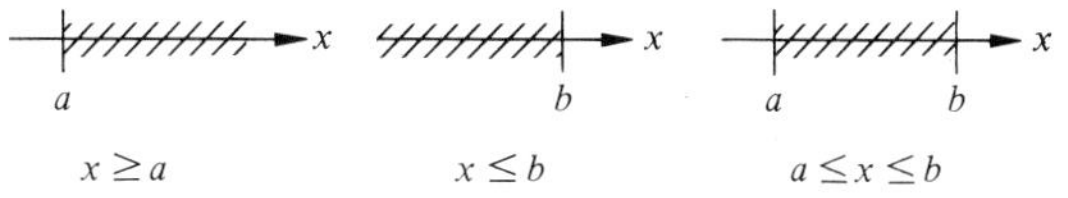

$x \geq a$ $\qquad$ $x \leq b$ $\qquad$ $a \leq x \leq b$

Figure 3 Intervals.

EXAMPLE 1 ● Express the following intervals algebraically.

(a) 3 x (b) −2 x

(c) −2 3 x

Express the following intervals geometrically.

(d) $x \leq \sqrt{7}$ (e) $x \geq -\frac{2}{3}$ (f) $0 \leq x \leq 6$

SOLUTION (a) $x \leq 3$ (b) $x \geq -2$ (c) $-2 \leq x \leq 3$

(d) $\sqrt{7}$ x (e) 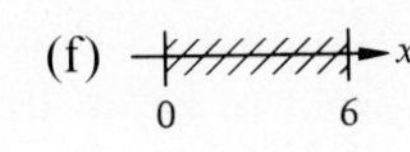

(f) 0 6 x

●

Rounding off Since calculators have become popular, it has become common to use numbers in decimal form. Many simple fractions are long and unwieldy when expressed in this way, and **rounding off** is necessary. To round off a number in decimal form is to shorten it by using only a limited number of places to the right of the decimal point. How much you round off depends on how much accuracy you need and how much space you have to display your data. For many calculations in this book, we have rounded off numbers to the "nearest hundredth," that is, using only two places to the right of the decimal point. We use the convention that if the leftmost digit to be eliminated is 5 or greater, then the preceding digit is increased by 1 in the rounded off number, and otherwise it remains the same. For example, 0.27164 rounded off to two decimal places becomes 0.27, since 1 is less than 5, while 0.27664 becomes 0.28, since 6 is greater than 5. This is further illustrated in Example 2.

EXAMPLE 2 ● Round off the following numbers to two decimal places.

(a) 0.17891 (b) 307.006 (c) 76.844 (d) 0.885 (e) 0.997

SOLUTION (a) 0.18 (b) 307.01 (c) 76.84 (d) 0.89 (e) 1.00 ●

Absolute value The **absolute value** of a real number is the nonnegative value, or magnitude of the number. Thus, the absolute value of a nonnegative number is the number itself, while the absolute value of a negative number is the number without the minus sign. To denote the absolute value of a real number x we write $|x|$. Here are some examples.

$$|3| = 3 \qquad |-27| = 27 \qquad |0| = 0 \qquad \text{and} \qquad |-247.18| = 247.18$$

The absolute value of a number can be interpreted geometrically as its distance from zero on the number line. Similarly, $|x - y|$ represents the distance between

x and y. This geometric interpretation can be used to simplify certain types of inequalities involving absolute values. This is illustrated in Example 3.

EXAMPLE 3 ● Simplify the inequality $|x - 3| \leq 2$.

SOLUTION In geometric terms this inequality says that the distance between x and 3 is less than or equal to 2. The situation is illustrated in Figure 4 and is clearly equivalent to the statement

$$1 \leq x \leq 5$$

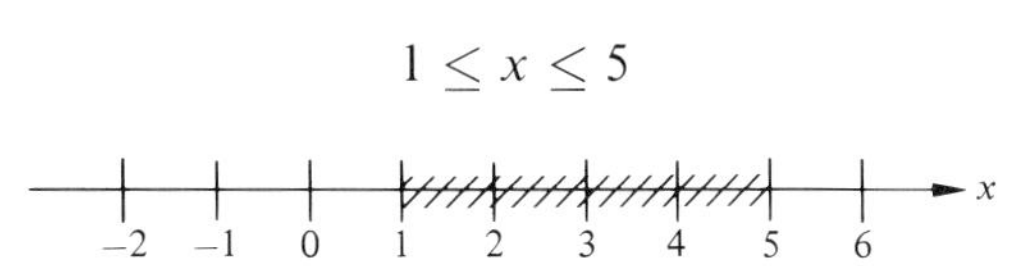

Figure 4 $|x - 3| \leq 2$. ●

Roots For any positive integer n and any positive real number x, the ***n*th root of *x***, denoted by $\sqrt[n]{x}$, is defined to be the number y such that $y^n = x$, where $y^n = y \cdot y \cdots y$ (n factors). For example,

$$\sqrt[4]{81} = 3 \qquad \sqrt[5]{32} = 2 \qquad \sqrt[3]{125} = 5 \qquad \text{and} \qquad \sqrt[2]{64} = 8$$

The root $\sqrt[3]{\ }$ is called the **cube root** and $\sqrt[2]{\ }$ is the **square root.** It is common to drop the "2" and denote the square root by $\sqrt{\ }$. For example, $\sqrt{64} = 8$ and $\sqrt{49} = 7$. Roots of negative numbers are defined in some situations but not in others. For example, since $(-3)^3 = -27$, we have $\sqrt[3]{-27} = -3$. However, there is no real number x such that $x^2 = -3$, so $\sqrt{-3}$ is not defined for real numbers. It is hard to obtain roots of most numbers without a calculator. We have used a calculator to obtain the following roots, and have rounded the answer off to three decimal places.

$$\sqrt{17} = 4.123 \qquad \sqrt[3]{11} = 2.224$$

Exponents For any positive integer n and any real number x, the ***n*th power of *x***, denoted by x^n, is

$$x^n = x \cdot x \cdots x$$

where the product contains n factors. For example,

$$3^2 = 9 \qquad 5^3 = 125 \qquad \text{and} \qquad 3^4 = 81$$

For any real number x, we define

$$x^0 = 1$$

For any positive real number x and any positive integers m and n, we define

$$x^{n/m} = (\sqrt[m]{x})^n$$

For example,

$$9^{3/2} = (\sqrt{9})^3 = 27 \qquad \text{and} \qquad \left(\frac{1}{16}\right)^{1/4} = \sqrt[4]{\frac{1}{16}} = \frac{1}{2}$$

For any positive rational number p and positive real number x we define

$$x^{-p} = \frac{1}{x^p}$$

For example,

$$2^{-3} = \frac{1}{8} \qquad \text{and} \qquad \left(\frac{1}{4}\right)^{-1/2} = 2$$

In the expression x^n, x is called the **base** and n is called the **exponent.**

Laws of exponents

Exponents obey the following laws for any rational numbers p and q.

Laws of exponents

1 $x^p x^q = x^{p+q}$

2 $x^p / x^q = x^{p-q}$

3 $(x^p)^q = x^{pq}$

4 $(xy)^p = x^p y^p$

Warning: $(x + y)^p \neq x^p + y^p$

Here are two examples to illustrate the use of these laws.

EXAMPLE 4 ● Evaluate the following expressions.

(a) $(2^{-3})^2$ (b) $\dfrac{7^{1/2}(7^{3/2})}{7^3}$ (c) $4^{2/5}(64^{1/5})$

SOLUTION (a) $(2^{-3})^2 = 2^{-3(2)} = 2^{-6} = \dfrac{1}{64}$

(b) $\dfrac{7^{1/2}(7^{3/2})}{7^3} = \dfrac{7^{(1/2+3/2)}}{7^3} = 7^{(1/2+3/2-3)} = 7^{-1} = \dfrac{1}{7}$

(c) $4^{2/5}(64^{1/5}) = 4^{2/5}[(4^3)^{1/5}] = 4^{2/5}(4^{3/5}) = 4^{(2/5+3/5)} = 4$ ●

EXAMPLE 5 ● Solve each of the following equations for n.

(a) $a^3/a^5 = a^n$ (b) $a^5 a^n = a^2$ (c) $(a^n)^2 = a^{12}$

SOLUTION (a) $a^3/a^5 = a^{3-5} = a^{-2}$, so $n = -2$
(b) $a^5 a^n = a^{5+n}$, and we get $a^{5+n} = a^2$, or $n = -3$
(c) $(a^n)^2 = a^{2n}$ and we get $a^{2n} = a^{12}$, or $n = 6$ ●

Factoring To **factor** an expression is to write it as a product of two or more terms. Complicated expressions can often be simplified by factoring. Factoring is based on the **distributive law,** which states that for any numbers a, b, and c,

$$ab + ac = a(b + c)$$

The use of the distributive law in factoring is illustrated in Example 6.

EXAMPLE 6 ● Factor the following expressions.

(a) $2x^2 + 11x$ (b) $x^2 - 9$ (c) $x^2 - 2x - 3$ (d) $x^3 - 8$

SOLUTION (a) We use the distributive law to factor out x and obtain

$$2x^2 + 11x = x(2x + 11)$$

(b) The distributive law implies that

$$(x + a)(x + b) = x^2 + (a + b)x + ab$$

Hence,

$$x^2 - 9 = (x + a)(x + b)$$

where

$$ab = -9 \qquad \text{and} \qquad a + b = 0$$

From the list

$$-1, 9 \qquad 1, -9 \qquad 3, -3$$

of pairs of integers whose product is -9, we choose $a = 3$ and $b = -3$ as the only pair whose sum is 0. Hence

$$x^2 - 9 = (x + 3)(x - 3)$$

(c) From the list

$$1, -3 \qquad -1, 3$$

of pairs of integers whose product is -3, we choose $a = -3$ and $b = 1$ as the only pair whose sum is -2. Hence,

$$x^2 - 2x - 3 = (x - 3)(x + 1)$$

(d) Since $2^3 = 8$, we begin by factoring out $(x - 2)$ to get

$$x^3 - 8 = (x - 2)(x^2 + ax + b)$$

Since

$$(x - 2)(x^2 + ax + b) = x^3 + (a - 2)x^2 + (b - 2a)x - 2b$$

we must have

$$a - 2 = 0 \qquad b - 2a = 0 \qquad \text{and} \qquad 2b = 8$$

This is possible only if $b = 4$ and $a = 2$. Hence,

$$x^3 - 8 = (x - 2)(x^2 + 2x + 4)$$

Convince yourself, by examining the pairs of integers whose product is 4, that the expression $x^2 + 2x + 4$ cannot be factored further. ●

Solving equations The **solutions** of an equation are the values of the variable that make the equation true. For example, the solution of the equation

$$3x + 2 = 14$$

is 4 because substituting 4 for the variable x gives

$$3(4) + 2 = 14$$

which is a true statement. The equation

$$x^2 - 9 = 0$$

has two solutions, 3 and -3, while the equation

$$x^3 - x^2 - 4x + 4 = 0$$

has three solutions, 2, -2, and 1.

EXAMPLE 7 ● Solve the equation $5x + 6 = 9$.

SOLUTION Subtracting 6 from both sides of the equation, we get

$$5x = 3$$

Dividing both sides of the equation by 5, we get the solution

$$x = \tfrac{3}{5}$$ ●

One technique that is often useful for the solution of equations is to combine factoring with the following property of multiplication: If the product of two terms equals zero, then at least one of the terms must equal zero. That is, if $ab = 0$, then either $a = 0$ or $b = 0$ (or both). In the following examples we shall use this fact to solve equations.

EXAMPLE 8 ● Solve the equation $x^2 = 2x$.

SOLUTION Subtracting $2x$ from both sides of the equation, we get

$$x^2 - 2x = 0$$

which we factor as

$$x(x - 2) = 0$$

Since the product $x(x - 2)$ can be zero only if one of its terms is zero, we conclude that the only solutions are $x = 0$ (which makes the first term zero) and $x = 2$ (which makes the second term zero). ●

EXAMPLE 9 ● Solve the equation $x^2 + 10x + 25 = 0$.

SOLUTION Factoring, we get

$$(x + 5)^2 = 0$$

from which we conclude that the only solution is $x = -5$. ●

EXAMPLE 10 ● Solve the equation $x^3 - x^2 - 4x + 4 = 0$.

SOLUTION Factoring, we get

$$x^3 - x^2 - 4x + 4 = (x - 1)(x^2 - 4)$$

which we can factor further as

$$x^3 - x^2 - 4x + 4 = (x - 1)(x + 2)(x - 2)$$

Hence our equation is

$$(x - 1)(x + 2)(x - 2) = 0$$

and its solutions are clearly $x = 1$, $x = -2$, and $x = 2$. ●

There are many equations that cannot be easily factored. The equation

$$x^2 + 3x + 1 = 0$$

is one. This equation is an example of a **quadratic equation,** that is, an equation of the form

$$ax^2 + bx + c = 0$$

Quadratic equations have at most two solutions, which can be found using a special formula called the **quadratic formula.**

The quadratic formula
The solutions of the quadratic equation

$$ax^2 + bx + c = 0$$

(for $a \neq 0$) are given by the formulas

$$x = \frac{-b + \sqrt{b^2 - 4ac}}{2a} \quad \text{and} \quad x = \frac{-b - \sqrt{b^2 - 4ac}}{2a}$$

The term $b^2 - 4ac$ is called the **discriminant** of a quadratic equation. If the discriminant is positive, the equation has two solutions, if it is zero, the equation has one solution, and if it is negative, the equation has no real solutions. (Do you see why?) If a quadratic equation can be easily factored, it is unnecessary to use the quadratic formula to obtain solutions (although the same results would be obtained). If the quadratic equation cannot be easily factored, the quadratic formula should be used. The next examples illustrate the use of the quadratic formula in solving equations.

EXAMPLE 11 ● Solve the equation $x^2 + 3x + 1 = 0$

SOLUTION This is a quadratic equation with $a = 1$, $b = 3$, and $c = 1$. Using the quadratic formula we get

$$x = \frac{-3 + \sqrt{5}}{2} \quad \text{and} \quad x = \frac{-3 - \sqrt{5}}{2}$$ ●

EXAMPLE 12 ● Solve the equation $x^2 + 18x + 81 = 0$.

SOLUTION This is a quadratic equation with $a = 1$, $b = 18$, and $c = 81$. In using the quadratic formula we find that the discriminant equals zero, and each of the two formulas for x gives

$$x = \frac{-b}{2a} = -9$$

Hence, $x = -9$ is the only solution of this equation. [Notice that this equation could have been factored as $(x + 9)^2 = 0$ and the solution $x = -9$ read off immediately.] ●

EXAMPLE 13 ● Solve the equation $x^2 + x + 1 = 0$.

SOLUTION This is a quadratic equation with $a = 1$, $b = 1$, and $c = 1$. Using the quadratic formula we get

$$x = \frac{-1 + \sqrt{-3}}{2} \quad \text{and} \quad x = \frac{-1 - \sqrt{-3}}{2}$$

Since there is no real number that is the square root of -3, this equation has no solution. ●

Summation notation

To describe a sum

$$X_1 + X_2 + \cdots + X_n$$

it suffices to specify the general term X_i and to indicate that n terms of this form are to be added, starting with the term in which $i = 1$ and ending with the term in which $i = n$. It is customary to use the Greek letter Σ (sigma) to denote the sum and to write

$$X_1 + X_2 + \cdots + X_n = \sum_{i=1}^{n} X_i$$

EXAMPLE 14 ● Use summation notation to express the following sums.

(a) $1 + 4 + 9 + 16 + 25 + 36 + 49 + 64$
(b) $X_1^2 + X_2^2 + \cdots + X_n^2$
(c) $(X_1 + X_2 + \cdots + X_n)^2$

SOLUTION (a) This is a sum of 8 terms of the form i^2, starting with $i = 1$ and ending with $i = 8$. Hence,

$$1 + 4 + 9 + 16 + 25 + 36 + 49 + 64 = \sum_{i=1}^{8} i^2$$

(b) This is a sum of n terms of the form X_i^2, starting with $i = 1$ and ending with $i = n$. Hence,

$$X_1^2 + X_2^2 + \cdots + X_n^2 = \sum_{i=1}^{n} X_i^2$$

(c) This is the *square* of a sum of n terms of the form X_i, starting with $i = 1$ and ending with $i = n$. Hence,

$$(X_1 + X_2 + \cdots + X_n)^2 = \left(\sum_{i=1}^{n} X_i\right)^2$$

●

If there is no ambiguity, we sometimes omit the index i from the symbol Σ and write ΣX_i instead of $\Sigma_{i=1}^{n} X_i$.

Problems

1 Express the following intervals geometrically.
(a) $x \geq -3$ (b) $x \leq 2$ (c) $-1 \leq x \leq 3$

2 Round off the following numbers to two decimal places.
(a) 3.1482 (b) -12.008 (c) 0.1346
(d) 3.998 (e) 0.105

3 Rewrite the following inequalities in the form $a \leq x \leq b$.
(a) $|x - 2| \leq 3$ (b) $|x + 2| \leq 3$ (c) $|x| \leq 1$

4 Rewrite each of the following expressions using exponential notation.
(a) $\sqrt[3]{x}$ (b) $1/\sqrt{x}$ (c) $\sqrt[5]{x^2}$ (d) $1/(\sqrt{x})^5$

5 Evaluate the following expressions.
(a) 5^3 (b) 2^{-3} (c) $16^{1/2}$ (d) $36^{-1/2}$ (e) $8^{2/3}$
(f) $27^{-4/3}$ (g) $(\frac{1}{4})^{1/2}$ (h) $(\frac{1}{4})^{-3/2}$ (i) $2^3(2^4)/(2^2)^3$ (j) $2(32^{3/5})/2^5$

6 Solve the following equations for n.
(a) $a^3a^5 = a^n$ (b) $a^2/a^n = a^7$
(c) $(a^2)^3 = a^n$ (d) $(a^2)^n = a^6$
(e) $\sqrt[n]{a}\,(a^{5/3}) = a^2$ (f) $a^{2/5}a^n = 1/a$

7 Factor the following expressions.
(a) $x^5 - 4x^4$ (b) $x^2 - 16$
(c) $x^2 - 4x - 5$ (d) $3x^4 - 3x^2$
(e) $6x^2 + 6x - 12$ (f) $x^3 - 27$

8 Solve the following equations.
(a) $3x - 2 = 7$ (b) $x^2 = 3x$
(c) $x^2 = 16$ (d) $x^2 + x - 6 = 0$
(e) $x^3 = 4x$ (f) $x^3 - 2x^2 - x + 2 = 0$

9 Use the quadratic formula to solve the following equations (if possible).
(a) $2x^2 + 3x + 1 = 0$ (b) $x^2 - 2x + 3 = 0$
(c) $x^2 - 2x + 1 = 0$ (d) $-x^2 + 3x - 1 = 0$

10 Use summation notation to express the following sums.

(a) $1 + \frac{1}{2} + \frac{1}{3} + \frac{1}{4} + \frac{1}{5} + \frac{1}{6}$

(b) $3 + 6 + 9 + 12 + 15 + 18 + 21 + 24 + 27 + 30$

(c) $2 + 4 + 8 + 16 + 32 + 64$

(d) $2x_1 + 2x_2 + 2x_3 + 2x_4 + 2x_5 + 2x_6$

11 Evaluate the following sums.

(a) $\sum_{i=1}^{5} i^2$

(b) $\sum_{i=1}^{4} (3i + 1)$

(c) $\sum_{i=1}^{5} 2^i$

(d) $\sum_{i=1}^{10} (-1)^i$

TABLES

Table I
Binomial probabilities:
$\binom{n}{k}p^k(1-p)^{n-k}$

k	.1	.2	.3	.4	.5	.6	.7	.8	.9
					$n = 5$				
0	.5905	.3277	.1681	.0778	.0312	.0102	.0024	.0003	.0000
1	.3280	.4096	.3602	.2592	.1562	.0768	.0284	.0064	.0004
2	.0729	.2048	.3087	.3456	.3125	.2304	.1323	.0512	.0081
3	.0081	.0512	.1323	.2304	.3125	.3456	.3087	.2048	.0729
4	.0004	.0064	.0284	.0768	.1562	.2592	.3602	.4096	.3280
5	.0000	.0003	.0024	.0102	.0312	.0778	.1681	.3277	.5905
					$n = 10$				
0	.3487	.1074	.0282	.0060	.0010	.0001	.0000	.0000	.0000
1	.3874	.2684	.1211	.0403	.0098	.0016	.0001	.0000	.0000
2	.1937	.3020	.2335	.1209	.0439	.0106	.0014	.0001	.0000
3	.0574	.2013	.2668	.2150	.1172	.0425	.0090	.0008	.0000
4	.0112	.0881	.2001	.2508	.2051	.1115	.0368	.0055	.0001
5	.0015	.0264	.1029	.2007	.2461	.2007	.1029	.0264	.0015
6	.0001	.0055	.0368	.1115	.2051	.2508	.2001	.0881	.0112
7	.0000	.0008	.0090	.0425	.1172	.2150	.2668	.2013	.0574
8	.0000	.0001	.0014	.0106	.0439	.1209	.2335	.3020	.1937
9	.0000	.0000	.0001	.0016	.0098	.0403	.1211	.2684	.3874
10	.0000	.0000	.0000	.0001	.0010	.0060	.0282	.1074	.3487
					$n = 15$				
0	.2059	.0352	.0047	.0005	.0000	.0000	.0000	.0000	.0000
1	.3432	.1319	.0305	.0047	.0005	.0000	.0000	.0000	.0000
2	.2669	.2309	.0916	.0219	.0032	.0003	.0000	.0000	.0000
3	.1285	.2501	.1700	.0634	.0139	.0016	.0001	.0000	.0000
4	.0428	.1876	.2186	.1268	.0417	.0074	.0006	.0000	.0000
5	.0105	.1032	.2061	.1859	.0916	.0245	.0030	.0001	.0000
6	.0019	.0430	.1472	.2066	.1527	.0612	.0116	.0007	.0000
7	.0003	.0138	.0811	.1711	.1964	.1181	.0348	.0035	.0000
8	.0000	.0035	.0348	.1181	.1964	.1771	.0811	.0138	.0003
9	.0000	.0007	.0116	.0612	.1527	.2066	.1472	.0430	.0019
10	.0000	.0001	.0030	.0245	.0916	.1859	.2061	.1032	.0105
11	.0000	.0000	.0006	.0074	.0417	.1268	.2186	.1876	.0428
12	.0000	.0000	.0001	.0016	.0139	.0634	.1700	.2501	.1285
13	.0000	.0000	.0000	.0003	.0032	.0219	.0916	.2309	.2669
14	.0000	.0000	.0000	.0000	.0005	.0047	.0305	.1319	.3432
15	.0000	.0000	.0000	.0000	.0000	.0005	.0047	.0352	.2059

k	.1	.2	.3	.4	.5	.6	.7	.8	.9
					$n = 20$				
0	.1216	.0115	.0008	.0000	.0000	.0000	.0000	.0000	.0000
1	.2702	.0576	.0068	.0005	.0000	.0000	.0000	.0000	.0000
2	.2852	.1369	.0278	.0031	.0002	.0000	.0000	.0000	.0000
3	.1901	.2054	.0716	.0123	.0011	.0000	.0000	.0000	.0000
4	.0898	.2182	.1304	.0350	.0046	.0003	.0000	.0000	.0000
5	.0319	.1746	.1789	.0746	.0148	.0013	.0000	.0000	.0000
6	.0089	.1091	.1916	.1244	.0370	.0049	.0002	.0000	.0000
7	.0020	.0545	.1643	.1659	.0739	.0146	.0010	.0000	.0000
8	.0004	.0222	.1144	.1797	.1201	.0355	.0039	.0001	.0000
9	.0001	.0074	.0654	.1597	.1602	.0710	.0120	.0005	.0000
10	.0000	.0020	.0308	.1171	.1762	.1171	.0308	.0020	.0000
11	.0000	.0005	.0120	.0710	.1602	.1597	.0654	.0074	.0001
12	.0000	.0001	.0039	.0355	.1201	.1797	.1144	.0222	.0004
13	.0000	.0000	.0010	.0146	.0739	.1659	.1643	.0545	.0020
14	.0000	.0000	.0002	.0049	.0370	.1244	.1916	.1091	.0089
15	.0000	.0000	.0000	.0013	.0148	.0746	.1789	.1746	.1319
16	.0000	.0000	.0000	.0003	.0046	.0350	.1304	.2182	.0898
17	.0000	.0000	.0000	.0000	.0011	.0123	.0716	.2054	.1901
18	.0000	.0000	.0000	.0000	.0002	.0031	.0278	.1369	.2852
19	.0000	.0000	.0000	.0000	.0000	.0005	.0068	.0576	.2702
20	.0000	.0000	.0000	.0000	.0000	.0000	.0008	.0115	.1216

Table II
Areas under the standard normal curve to the left of positive Z values: $P(Z \leq z)$

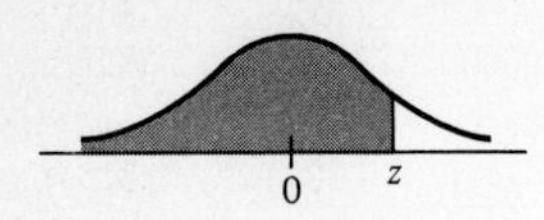

z	.00	.01	.02	.03	.04	.05	.06	.07	.08	.09
.0	.5000	.5040	.5080	.5120	.5160	.5199	.5239	.5279	.5319	.5359
.1	.5398	.5438	.5478	.5517	.5557	.5596	.5636	.5675	.5714	.5753
.2	.5793	.5832	.5871	.5910	.5948	.5987	.6026	.6064	.6103	.6141
.3	.6179	.6217	.6255	.6293	.6331	.6368	.6406	.6443	.6480	.6517
.4	.6554	.6591	.6628	.6664	.6700	.6736	.6772	.6808	.6844	.6879
.5	.6915	.6950	.6985	.7019	.7054	.7088	.7123	.7157	.7190	.7224
.6	.7257	.7291	.7324	.7357	.7389	.7422	.7454	.7486	.7517	.7549
.7	.7580	.7611	.7642	.7673	.7704	.7734	.7764	.7794	.7823	.7852
.8	.7881	.7910	.7939	.7967	.7995	.8023	.8051	.8078	.8106	.8133
.9	.8159	.8186	.8212	.8238	.8264	.8289	.8315	.8340	.8365	.8389
1.0	.8413	.8438	.8461	.8485	.8508	.8531	.8554	.8577	.8599	.8621
1.1	.8643	.8665	.8686	.8708	.8729	.8749	.8770	.8790	.8810	.8830
1.2	.8849	.8869	.8888	.8907	.8925	.8944	.8962	.8980	.8997	.9015
1.3	.9032	.9049	.9066	.9082	.9099	.9115	.9131	.9147	.9162	.9177
1.4	.9192	.9207	.9222	.9236	.9251	.9265	.9279	.9292	.9306	.9319
1.5	.9332	.9345	.9357	.9370	.9382	.9394	.9406	.9418	.9429	.9441
1.6	.9452	.9463	.9474	.9484	.9495	.9505	.9515	.9525	.9535	.9545
1.7	.9554	.9564	.9573	.9582	.9591	.9599	.9608	.9616	.9625	.9633
1.8	.9641	.9649	.9656	.9664	.9671	.9678	.9686	.9693	.9699	.9706
1.9	.9713	.9719	.9726	.9732	.9738	.9744	.9750	.9756	.9761	.9767
2.0	.9772	.9778	.9783	.9788	.9793	.9798	.9803	.9808	.9812	.9817
2.1	.9821	.9826	.9830	.9834	.9838	.9842	.9846	.9850	.9854	.9857
2.2	.9861	.9864	.9868	.9871	.9875	.9878	.9881	.9884	.9887	.9890
2.3	.9893	.9896	.9898	.9901	.9904	.9906	.9909	.9911	.9913	.9916
2.4	.9918	.9920	.9922	.9925	.9927	.9929	.9931	.9932	.9934	.9936
2.5	.9938	.9940	.9941	.9943	.9945	.9946	.9948	.9949	.9951	.9952
2.6	.9953	.9955	.9956	.9957	.9959	.9960	.9961	.9962	.9963	.9964
2.7	.9965	.9966	.9967	.9968	.9969	.9970	.9971	.9972	.9973	.9974
2.8	.9974	.9975	.9976	.9977	.9977	.9978	.9979	.9979	.9980	.9981
2.9	.9981	.9982	.9982	.9983	.9984	.9984	.9985	.9985	.9986	.9986
3.0	.9987	.9987	.9987	.9988	.9988	.9989	.9989	.9989	.9990	.9990
3.1	.9990	.9991	.9991	.9991	.9992	.9992	.9992	.9992	.9993	.9993
3.2	.9993	.9993	.9994	.9994	.9994	.9994	.9994	.9995	.9995	.9995
3.3	.9995	.9995	.9995	.9996	.9996	.9996	.9996	.9996	.9996	.9997
3.4	.9997	.9997	.9997	.9997	.9997	.9997	.9997	.9997	.9997	.9998

From *Introduction to the Theory of Statistics,* 3d ed., by A. M. Mood, F. A. Graybill, and D. C. Boes.

Table III
Powers of e

x	e^x	e^{-x}	x	e^x	e^{-x}	x	e^x	e^{-x}
0.00	1.0000	1.00000	0.50	1.6487	.60653	1.00	2.7183	.36788
0.01	1.0101	0.99005	0.51	1.6653	.60050	1.20	3.3201	.30119
0.02	1.0202	.98020	0.52	1.6820	.59452	1.30	3.6693	.27253
0.03	1.0305	.97045	0.53	1.6989	.58860	1.40	4.0552	.24660
0.04	1.0408	.96079	0.54	1.7160	.58275	1.50	4.4817	.22313
0.05	1.0513	.95123	0.55	1.7333	.57695	1.60	4.9530	.20190
0.06	1.0618	.94176	0.56	1.7507	.57121	1.70	5.4739	.18268
0.07	1.0725	.93239	0.57	1.7683	.56553	1.80	6.0496	.16530
0.08	1.0833	.92312	0.58	1.7860	.55990	1.90	6.6859	.14957
0.09	1.0942	.91393	0.59	1.8040	.55433	2.00	7.3891	.13534
0.10	1.1052	.90484	0.60	1.8221	.54881	3.00	20.086	.04979
0.11	1.1163	.89583	0.61	1.8404	.54335	4.00	54.598	.01832
0.12	1.1275	.88692	0.62	1.8589	.53794	5.00	148.41	.00674
0.13	1.1388	.87809	0.63	1.8776	.53259	6.00	403.43	.00248
0.14	1.1503	.86936	0.64	1.8965	.52729	7.00	1096.6	.00091
0.15	1.1618	.86071	0.65	1.9155	.52205	8.00	2981.0	.00034
0.16	1.1735	.85214	0.66	1.9348	.51685	9.00	8103.1	.00012
0.17	1.1853	.84366	0.67	1.9542	.51171	10.00	22026.5	.00005
0.18	1.1972	.83527	0.68	1.9739	.50662			
0.19	1.2092	.82696	0.69	1.9937	.50158			
0.20	1.2214	.81873	0.70	2.0138	.49659			
0.21	1.2337	.81058	0.71	2.0340	.49164			
0.22	1.2461	.80252	0.72	2.0544	.48675			
0.23	1.2586	.79453	0.73	2.0751	.48191			
0.24	1.2712	.78663	0.74	2.0959	.47711			
0.25	1.2840	.77880	0.75	2.1170	.47237			
0.26	1.2969	.77105	0.76	2.1383	.46767			
0.27	1.3100	.76338	0.77	2.1598	.46301			
0.28	1.3231	.75578	0.78	2.1815	.45841			
0.29	1.3364	.74826	0.79	2.2034	.45384			
0.30	1.3499	.74082	0.80	2.2255	.44933			
0.31	1.3634	.73345	0.81	2.2479	.44486			
0.32	1.3771	.72615	0.82	2.2705	.44043			
0.33	1.3910	.71892	0.83	2.2933	.43605			
0.34	1.4049	.71177	0.84	2.3164	.43171			
0.35	1.4191	.70469	0.85	2.3396	.42741			
0.36	1.4333	.69768	0.86	2.3632	.42316			
0.37	1.4477	.69073	0.87	2.3869	.41895			
0.38	1.4623	.68386	0.88	2.4109	.41478			
0.39	1.4770	.67706	0.89	2.4351	.41066			
0.40	1.4918	.67032	0.90	2.4596	.40657			
0.41	1.5068	.66365	0.91	2.4843	.40252			
0.42	1.5220	.65705	0.92	2.5093	.39852			
0.43	1.5373	.65051	0.93	2.5345	.39455			
0.44	1.5527	.64404	0.94	2.5600	.39063			
0.45	1.5683	.63763	0.95	2.5857	.38674			
0.46	1.5841	.63128	0.96	2.6117	.38298			
0.47	1.6000	.62500	0.97	2.6379	.37908			
0.48	1.6161	.61878	0.98	2.6645	.37531			
0.49	1.6323	.61263	0.99	2.6912	.37158			

Table IV
Natural logarithms (base e)

x	$\ln x$	x	$\ln x$	x	$\ln x$	x	$\ln x$
.01	−4.60517	0.50	−0.69315	1.00	0.00000	1.5	0.40547
.02	−3.91202	.51	.67334	1.01	.00995	1.6	7000
.03	.50656	.52	.65393	1.02	.01980	1.7	0.53063
.04	.21888	.53	.63488	1.03	.02956	1.8	8779
		.54	.61619	1.04	.03922	1.9	0.64185
.05	−2.99573	.55	.59784	1.05	.04879	2.0	9315
.06	.81341	.56	.57982	1.06	.05827	2.1	0.74194
.07	.65926	.57	.56212	1.07	.06766	2.2	8846
.08	.52573	.58	.54473	1.08	.07696	2.3	0.83291
.09	.40795	.59	.52763	1.09	.08618	2.4	7547
0.10	−2.30259	0.60	−0.51083	1.10	.09531	2.5	0.91629
.11	.20727	.61	.49430	1.11	.10436	2.6	5551
.12	.12026	.62	.47804	1.12	.11333	2.7	9325
.13	.04022	.63	.46204	1.13	.12222	2.8	1.02962
.14	−1.96611	.64	.44629	1.14	.13103	2.9	6471
.15	.89712	.65	.43078	1.15	.13976	3.0	9861
.16	.83258	.66	.41552	1.16	.14842	4.0	1.38629
.17	.77196	.67	.40048	1.17	.15700	5.0	1.60944
.18	.71480	.68	.38566	1.18	.16551	10.0	2.30258
.19	.66073	.69	.37106	1.19	.17395		
0.20	−1.60944	0.70	−0.35667	1.20	.18232		
.21	.56065	.71	.34249	1.21	.19062		
.22	.51413	.72	.32850	1.22	.19885		
.23	.46968	.73	.31471	1.23	.20701		
.24	.42712	.74	.30111	1.24	.21511		
.25	.38629	.75	.28768	1.25	.22314		
.26	.34707	.76	.27444	1.26	.23111		
.27	.30933	.77	.26136	1.27	.23902		
.28	.27297	.78	.24846	1.28	.24686		
.29	.23787	.79	.23572	1.29	.25464		
0.30	−1.20397	0.80	−0.22314	1.30	.26236		
.31	.17118	.81	.21072	1.31	.27003		
.32	.13943	.82	.19845	1.32	.27763		
.33	.10866	.83	.18633	1.33	.28518		
.34	.07881	.84	.17435	1.34	.29267		
.35	−1.04982	.85	−0.16252	1.35	.30010		
.36	.02165	.86	.15032	1.36	.30748		
.37	−0.99425	.87	.13926	1.37	.31481		
.38	.96758	.88	.12783	1.38	.32208		
.39	.94161	.89	.11653	1.39	.32930		
0.40	−0.91629	0.90	−0.10536	1.40	.33647		
.41	.89160	.91	.09431	1.41	.34359		
.42	.86750	.92	.08338	1.42	.35066		
.43	.84397	.93	.07257	1.43	.35767		
.44	.82098	.94	.06188	1.44	.36464		
.45	.79851	.95	.05129	1.45	.37156		
.46	.77653	.96	.04082	1.46	.37844		
.47	.75502	.97	.03046	1.47	.38526		
.48	.73397	.98	.02020	1.48	.39204		
.49	.71335	.99	.01005	1.49	.39878		

ANSWERS TO ODD-NUMBERED PROBLEMS

Chapter 1 Section 1 (page 11)

1 (a) \$1,300 (b) \$30
3 (a) 140 million gallons (b) 200 million gallons
(c) 4 million gallons (d) 4 million gallons
5 (a) and (d) are linear. **7** 1 **9** 0 **11** $m = 3, b = 0$
13 $m = 1, b = 0$ **15** $m = -\frac{3}{2}, b = 3$
17 $m = \frac{3}{5}, b = \frac{4}{5}$ **19** $m = -\frac{5}{2}, b = 5$
21 m is undefined; there is no y intercept. **23** $y = \frac{2}{3}x + \frac{8}{3}$
25 $y = 5x$ **27** $x = 2$ **29** $y = 7x - 9$ **31** $y = 5$

Chapter 1 Section 2 (page 15)

1 (a) $R(x) = 6{,}000x$ (b) \$36,000 **3** $C(x) = 280x$
5 (a) $f(x) = -12.5x + 150$ (b) \$87.50
7 (a) $V(x) = 248 - 4x$ million gallons after x days.
(b) 216 million gallons
9 (a) $f(x) = \frac{9}{5}x + 32$ (b) 59°F (c) 20°C
11 (a) $C(x) = \begin{cases} 0.12x + 2.95 \text{ if } x \geq 15 \\ 4.75 \text{ if } 0 \leq x < 15 \end{cases}$ (b) \$10.15
13 (a) \$24, \$48 (b) No. The rate of change is not constant.

Chapter 1 Section 3 (page 22)

1 $(-\frac{1}{2}, \frac{7}{2})$ **3** $(1, -1)$ **5** None **7** (a) 4 (b) 7
9 Choose the first club if you are planning to play more than 80 hours a year and the second if you are planning to play less than 80 hours a year.
11 (a) 20 units at \$30 **13** Of course!

Chapter 2 Section 1 (page 29)

1 (a) 3×4 (b) $a_{12} = 0$ $a_{22} = -2$ $a_{14} = 1$ $a_{34} = -4$
(c) $9 = a_{23}$ $-3 = a_{31}$ $0 = a_{12}$ $5 = a_{33}$
3 $\begin{bmatrix} 2 & 3 & 4 & 5 \\ 3 & 4 & 5 & 6 \\ 4 & 5 & 6 & 7 \end{bmatrix}$
5 (a) Market 1
(b) The total cost at market 2 of one unit of each of items A, B, C, and D.
(c) \$14.38 at market 1; \$14.35 at market 2; \$14.03 at market 3
7 (a)

	Eric	Gabriel	Kirk
Eric	0	2	1
Gabriel	1	0	6
Kirk	5	3	0

where a_{ij} = number of letters from i received by j

(b) Kirk; 8 letters
9 Impossible
11 $\begin{bmatrix} -5 & 5 & 5 \\ -1 & -2 & -3 \end{bmatrix}$ **13** $\begin{bmatrix} 11 & -15 & -10 \\ 5 & 4 & 8 \end{bmatrix}$

15 $\begin{bmatrix} 2 & -1 & 3 \\ -7 & 3 & 2 \\ 0 & -5 & -2 \end{bmatrix}$ **17** $\begin{bmatrix} -1 & 2 & 3 \\ 8 & -3 & 11 \\ 0 & -2 & 10 \end{bmatrix}$

19 (a)

	Guitar	Case
X	200	45
Y	260	40
Z	320	50

(b) $A - 0.15A = 0.85A = \begin{bmatrix} 170 & 38.25 \\ 221 & 34 \\ 272 & 42.50 \end{bmatrix}$

21

	Prod. ed.	Designer	Copy ed.
N.Y.	430	52	270
S.F.	140	14	90

Chapter 2 Section 2 (page 36)

1 (a) 1×3 (b) Undefined (c) Undefined (d) 3×4
(e) 2×4 (f) Undefined

3 $\begin{bmatrix} 3 & 1 \\ -2 & 1 \\ 0 & -1 \end{bmatrix}$ **5** Undefined

7 $\begin{bmatrix} -8 & -2 & -73 & 38 \\ -15 & 6 & -15 & 3 \\ 21 & 2 & 151 & -77 \end{bmatrix}$ **9** $\begin{bmatrix} 15 & -11 \\ 62 & -3 \\ -20 & 19 \end{bmatrix}$

11 $AB + CB = (A + C)B = \begin{bmatrix} 37 & 2 & 12 \\ 19 & -6 & 4 \\ -6 & 0 & -2 \end{bmatrix}$

13 False **15** False **17** $-3x_1 + 2x_2 + x_3 = 1$
$5x_1 - x_2 = 0$
$4x_1 + x_2 + 2x_3 = 3$

19 (a) The number of hours of each type of labor needed in each state in June to fill the orders.
(b) The labor cost per item.
(c) The total labor cost in each state in June.

21 (a)

		P.K.	C.P.	J's
$B^*A =$	Club	10.70	10.55	10.45
	Senate	31.25	30.50	30.25
	Team	35.90	33.85	34.95

where B^* is obtained from B by interchanging the rows and columns.
(b) Math club at Joe's; senate at Joe's; rugby team at Campus Pizza.

Chapter 2 Section 3 (page 44)

1 $x_1 = 0$ $x_2 = -1$ $x_3 = 3$
3 $x_3 = 3$ $x_2 =$ anything $x_1 = 3 - 2x_2$
5 $x_1 =$ anything $x_2 = 7$ $x_3 = 9$ $x_4 = -8$

7 $x_1 = 0 \quad x_2 = 3$
9 No solution
11 $x_2 =$ anything $\quad x_1 = 3x_2 - 10$
13 $x_2 =$ anything $\quad x_3 =$ anything $\quad x_1 = 5 + 2x_2 - 3x_3$
15 $x_1 = -4 \quad x_2 = 3 \quad x_3 = 4$
17 $x_2 = 0 \quad x_3 =$ anything $\quad x_1 = 1 + 2x_3$
19 $x_1 = 1 \quad x_2 = 1 \quad x_3 = 2 \quad x_4 = 0$
21 15 and 40
23 90 grams of the first, 30 of the second, and 60 of the third.
25 \$30,000 at 4 percent, \$10,000 at 5 percent, \$50,000 at 8 percent

Chapter 2 Section 4 (page 51)

1 Yes **3** No **5** $\begin{bmatrix} 9 & 2 \\ 5 & 1 \end{bmatrix}$

7 Not invertible **9** $\begin{bmatrix} \frac{23}{2} & -\frac{3}{2} & \frac{5}{2} \\ -4 & \frac{1}{2} & -1 \\ -\frac{3}{2} & 0 & -\frac{1}{2} \end{bmatrix}$ **11** $\begin{bmatrix} 7 & 1 & 3 \\ \frac{3}{2} & \frac{1}{2} & 0 \\ -1 & 0 & -1 \end{bmatrix}$

13 $x_1 = -23 \quad x_2 = 5$ **15** $x_1 = 3 \quad x_2 = 5 \quad x_3 = 2$
17 $x_1 = 0 \quad x_2 = -6 \quad x_3 = -4$ **19** $\begin{bmatrix} -2 & \frac{15}{7} \\ 0 & -\frac{4}{7} \end{bmatrix}$

21 $\begin{bmatrix} -29 & -29 & -29 \\ 9 & 9 & 9 \\ 5 & 5 & 5 \end{bmatrix}$

23 (b) 5 of brand I; 2 of brand II; 1 of brand III
25 The rows represent the two types of furniture, the columns the two plants, and the entries are the numbers of pieces of each type of furniture produced at each plant in July.

Chapter 2 Section 5 (page 58)

1 (a)

	p_1	p_2	p_3	p_4	p_5
p_1	0	1	0	0	1
p_2	1	0	0	1	0
p_3	0	0	0	0	1
p_4	0	1	0	0	1
p_5	1	0	1	1	0

(b)

	p_1	p_2	p_3	p_4	p_5
p_1	2	0	1	2	0
p_2	0	2	0	0	2
p_3	1	0	1	1	0
p_4	2	0	1	2	0
p_5	0	2	0	0	3

(c) $p_1 \leftrightarrow p_2 \leftrightarrow p_4 \quad p_1 \leftrightarrow p_5 \leftrightarrow p_4$

3

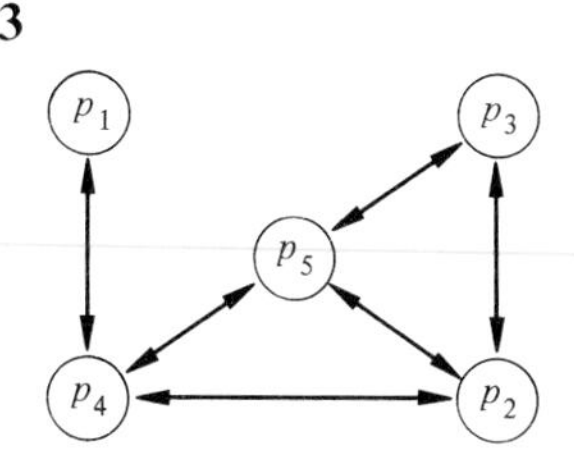

5 (a) The number of ways each pair of individuals can communicate using no more than $n - 2$ intermediaries.
(b) p_i and p_j cannot communicate with each other if and only if the entry in row i, column j is zero.
7 (a) 0.1 (b) M (c) 3.6
9 S_1 must produce 372; S_2 must produce 122

Chapter 3 Section 1 (page 67)

1 Maximize $f = 15x_1 + 20x_2$ subject to $x_1 + x_2 \leq 7$, $x_1 + 2x_2 \leq 11$, $x_1 \geq 0$, $x_2 \geq 0$
3 Maximize $f = 15x_1 + 20x_2$ subject to $x_1 + x_2 \leq 7$, $x_1 + 2x_2 \leq 11$, $0.2x_1 + 0.6x_2 \leq 1.8$, $x_2 \geq 2$, $x_1 \geq 0$
5 Maximize $f = 1.2x_1 + x_2 + 1.2x_3$ subject to $x_1 + 0.4x_3 \leq 400$, $x_2 + 0.6x_3 \leq 480$, $x_1 \geq 0$, $x_2 \geq 0$, $x_3 \geq 0$
7 Minimize $f = 60x_1 + 140x_2 + 120x_3$ subject to $5x_1 + 10x_2 + 15x_3 \geq 30$, $2x_1 + 3x_2 + 2x_3 \geq 8$, $3x_1 + 6x_2 + x_3 \geq 10$, $x_1 \geq 0$, $x_2 \geq 0$, $x_3 \geq 0$
9 Minimize $f = 20x_1 + 26x_2$ subject to $3x_1 + 12x_2 \geq 396$, $6x_1 + 3x_2 \geq 288$, $4x_1 + 3x_2 \geq 255$, $x_1 \geq 0$, $x_2 \geq 0$
11 Minimize $f = 50x_{11} + 15x_{12} + 25x_{21} + 45x_{22}$ subject to $x_{11} + x_{12} \leq 100$, $x_{21} + x_{22} \leq 120$, $x_{11} + x_{21} \geq 80$, $x_{12} + x_{22} \geq 125$, $x_{11} \geq 40$, $x_{12} \leq 50$, $x_{12} \geq 0$, $x_{21} \geq 0$, $x_{22} \geq 0$
13 Minimize $f = c_{11}x_{11} + c_{12}x_{12} + c_{21}x_{21} + c_{22}x_{22}$ subject to $x_{11} + x_{12} \leq s_1$, $x_{21} + x_{22} \leq s_2$, $x_{11} + x_{21} \geq d_1$, $x_{12} + x_{22} \geq d_2$, $x_{11} \geq 0$, $x_{12} \geq 0$, $x_{21} \geq 0$, $x_{22} \geq 0$

Chapter 3 Section 2 (page 75)

1 $f(6, 0) = 30$ **3** $f(1, 1) = 16$ **5** $f(1, 4) = f(4, 2) = 140$
7 $f(7, 3) = 17$ **9** $f(4, 5) = 23$ **11** $f(2, 3) = 9$ **13** $f(1, 7) = 10$
15 $f(3, 4) = 125$ **17** $f(3, 2) = 85$ **19** $f(48, 21) = 1{,}506$
21 There is no solution. f has no lower bound subject to these constraints.

Chapter 3 Section 3 (page 81)

1 $f(6, 0) = 6$ **3** $f(3, 4) = 15$ **5** $f(0, 4) = -12$
7 $f(120, 60) = 204$ **9** $f(180, 10, 0, 140) = 18{,}100$
11 $f(3, 0, 1) = 300$ **13** $f(48, 21) = 1{,}506$ **15** $f(40, 50, 40, 75) = 7,125$

Chapter 3 Section 4 (page 88)

1 $x_{11} = 75$ $x_{21} = 125$ $x_{31} = 0$ $x_{12} = 0$ $x_{22} = 0$
$x_{32} = 150$ $y_{11} = 150$ $y_{21} = 0$ $y_{12} = 50$ $y_{22} = 150$
3 $a_1 \leftrightarrow a_2 \leftrightarrow a_3 \leftrightarrow a_5$ **5** Four semesters
7 $x_{12} = 10$ $x_{13} = 10$ $x_{14} = 20$ $x_{25} = 0$ $x_{32} = 0$
$x_{34} = 0$ $x_{35} = 20$ $x_{45} = 0$
9 54,000

Chapter 4 Section 1 (page 100)

1 (a) T (b) F (c) F (d) F
(e) T (f) T (g) T (h) F
(i) F (j) F (k) F (l) T

3 (a) $\{x \mid x$ is a star in the Milky Way$\}$
(b) $\{x \mid x$ is a U.S. city west of the Mississippi$\}$
(c) $\{x \mid x$ is a positive integer divisible by 11$\}$

5 $\varnothing$ $\{w\}$ $\{x\}$ $\{y\}$ $\{z\}$ $\{w, x\}$ $\{w, y\}$ $\{w, z\}$ $\{x, y\}$ $\{x, z\}$ $\{y, z\}$ $\{w, x, y\}$ $\{w, x, z\}$ $\{w, y, z\}$ $\{x, y, z\}$ $\{w, x, y, z\}$

13 (a) F (b) T (c) F (d) T (e) T

15 (a) 12 (b) 4 (c) 16 (d) 4 (e) 4

17 (a)

(b)

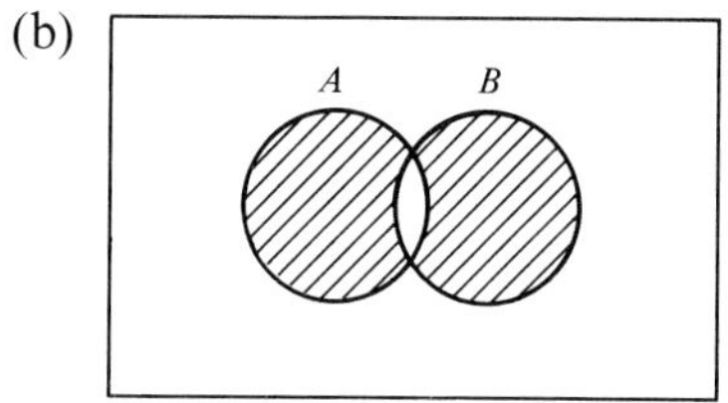

(c)

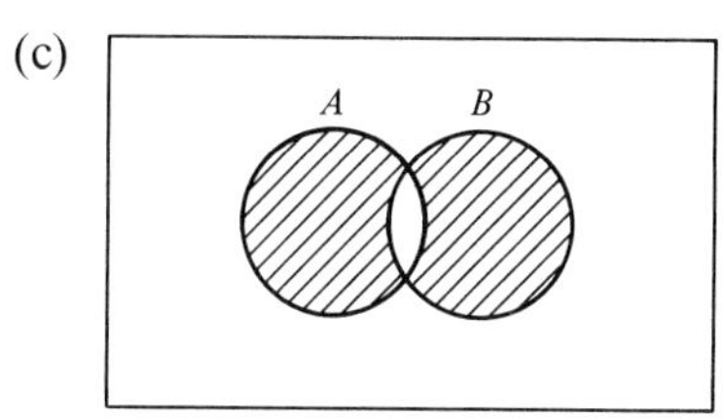

(d)

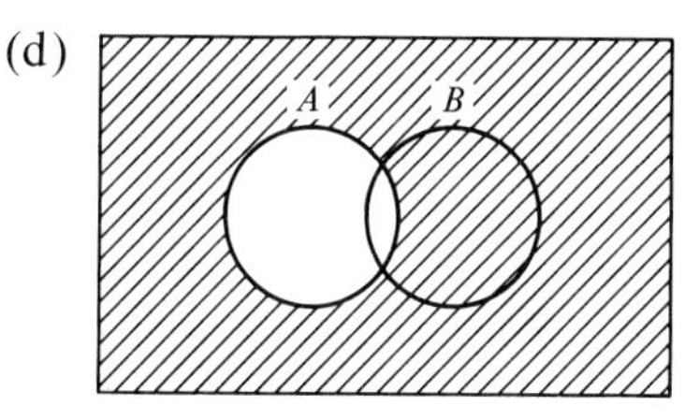

19 (a) The set of all even integers
(b) The set of all perfect squares
(c) The set of all positive multiples of 8
(d) The set of all integers greater than 18

21 (a) $(0, 1)$ $(0, 0)$ $(1, 0)$ $(1, 1)$
(b) $(x, 5)$ $(x, 6)$ $(y, 5)$ $(y, 6)$ $(z, 5)$ $(z, 6)$
(c) (r, r) (r, s) (r, t) (s, r) (s, s) (s, t) (t, r) (t, s) (t, t)

23 Express the record as a 4-tuple $(a_1, a_2, a_3, a_4) \in A \times A \times A \times A$ where $A = \{\text{W}, \text{L}\}$ and a_i is the outcome of the ith game.

25 $\#(A_1 \times A_2 \times \cdots \times A_n) = \#(A_1)\#(A_2) \cdots \#(A_n)$

Chapter 4 Section 2 (page 111)

1 12

3 (a) 8
(b) HHH HTH THH TTH HHT HTT THT TTT
(d) 1,024

5 3,486,784,401

7 (a) 8 (b) 16 (c) 64 (d) 1,024 (e) 2^n

9 20,274,183,400,000,000 **11** (a) 120 (b) 625

13 121,399,651,100 **17** 2^n **19** (a) 225 (b) 24

21 (a) 24 (b) 60 (c) 30 (d) 495

23 (a) 24 (b) 4 (c) 1,287 (d) 4,512

Chapter 5 Section 1 (page 125)

1 (a) 11 12 13 14 15 16 21 22 23 24 25 26 31 32 33 34 35 36 41 42 43 44 45 46
(b) 11 13 15 22 24 26 31 33 35 42 44 46
(c) 11 22 33 44

7 0.1

9 (a) HHHH HHHT HHTH HHTT HTHH HTHT HTTH HTTT THHH THHT THTH THTT TTHH TTHT TTTH TTTT
(b) $\frac{1}{16}$ (c) $\frac{1}{4}$ (d) $\frac{5}{16}$

11 $\frac{1}{9}$ **13** (a) $\frac{1}{2}$ (b) $\frac{1}{4}$ (c) $\frac{1}{10}$ **15** (a) 0.4 (b) 0.6 (c) 0.65

17 (a) 14 15 16 24 25 26 34 35 36 44 45 46 54 55 56
(b) $\frac{2}{15}$ (c) $\frac{13}{15}$ (d) $\frac{8}{15}$ (e) $\frac{6}{15}$

19 (a)

	Smoker	Nonsmoker	
High blood pressure	180	90	270
Normal or low blood pressure	120	360	480
	300	450	750

(b) 0.36 (c) 0.48 (d) 0.24 (e) no change

21 0.8 **23** The theoretical probability is $\frac{1}{3}$.

25 The theoretical probability is 0.167

Chapter 5 Section 2 (page 131)

1 (a) 58 59 68 69 78 79 81 88 89 90 91 98 99
(b) 50 51 58 59 70 71 78 79 81 90 91 98 99
(c) 81 90
(d) 50 51 58 59 68 69 70 71 78 79 81 88 89 90 91 98 99
(e) 50 51 60 61 70 71 80
(f) 60 61 80

3 (a) 0.75 (b) 0.55 (c) 0

5 (a) 0.35 (b) 0.6 (c) 0.65

7 (a) H1 H2 H3 H4 H5 H6 T1 T2 T3 T4 T5 T6
(b) $\frac{1}{4}$ (c) $\frac{3}{4}$ (d) $\frac{1}{2}$

Chapter 5 Section 3 (page 136)

1 (a) 0.07 (b) 0.05

3 (a) 0.35 (b) 0.26 (c) 0.33 (d) 0.35 (e) no

5 (a) 0.39 (b) 0.22 (c)

	C	C′	
M	900	600	1,500
M′	1,400	2,100	3,500
	2,300	2,700	5,000

7 (a) 0.03 (b) almost 1 **9** (a) 0.50 (b) 0.02 **11** 0.51

Chapter 5 Section 4 (page 145)

1 (a) 0.004 (b) 0.491 (c) 0.509 (d) 0.421
3 (a) 0.011 (b) 0.725 (c) 0.275 **5** 0.000016 **7** 0.533
9 (a) 0.484 (b) 0.089
11 (a) 0.817 (b) 0.183 (c) 0.012 (d) 0.000003
13 (a) 0.012 (b) 0.0006

Chapter 5 Section 5 (page 152)

1 0.999
3 (a) 0.063 (b) 0.438 (c) 0.375
5 0.034
7 (a) 0.240 (b) 0.992 (c) 0.265
9 (a) 0.251 (b) 0.633 (c) 0.618
11 (a) 0.797 (b) 0.001 (c) 0.168
13 (a) 0.133 (b) 0.236 (c) 0.177
15 (a) 0.044 (b) 0.999

Chapter 5 Section 6 (page 164)

1
$$\begin{array}{c} \\ D \\ R \end{array}\begin{array}{c} \begin{array}{cc} D & R \end{array} \\ \begin{bmatrix} 0.95 & 0.05 \\ 0.10 & 0.90 \end{bmatrix} \end{array}$$

3 (a)
$$\begin{array}{c} \\ S \\ C \end{array}\begin{array}{c} \begin{array}{cc} S & C \end{array} \\ \begin{bmatrix} 0.95 & 0.05 \\ 0.03 & 0.97 \end{bmatrix} \end{array}$$
(b) 0.058 (c)
$$\begin{array}{c} \\ S \\ C \end{array}\begin{array}{c} \begin{array}{cc} S & C \end{array} \\ \begin{bmatrix} 0.904 & 0.096 \\ 0.058 & 0.942 \end{bmatrix} \end{array}$$

5 (a)
$$\begin{array}{c} \\ \text{Tuned} \\ \text{Untuned} \end{array}\begin{array}{c} \begin{array}{cc} \text{Tuned} & \text{Untuned} \end{array} \\ \begin{bmatrix} 0.9 & 0.1 \\ 0.8 & 0.2 \end{bmatrix} \end{array}$$
(b) 0.889

(c)
$$\begin{array}{c} \\ \text{Tuned} \\ \text{Untuned} \end{array}\begin{array}{c} \begin{array}{cc} \text{Tuned} & \text{Untuned} \end{array} \\ \begin{bmatrix} 0.889 & 0.111 \\ 0.888 & 0.112 \end{bmatrix} \end{array}$$

7 (a)
$$\begin{array}{c} \\ 1 \\ 2 \\ 3 \end{array}\begin{array}{c} \begin{array}{ccc} 1 & 2 & 3 \end{array} \\ \begin{bmatrix} 0.7 & 0.2 & 0.1 \\ 0.3 & 0.5 & 0.2 \\ 0.2 & 0.4 & 0.4 \end{bmatrix} \end{array}$$
(b) 0.15 (c) 0.203
9 Regular **11** Regular
13 The matrices in Problems 1, 2, 3, and 5.
15 88.89 percent

Chapter 6 Section 1 (page 171)

1

x	0	1	2	3
$P(X = x)$	$\frac{1}{8}$	$\frac{3}{8}$	$\frac{3}{8}$	$\frac{1}{8}$

3 (a)

x	1	2	3	4	5	6
$P(X_i = x)$	$\frac{1}{6}$	$\frac{1}{6}$	$\frac{1}{6}$	$\frac{1}{6}$	$\frac{1}{6}$	$\frac{1}{6}$

(b)

x	2	3	4	5	6	7	8	9	10	11	12
$P(X_1 + X_2 = x)$	$\frac{1}{36}$	$\frac{2}{36}$	$\frac{3}{36}$	$\frac{4}{36}$	$\frac{5}{36}$	$\frac{6}{36}$	$\frac{5}{36}$	$\frac{4}{36}$	$\frac{3}{36}$	$\frac{2}{36}$	$\frac{1}{36}$

5 (a)

x	1	2	3
$P(X_i = x)$	0.2	0.3	0.5

(b)

x	2	3	4	5	6
$P(X_1 + X_2 = x)$	0.04	0.12	0.29	0.30	0.25

7 (a) 0.285 (b) 0.677 **9** (a) 0.420 (b) 0.157

11 $P(X = k) = \binom{15}{k}\left(\frac{1}{3}\right)^k\left(\frac{2}{3}\right)^{15-k}$ **13** $P(X = k) = (0.1)(0.9)^{k-1}$

15 (a) 0.035 (b) $P(X = k) = \binom{k-1}{3}(0.15)^4(0.85)^{k-4}$

Chapter 6 Section 2 (page 179)

1 5 **3** 24.625 **5** 98.25 cents **7** 6 months **9** −5.3 cents
11 −7.9 cents **13** 1.875 **15** (a) 10 (b) 8 (c) 61
17 (a) $E(X_1) = E(X_2) = 3$ (b) 6 (c) 9 **19** 10 **21** 80
23 $\frac{3}{2}$ to 1 **25** $\frac{1}{11}$ **27** 0.374 **29** 1.028 to 1; odds favor casino

Chapter 6 Section 3 (page 188)

1 (a) 0.8925 (b) 0.4247 (c) 0.1151 (d) 0.6826 (e) 1.0000
3 (a) 0.6 (b) 0 (c) 4.6 (d) −1 (e) −5
5 (a) 4.42 (b) 4.37 (c) 3.9 (d) 4

Chapter 6 Section 4 (page 191)

1 0.8664
3 (a) 0.2112 (b) 0.9876
5 0.0062 **7** 0.0228
9 (a) 0.9332 (b) 69.8 (c) 87.8
11 (a) 0.68 (b) 0.95 (c) 0.997
13 68 percent between 71 and 85; 95 percent between 64 and 92; 99.7 percent between 57 and 99.
15 (a) 0.1587 (b) 0.6826 (c) 1.0000 **17** 0.0516
19 (a) 0.9177 (b) 319 **21** B offers the better deal.

Chapter 7 Section 1 (page 200)

1 (a) $\bar{X} = 0$ $s = 2.37$ (b) $\bar{X} = 2.83$ $s = 1.72$
(c) $\bar{X} = 14.36$ $s = 2.55$ (d) $\bar{X} = 8$ $s = 0$
3 (a) 0.3085 (b) 0.5 (c) 0.6826
5 (a) 0 (b) 0.5 (c) 1

Chapter 7 Section 2 (page 207)

1 2288.4 to 2291.6; 2287.9 to 2292.1 **3** 0.91 to 1.29; 0.86 to 1.34
5 97 **9** 0.38 to 0.46; 0.36 to 0.48
11 (a) 0.07 to 0.18 (b) Double-parking
13 (a) 4,161 (b) 0.12 to 0.14

Chapter 7 Section 3 (page 210)

1 $Z = 3$; reject the null hypothesis
3 $Z = 5.025$; yes
5 (a) $Z = 4$; yes (b) $Z = -4$; yes
7 $Z = 10.29$; The evidence is sufficient to imply a relationship between the chemical and liver cancer.
9 2.575
11 $Z = -1.37$; results could be due to chance.

Chapter 7 Section 4 (page 219)

1 $r = 0.92$; $Y = 1.13X + 0.83$
3 (a) $r = 1$; $Y = 2X + 1$ (b) $r = -1$; $Y = -X + 5$
(c) $r = 0$; $Y = 10$
5 (b) $r = 0.97$ (c) $Y = 0.32X - 0.34$ (d) 1,196
7 (b) $r = 0.645$ (c) $Y = 0.95X + 3.12$ (d) \$5,020

Chapter 8 Section 1 (page 228)

1 Produce cannon; $E = 4$ **3** Do not immunize; $E = 2.7$
5 (a) Drill; $E = \$10,000$ (b) Drill if + and not if −.
(c) \$9,000; yes
7 (a) $E = 80¢$ for either guess
(b) Guess weighted if heads comes up and balanced if tails does.
(c) $E = \$2.07$; worth $= \$1.27$.

Chapter 8 Section 2 (page 235)

1 Miles' payoff matrix is

	1	2	3	4	
1	−2	3	−4	5	(−4)
2	3	−4	5	−6	−6
3	−4	5	−6	7	−6
4	5	−6	7	−8	−8
	(5)	(5)	7	7	

3

	C	S	
C	45	55	(45)
S	65	40	40
	65	(55)	

5

	1	2	3	4	5	
1	0	1	2	3	4	0
2	1	0	1	2	3	0
3	2	1	0	1	2	0
4	3	2	1	0	1	0
5	4	3	2	1	0	0
	4	3	(2)	3	4	

7

	0	1	2	3	4	
0	5	3	2	1	0	(0)
1	1	4	1	0	−1	−1
2	−3	2	3	−1	−2	−3
3	−2	−1	3	2	−3	−3
4	−1	0	1	4	1	−1
5	0	1	2	3	5	(0)
	5	4	(3)	4	5	

where the rows and columns represent numbers of regiments sent to A.

Chapter 8 Section 3 (page 244)

1 Yes; the entry 1 in row 1, column 1. **3** No
5 (a) $x = 6$ (b) $x = 13$ **7** 0.5 **9** 3.675
11 $P = [\frac{3}{8} \quad \frac{5}{8}]$ $Q = \begin{bmatrix} \frac{3}{8} \\ \frac{5}{8} \end{bmatrix}$ $E = 1.875$
13 $P = [0.48 \quad 0.52]$ $Q = \begin{bmatrix} 0.32 \\ 0.68 \end{bmatrix}$ $E = 12.16$
17 (a) $P = [0.44 \quad 0.56]$ $E = 0.11$ (b) 11 cents for each $1 bet

Chapter 8 Section 4 (page 250)

1 (a) 42 percent in A; 58 percent in B (b) 8.37 percent
3 If rainy, invest in A for an expected gain of 12.75 percent. If dry, invest in B for an expected gain of 8.64 percent.
5 (a) Extend two fingers for an expected payoff of $\frac{1}{3}$ dollar.
(b) Extend one finger if opponent touches his chin and two fingers if he does not.

Chapter 9 Section 1 (page 259)

1 $550 **3** (a) $135 (b) $270 (c) $810
5 5 years ago **7** 240 percent **9** $r = i/nP$
11 (a) 20 years (b) 20 years (c) $1/r$ years (d) $(q - 1)/r$ years
13 (a) $1,346.86 (b) $2,391.24 (c) $3,121.20
15 8 percent **17** (a) 6.17 percent (b) 6.14 percent (c) 6.09 percent

19 (a) \$3,345.55 (b) \$6,139.13 (c) \$20,396.87
21 \$18,308.13

Chapter 9 Section 2 (page 266)

1 (a) \$6,977.00 (b) \$6,897.78 (c) \$359,497.30
3 \$7,952.98 **5** \$152.74 **7** \$7,065.63

Chapter 9 Section 3 (page 270)

1 \$3,372.44; \$775.56 **3** \$349.47; \$1,387.28 **5** \$28,759.39
7 \$36,090.75 **9** \$2,659.22

Chapter 10 Section 1 (page 281)

1 (a) 46 (b) 26
3 (a) $33\frac{1}{3}°$ (b) 7.5° decrease
5 (a) 25.344 centimeters per second (b) 19.008 centimeters per second
7 (a) $0 \leq x \leq 100$ (b) 120 (c) 300 (d) 60 percent

9

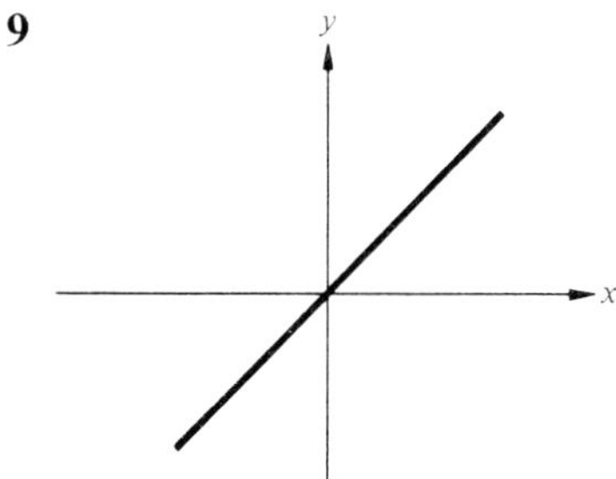

11

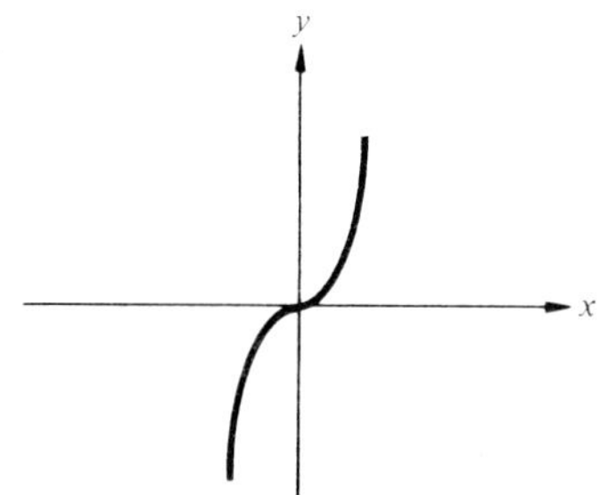

13

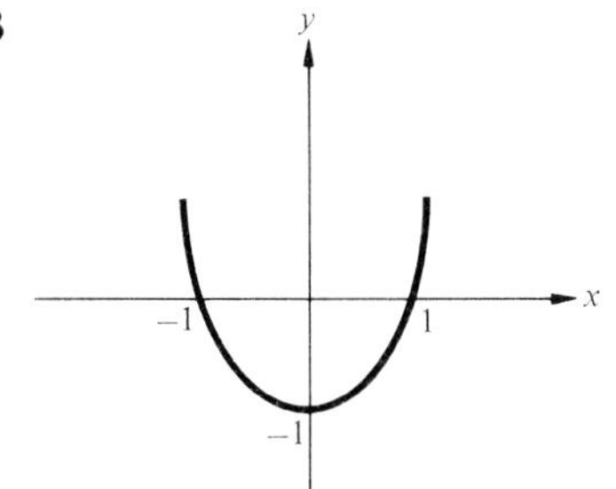

15

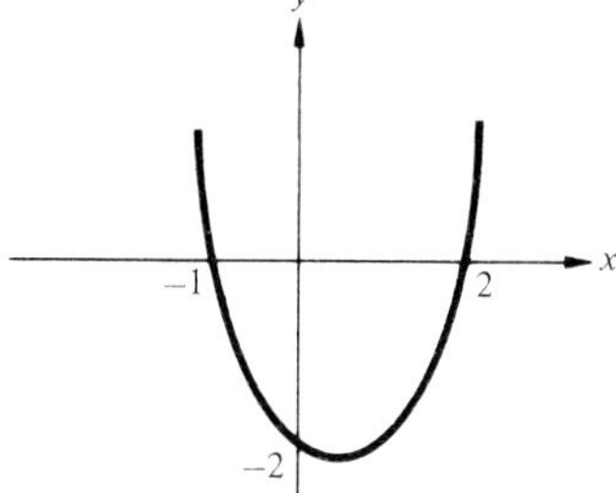

17

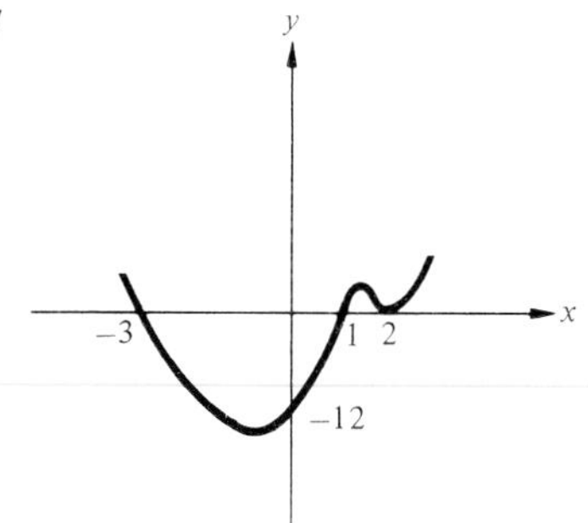

19

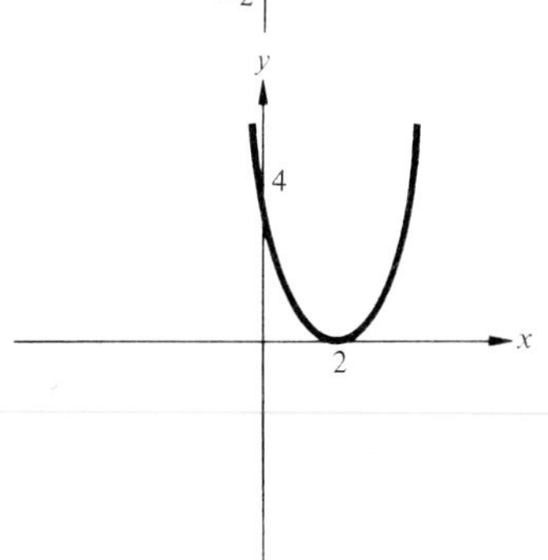

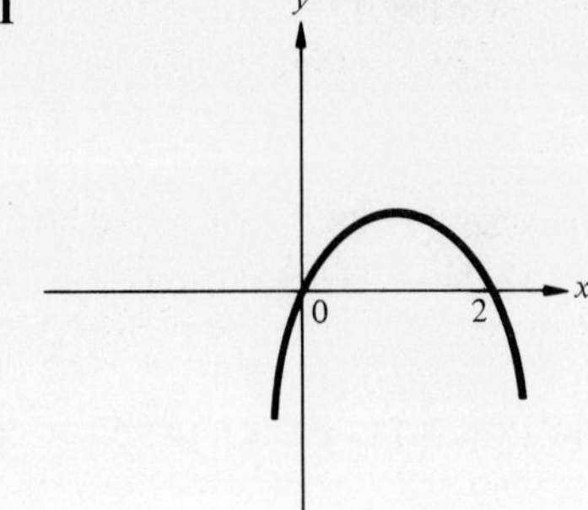

23

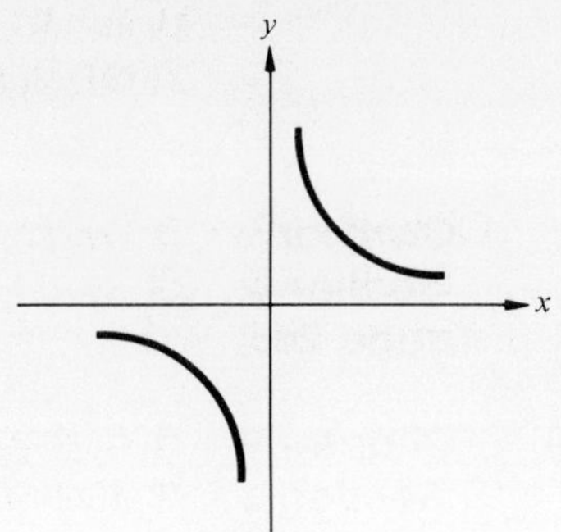

25

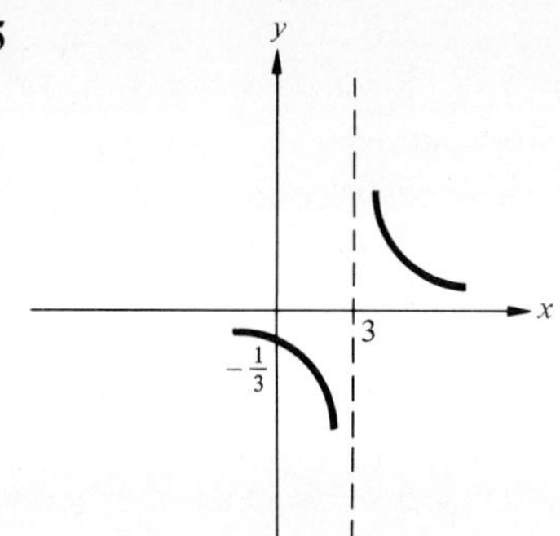

27

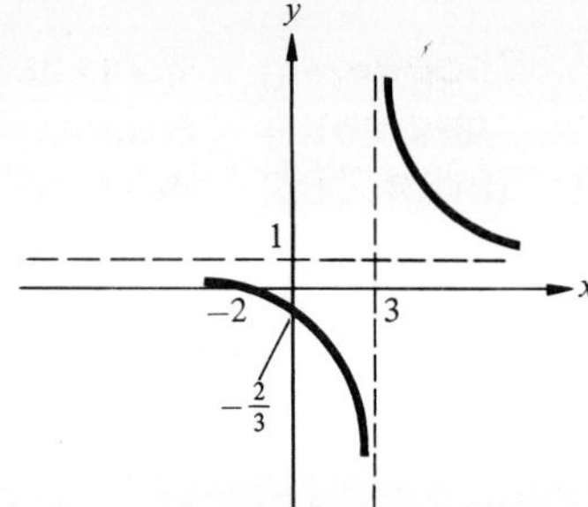

29

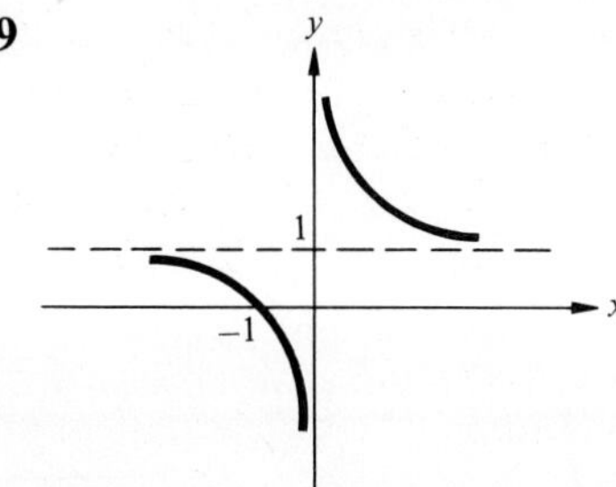

31 (a) (d)

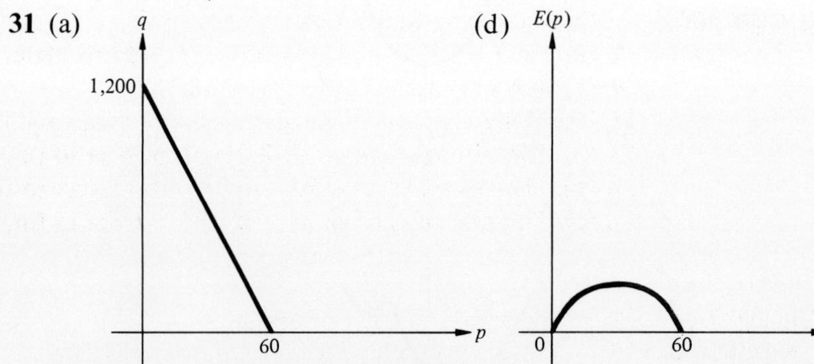

(b) $p = 60$ (c) $E(p) = -200p(p - 60)$

(e) $E(0) = 0$ since the price is zero; $E(60) = 0$ since the demand is zero.

33 (a) $M(x) = 2x^2 + 50x + 600$ (b) 1,300
35 (a) $D(p) = -10p^2 + 250$ (b) 90 pounds

Chapter 10 Section 2 (page 286)

1 (a) $P(x) = 20(25 - x)(x - 3)$ (b) \$2,100
3
$$R(x) = \begin{cases} 2{,}400 & \text{if } 0 < x \le 40 \\ 0.5x(160 - x) & \text{if } 40 < x < 80 \\ 40x & \text{if } x \ge 80 \end{cases}$$
5 $C(x) = 4x^2 + 1{,}000/x$ **7** $D(x) = \sqrt{(975 - 60x)^2 + (90x)^2}$
9 $A(x) = 57 + 8x + 100/x$ **11** $Y(x) = 4(60 + x)(100 - x)$; 80 trees

Chapter 11 Section 1 (page 297)

1 $f'(x) = 5$; slope $= 5$ **3** $dy/dx = 4x - 3$; slope $= -3$
5 $dy/dx = -1/x^2$; slope $= -4$ **7** (a) -3.9 (b) -4
9 (a) $dy/dx = 3$ (b) $y = 3x - 2$
(c) The line $y = 3x - 2$ is its own tangent.
11 (0, 1) **13** The graph rises as x increases. **15** \$70
17 (d) $f' = g' + h'$

Chapter 11 Section 2 (page 302)

1 $dy/dx = 15x^4 - 12x^2 + 9$ **3** $g'(u) = 2 + 6u^{-3} + \frac{5}{3}u^{-2/3}$
5 $h'(t) = -16 - \frac{1}{2}t^{-3/2} - \frac{3}{2}t^{1/2} - \frac{1}{3}t^{-2} + \frac{1}{3}$
7 $g'(x) = \frac{1}{3}(6x^5 - 6x^2)$ **9** $g'(x) = -(x^2 + 2)/(x^2 - 2)^2$
11 $h'(u) = (u^4 + 6u^2 - 3)/(3u - u^3)^2$ **13** $dy/dx = \frac{1}{3}(2x + 2)$
15 $f'(x) = -16x^7 - 91x^6 + 15x^4 + 6x^2 + 26x$ **17** $y = -9x + 4$
19 $y = -2x - 1$ **21** $y = 4x - 14$ **23** $y = 8x + 5$ **29** $(-1, 4)$
31 (a) $E(p) = -200p^2 + 12{,}000p$ (b) \$30

Chapter 11 Section 3 (page 306)

1 (a) 9.8 meters per minute (b) 9.83 meters
3 (a) $C'(t) = 200t + 400$
(b) Increasing at a rate of 1,400 per year. (c) 1,300
7 (a) $f'(x) = -3x^2 + 12x + 15$ (b) 24 per hour (c) 26
9 (a) approximately \$248 (b) \$248.05
11 The daily output will increase by approximately 10 units.
13 (a) The rate of change of cost with respect to output; dollars per unit
(b) The rate of change of output with respect to time; units per hour
(c) The rate of change of cost with respect to time; dollars per hour

Chapter 11 Section 4 (page 312)

1 $dy/dx = 6(3x - 2)$ **3** $dy/dx = (x + 1)/\sqrt{x^2 + 2x - 3}$
5 $dy/dx = -4x/(x^2 + 1)^3$ **7** $dy/dx = -x(x^2 - 9)^{-3/2}$
9 $dy/dx = -2x/(x^2 - 1)^2$
11 $f'(x) = 10x(3x^4 - 7x^2 + 9)^4(6x^2 - 7)$
13 $dy/dx = 15x^5(5x^6 - 12)^{-1/2}$

15 $h'(u) = -\frac{15}{2}u^2(5u^3 + 2)^{-3/2}$
17 $g'(u) = 2(2u - 1)(u^2 + 1)^2(8u^2 - 3u + 2)$
19 $h'(x) = (x + 1)^4(9 - x)/(1 - x)^5$
21 $y = 594x - 1{,}161$ **23** $y = -6x + 26$
25 \$2,025 per year **27** 0.31 parts per million per year
29 6 pounds per week; decreasing

Chapter 11 Section 5 (page 320)

1 Increasing for $-3 < x < 3$; decreasing for $x < -3$ and $x > 3$.
3 Increasing for $x < -4$ and $0 < x < 2$; decreasing for $-4 < x < -2$, $-2 < x < 0$, and $x > 2$

5

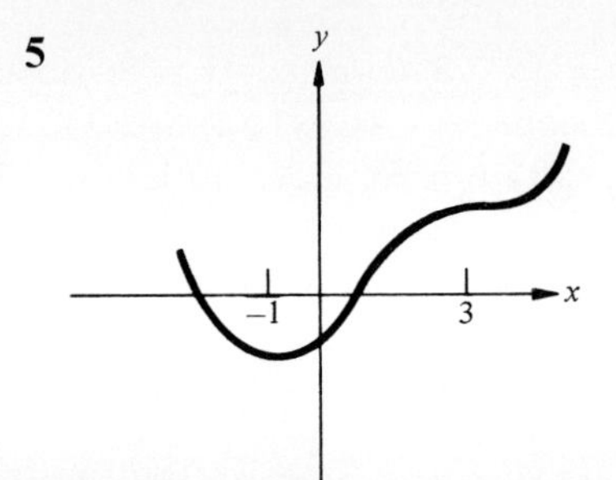

7

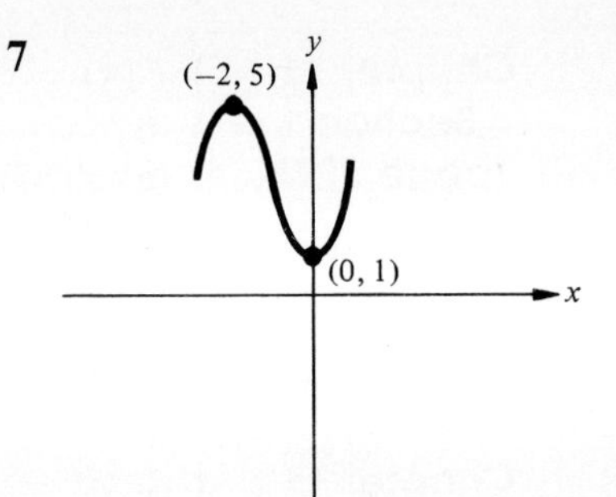

9

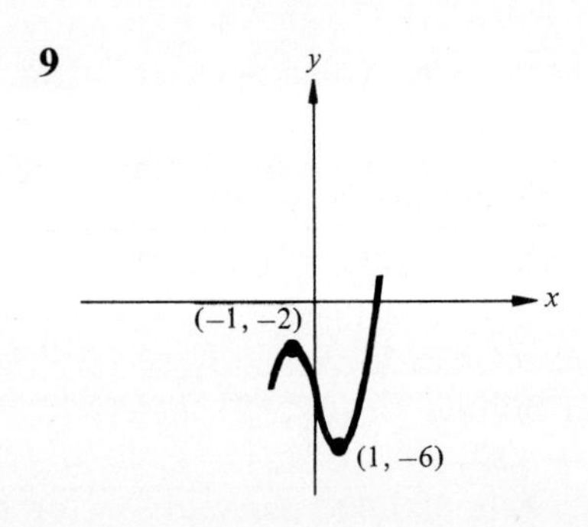

11

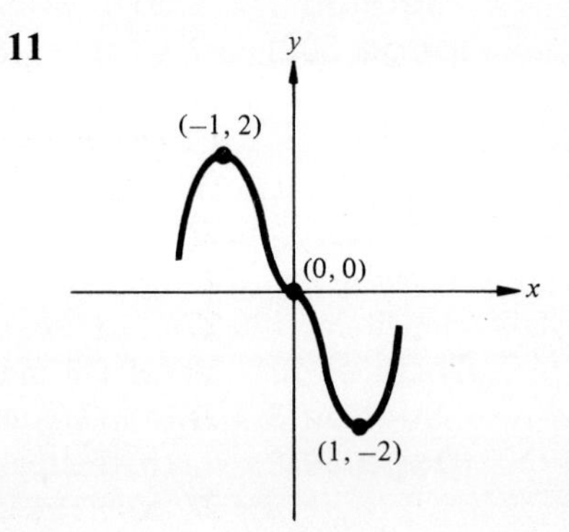

13

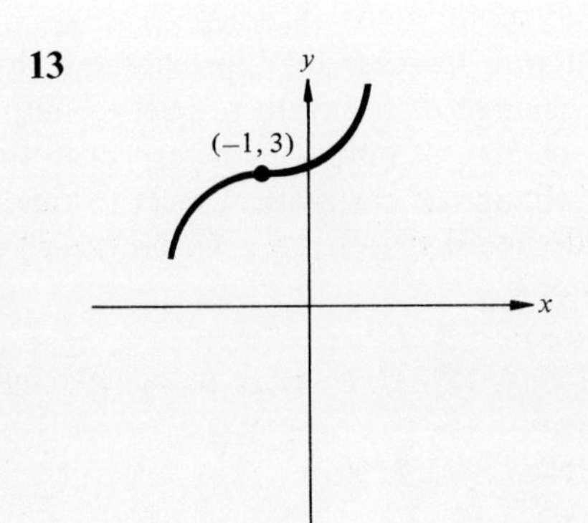

15

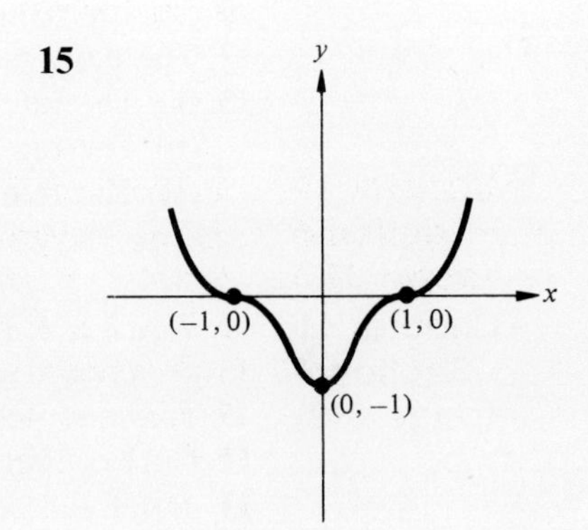

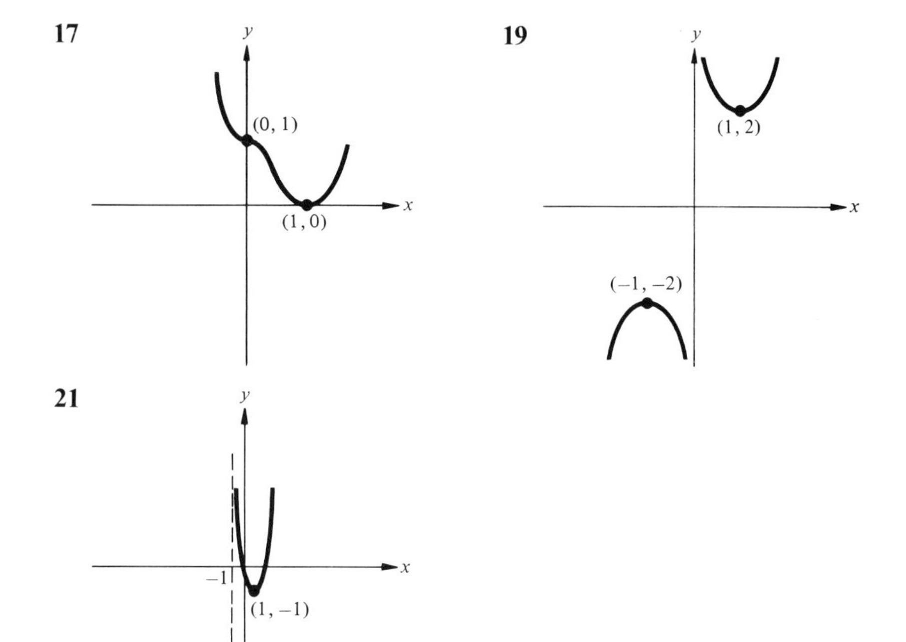

Problem	Absolute maximum	Absolute minimum
23	$f(2) = 21$	$f(0) = f(-3) = 1$
25	$f(0) = f(5) = 1$	$f(4) = -255$
27	$f(1) = 52$	$f(0) = 3$
29	$f(-\frac{1}{2}) = -\frac{1}{6}$	$f(-2) = -\frac{4}{3}$
31	$f(0) = -\frac{1}{9}$	$f(2) = -\frac{1}{5}$

33 $f(x) = (x - 2)^3$
37 (a) 5:00 P.M. (b) 8:00 P.M.

Chapter 11 Section 6 (page 326)

1 The field should be a square, 80 meters by 80 meters.
3 The playground should be a square, 60 meters by 60 meters.
5 \$14 **7** 80 people
9 $2 \times 2 \times \frac{4}{3}$ meters
11 A straight line from the power plant to the factory.
13 5.28 minutes
15 (a) 8 (b) \$160 (c) \$160
17 11:00 A.M.

19 (a) 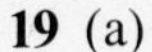

(b) $E(p) = p(mp + b)$

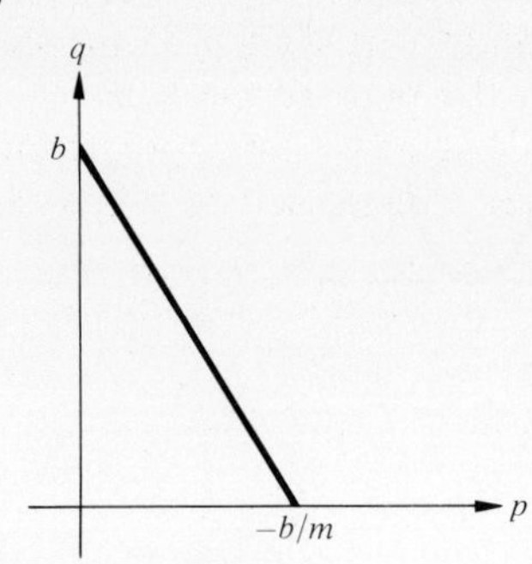

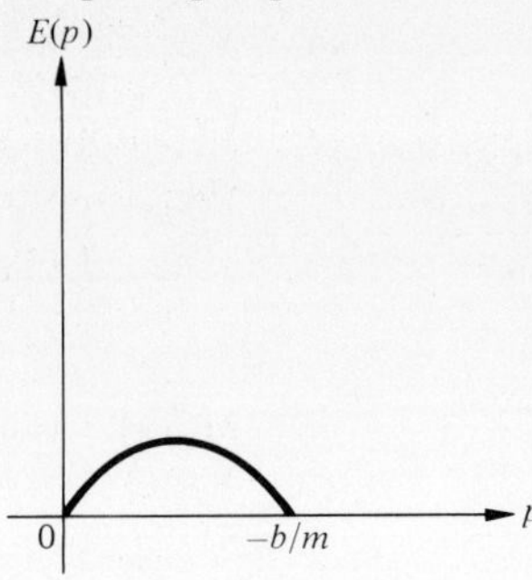

(c) $p = -b/2m$

21 (a) $A(q) = (3q^2 + q + 48)/q$ (b) 4 (c) 4

(d)

Chapter 11 Section 7 (page 332)

1 $f''(x) = 450x^8 - 120x^3$ **3** $d^2y/dx^2 = 2 - 6/x^4$

5 $f''(x) = 30x(x^3 + 1)^3(7x^3 + 1)$ **7** $d^2y/dx^2 = 18/(x - 3)^3$

Problem	Relative maxima	Relative minima
9	$(0, 3)$	$(1, 2)$ and $(-1, 2)$
11	$(0, 81)$	$(3, 0)$ and $(-3, 0)$
13	$(-1, -2)$	$(1, 2)$
15	$(0, 0)$	$(4, 8)$

17 $(-1, \frac{7}{3})$; slope $= -5$

Chapter 11 Section 8 (page 335)

1 -5; 0 **3** 1; $2\sqrt{2}$

5 (a) $C(x, y) = 80x + 20y$ (b) \$56,000
(c) 400 fewer manual typewriters should be produced.

7 (a) 160,000 (b) 16,400 (c) 4,000 (d) 20,810

9 $\partial z/\partial x = \sqrt{y} - 2x^{-5/3}y^5$; $\partial z/\partial y = \frac{1}{2}xy^{-1/2} + 15x^{-2/3}y^4 + 6$

11 $f_x = 15(3x + 2y)^4$ $f_y = 10(3x + 2y)^4$

13 $\partial z/\partial u = -2v/(u - v)^2$ $\partial z/\partial v = 2u/(u - v)^2$

15 $f_x \approx$ change in the monthly demand due to an increase of \$1,000 for monthly advertising; $f_y \approx$ change in the monthly demand due to an increase of \$1 in the price of the toasters.

17 Monthly sales will increase by approximately 18 bicycles.

19 (a) As x increases, f decreases; as y increases, f increases; as z increases, f increases.
(b) $\partial f/\partial x < 0$ $\quad \partial f/\partial y > 0$ $\quad \partial f/\partial z > 0$
(c) $b < 0$ $\quad c > 0$ $\quad d > 0$

21 (a) cameras and film (b) $\partial Q_1/\partial p_2 < 0$ and $\partial Q_2/\partial p_1 < 0$
(c) yes

Chapter 12 Section 1 (page 345)

3 $e^2 = 7.3891$ $\quad e^{-2} = 0.1353$ $\quad e^{0.05} = 1.0513$ $\quad e^{-0.05} = 0.9512$
$e^0 = 1$ $\quad \sqrt{e} = 1.6487$ $\quad 1/\sqrt{e} = 0.6065$

5 The balance will be 4 times the initial deposit.

7 **9** **11** **13**

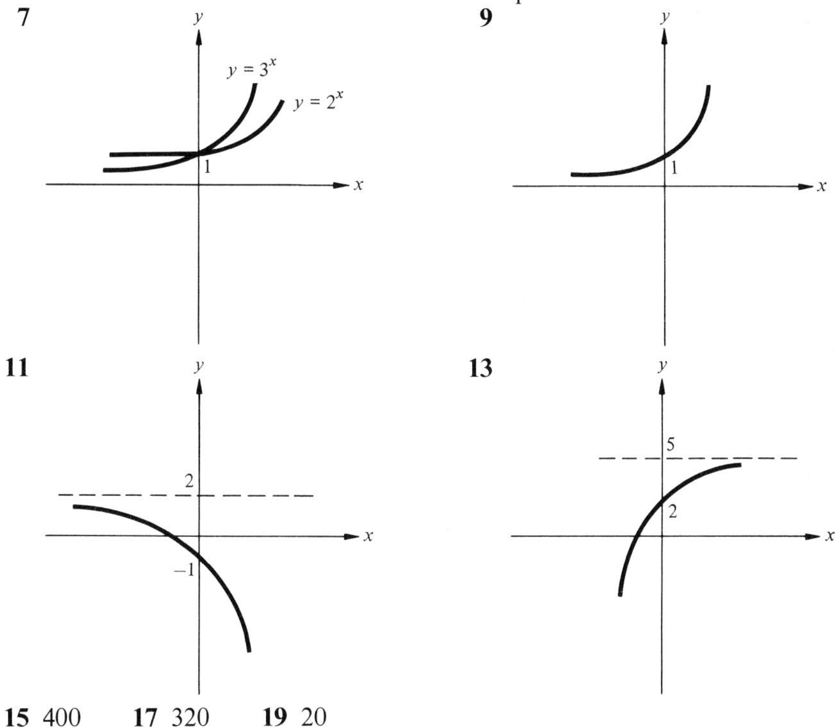

15 400 **17** 320 **19** 20

Chapter 12 Section 2 (page 350)

1 (a) 50 million (b) 91,105,940

3 202.5 million

5 324 billion dollars

7 3,297.44 grams

9 (a) 0.55 (b) 0.12 (c) 0.18

11 (a)

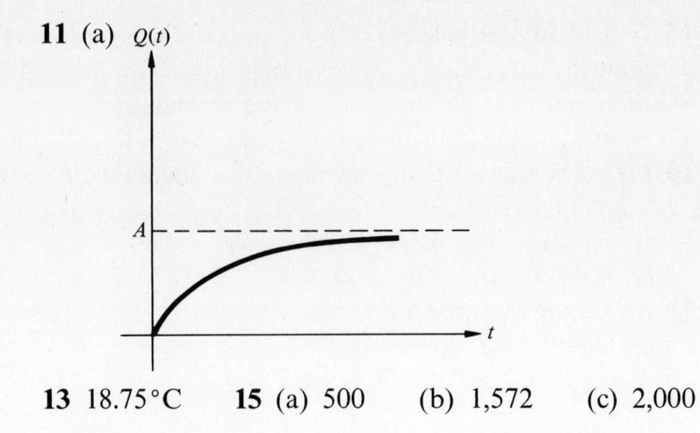

13 18.75°C **15** (a) 500 (b) 1,572 (c) 2,000

Chapter 12 Section 3 (page 357)

1 $\ln 1 = 0$ $\ln 2 = 0.6931$ $\ln e = 1$ $\ln 5 = 1.6094$ $\ln \frac{1}{5} = -1.6094$ $\ln e^2 = 2$ $\ln 0$ and $\ln -2$ are undefined
3 $\frac{1}{2}$ **5** 9 **7** $\frac{19}{6}$ **9** 0.5776 **11** $e^{-b/2}$ **13** e^2 **15** $\frac{9}{25}$
17 $\ln a$ **19** e and $1/e$ **21** 5 **23** 5.33 percent **25** 5.83 percent
27 Approximately 5,614 years **29** 3,143
31 (a) 268 billion (b) 343.62 billion **33** 9,082 years old

Chapter 12 Section 4 (page 363)

1 $f'(x) = 5e^{5x}$ **3** $f'(x) = (2x + 2)e^{x^2+2x-1}$ **5** $f'(x) = -0.5e^{-0.05x}$
7 $f'(x) = (1 - 2x^2)e^{-x^2}$ **9** $f'(x) = 2e^x(1 + e^x)$
11 $f'(x) = (4x + 1)/(2x^2 + x - 3)$ **13** $f'(x) = x(1 + 2 \ln x)$
15 $f(x) = (-1 + \ln x)/(\ln x)^2$ **17** $f'(x) = e^x(1/x + \ln x)$
19 $f'(x) = (x + 1)e^x$
21 (a) 1.22 million per year (b) 1.23 million per year
23 (a) Approximately 406 copies (b) 368 copies
27 \$15
29 $f'(x) = x^x(1 + \ln x)$

Chapter 13 Section 1 (page 371)

1 $\frac{4}{7}x^{7/4} + C$ **3** $\frac{2}{3}u^{3/2} + C$ **5** $t^3 - \frac{5}{2}t^2 + 2t + C$
7 $\frac{2}{3}x^{3/2} - \frac{9}{5}x^{5/3} + 6x + C$ **9** $\frac{a}{3}x^3 + \frac{b}{2}x^2 + cx + d$
11 $\frac{1}{5}x^5 + \frac{3}{4}x^4 + \frac{2}{3}x^3 + C$ **13** $\frac{2}{5}x^5 + \frac{1}{3}x^3 + C$ **15** 10,128
17 \$1,000 **19** \$436 **21** \$2,300 **23** $y = e^x + 1$

Chapter 13 Section 2 (page 375)

1 $\frac{1}{12}(2x + 6)^6 + C$ **3** $\frac{1}{3}\ln|3x + 5| + C$
5 $\frac{1}{6}(x - 1)^6 + (x - 1)^3 + 3(x - 1)^2 + 5x + C$
7 $\frac{1}{12}(x^2 + 1)^6 + C$ **9** $\frac{4}{21}(x^3 + 1)^{7/4} + C$ **11** $-\frac{1}{6}e^{1-x^6} + C$
13 $-\frac{1}{3}(x^3 + 5)^{-1} + C$ **15** $e^{x^3-x} + C$ **17** $5(x^4 - x^2 + 6)^{1/2} + C$
19 $\ln|\ln x| + C$ **21** $\frac{2}{5}(x + 1)^{5/2} - \frac{2}{3}(x + 1)^{3/2} + C$

23 $\frac{1}{11}(x-2)^{11}+\frac{3}{10}(x-2)^{10}+C$ **25** $y=\frac{1}{3}(x^2+5)^{3/2}+1$
27 $\frac{7}{3}$ meters

Chapter 13 Section 3 (page 380)

3 $dQ/dt=kQ$ **5** $dQ/dt=-kQ \quad (k>0)$ **7** $dQ/dt=0.07Q$
9 $y=x^3+\frac{5}{2}x^2-6x+C$ **11** $V=2\ln|x+1|+C$
13 $y=\frac{1}{5}e^{5x}+\frac{4}{5}$ **15** $V=2(t^2+1)^4-1$
17 (a) $C(q)=(q-4)^3+64+C_0$ (b) \$1,500
19 \$3.52 per kilogram **21** $y=-1/(x+C)$ **23** $y=-\ln(C-e^x)$
25 $y=Cx$ **27** $y=500e^{0.05x}$ **29** $y=8-2e^{-5x}$ **31** 83,886
33 (a) $Q(t)=1{,}600e^{-t/50}$ (b) 69.3 days

Chapter 13 Section 4 (page 388)

1 $\frac{9}{20}$ **3** 144 **5** $\frac{8}{3}+\ln 3$ **7** 777.6 **9** $\frac{4}{3}$ **11** $\frac{7}{234}$ **13** $-\frac{23}{110}$
15 98 people **17** \$1,870 **19** \$176,256 **21** 18 **23** 14
25 16 **27** $\frac{32}{3}$ **29** $\frac{4}{3}$

Chapter 13 Section 5 (page 393)

1 30 meters **3** $\int_0^5 r(t)\,dt$ **5** \$7,040,000 **7** $\int_0^{12} n(x)p(x)\,dx$
9 \$453,600 **11** $Pf(10)+\int_0^{10} r(t)f(10-t)\,dt$

Algebra review (page 408)

1 (a) 

(b) 

(c) 

2 (a) 3.15 (b) −12.01 (c) 0.13 (d) 4.00 (e) 0.11
3 (a) $-1\le x\le 5$ (b) $-5\le x\le 1$ (c) $-1\le x\le 1$
4 (a) $x^{1/3}$ (b) $x^{-1/2}$ (c) $x^{2/5}$ (d) $x^{-5/2}$
5 (a) 125 (b) $\frac{1}{8}$ (c) 4 (d) $\frac{1}{6}$ (e) 4 (f) $\frac{1}{81}$ (g) $\frac{1}{2}$
(h) 8 (i) 2 (j) $\frac{1}{2}$
6 (a) 8 (b) −5 (c) 6 (d) 3 (e) 3 (f) $-\frac{7}{5}$
7 (a) $x^4(x-4)$ (b) $(x-4)(x+4)$ (c) $(x+1)(x-5)$
(d) $3x^2(x+1)(x-1)$ (e) $6(x-1)(x+2)$
(f) $(x-3)(x^2+3x+9)$
8 (a) $x=3$ (b) $x=0$ and $x=3$ (c) $x=\pm 4$
(d) $x=2$ and $x=-3$ (e) $x=0$ and $x=\pm 2$
(f) $x=2$ and $x=\pm 1$
9 (a) $x=-1$ and $x=-\frac{1}{2}$ (b) No solution (c) $x=1$
(d) $x=\frac{3}{2}\pm\frac{1}{2}\sqrt{5}$
10 (a) $\sum_{i=1}^{6} 1/i$ (b) $\sum_{i=1}^{10} 3i$ (c) $\sum_{i=1}^{6} 2^i$ (d) $\sum_{i=1}^{6} 2x_i$
11 (a) 55 (b) 34 (c) 62 (d) 0

INDEX

Conditional probability:
  definition, 132

Differential equation: